Earth Surface: Processes, Emerging Trends and Concerns

Earth Surface: Processes, Emerging Trends and Concerns

Editor: Leon Cameron

Callisto Reference
www.callistoreference.com

Callisto Reference,
118-35 Queens Blvd., Suite 400,
Forest Hills, NY 11375, USA

Visit us on the World Wide Web at:
www.callistoreference.com

ISBN: 978-1-64116-270-8 (Hardback)

The publisher's policy is to use permanent paper from mills that operate a sustainable forestry policy. Furthermore, the publisher ensures that the text paper and cover boards used have met acceptable environmental accreditation standards.

Printed in China.

Cataloging-in-Publication Data

Earth surface : processes, emerging trends and concerns / edited by Leon Cameron.
p. cm.
Includes bibliographical references and index.
ISBN 978-1-64116-270-8
1. Earth (Planet)--Surface. 2. Earth (Planet)--Crust. 3. Earth sciences. I. Cameron, Leon.
QE511 .E27 2020
550--dc23

Table of Contents

Preface

The surface of the Earth as well as its structure and processes are studied under the domain of Earth science. The Earth consists of multiple layers such as crust, mantle and core. The thin layer or shell which covers the outside of the Earth is known as crust. It is the topmost constituent of the Earth's lithosphere. The lithosphere comprises of tectonic plates, and the movement of tectonic plates allows the Earth's interior heat to escape into space. The crust is placed above the mantle. The Earth's mantle is made up of silicate rocks. Partial melting of the mantle is responsible for the formation of oceanic as well as continental crust. The topics included in this book on Earth surface are bound to provide incredible insights to readers. It includes some of the vital pieces of work being conducted across the world, on various topics related to the surface of the Earth. The extensive content of this book provides the readers with a thorough understanding of the subject.

This book is the end result of constructive efforts and intensive research done by experts in this field. The aim of this book is to enlighten the readers with recent information in this area of research. The information provided in this profound book would serve as a valuable reference to students and researchers in this field.

At the end, I would like to thank all the authors for devoting their precious time and providing their valuable contribution to this book. I would also like to express my gratitude to my fellow colleagues who encouraged me throughout the process.

Editor

1

Late Holocene channel pattern change from laterally stable to meandering – a palaeohydrological reconstruction

Jasper H. J. Candel[1], Maarten G. Kleinhans[2], Bart Makaske[1], Wim Z. Hoek[2], Cindy Quik[1], and Jakob Wallinga[1]

[1]Soil Geography and Landscape Group, Wageningen University & Research, Wageningen, P.O. Box 47, 6700AA, the Netherlands
[2]Department of Physical Geography, Utrecht University, Utrecht, P.O. Box 80125, 3508TC, the Netherlands

Correspondence: Jasper H. J. Candel (jasper.candel@wur.nl)

Abstract. River channel patterns may alter due to changes in hydrological regime related to changes in climate and/or land cover. Such changes are well documented for transitions between meandering and braiding rivers, whereas channel pattern changes between laterally stable and meandering rivers are poorly documented and understood. We hypothesize that many low-energy meandering rivers had relatively low peak discharges and were laterally stable during most of the Holocene, when climate was relatively stable and human impact was limited. Our objectives in this work are to identify a Late Holocene channel pattern change for the low-energy Overijsselse Vecht river, to develop and apply a novel methodology to reconstruct discharge as a function of time following a stochastic approach, and to relate this channel pattern change to reconstructed hydrological changes. We established that the Overijsselse Vecht was laterally virtually stable throughout the Holocene until the Late Middle Ages, after which large meanders formed at lateral migration rates of about 2 m yr^{-1}. The lateral stability before the Late Middle Ages was constrained using a combination of coring information, ground-penetrating radar (GPR), radiocarbon (^{14}C) dating, and optically stimulated luminescence (OSL) dating. We quantified bankfull palaeodischarge as a function of time based on channel dimensions that were reconstructed from the scroll bar sequence and channel cut-offs using coring information and GPR data, combined with chronological constraints from historical maps and OSL dating. We found that the bankfull discharge was significantly greater during the meandering phase compared to the laterally stable phase. Empirical channel and bar pattern models showed that this increase can explain the channel pattern change. The bankfull discharge increase likely reflects climate changes related to the Little Ice Age and/or land use changes in the catchment, in particular as a result of peat reclamation and exploitation.

1 Introduction

Channel patterns describe the planform of a river, which reflects the interaction of the river channel with its floodplain. Channel patterns are classically distinguished: laterally inactive channels consist of straight and sinuous stable planforms, whereas laterally active channels consist of meandering and braiding planforms (Leopold and Wolman, 1957; Nanson and Knighton, 1996). Flume experiments and field data have shown that the channel pattern depends on several variables (Kleinhans, 2010). The first is the potential specific stream power, which is the product of the channel-forming discharge and valley slope (Kleinhans and Van den Berg, 2011; Nanson and Croke, 1992). The second is the bank erodibility (Ferguson, 1987; Friedkin, 1945), which is determined by the presence of bedrock in the valley side (Turowski et al., 2008), the bank cohesiveness (Peakall et al., 2007), and vegetation (Gurnell, 2014; Millar, 2000). The third is the type and amount of sediment supply (Gibling and Davies, 2012; Nanson and Croke, 1992).

Channel patterns can change in response to environmental variations (Ferguson, 1987). Many examples of channel pattern changes from braiding to meandering and vice versa are known to be associated with glacial–interglacial oscillations (Vandenberghe, 1995, 2002). Studies on the last glacial–interglacial transition have especially shown the simultaneous occurrence of channel pattern changes with a changing climate (Kasse et al., 2016; Vandenberghe et al., 1994). Climate change affects the vegetation, sediment availability, and discharge regime and consequently the bank stability, sediment transport, and potential specific stream power, resulting in different channel patterns.

Within the Holocene, several examples of channel pattern changes are documented from braiding to meandering rivers and vice versa (Brewer and Lewin, 1998; Lewin et al., 1977; Passmore et al., 1993; Słowik, 2015). However, channel pattern changes between laterally stable and meandering rivers have rarely been reported (Lewin and Macklin, 2010), except where human intervention transforms meandering rivers into heavily regulated and laterally stable rivers by introducing weirs, dams, groynes, and bank protection measures (Hesselink et al., 2003; Hobo et al., 2014; Słowik, 2013; Surian and Rinaldi, 2003). The partial abandonment of former meandering valleys may also result in underfit, laterally stable rivers like the former Rhine branches in the Niers and Oude IJssel valley (Janssens et al., 2012; Kasse et al., 2005).

Many studies have reported increased fluvial activity (e.g. increased discharge, sediment transport and deposition, and bank erosion rates) in relation to human, environmental, and climatic pressures during the Holocene (e.g. Hoffmann et al., 2008; Lespez et al., 2015; Macklin et al., 2010; Notebaert et al., 2018; Notebaert and Verstraeten, 2010). An example of increased fluvial activity is known from the Pine Creek (Idaho, USA), where mining and deforestation combined with intensive grazing resulted in an increase in discharge and sediment input, followed by river widening and an increase in bank erosion (Kondolf et al., 2002). The reverse change has been observed in settings as a result of afforestation (Kondolf et al., 2002; Liébault and Piégay, 2001) or increases in riparian vegetation fixing the channel banks (Eekhout et al., 2014; Vargas-Luna et al., 2016). The increase in fluvial activity during the Holocene was corroborated by an extensive review of existing studies concerning sediment accumulation in west and central European river floodplains by Notebaert and Verstraeten (2010). They concluded that sedimentation rates increased during the Middle and Late Holocene due to environmental changes. However, it is unknown whether the channel pattern changed simultaneously with the floodplain because no Early Holocene channel deposits representing a stable phase were identified. De Moor et al. (2008) hypothesized that the Geul River in southern Netherlands may have been relatively laterally stable during the Early and Middle Holocene, while it was actively meandering during the past 2000 years. Most of the floodplain deposits from the laterally stable phase have not been preserved, but De Moor et al. (2008) were able to reconstruct the bankfull depth for both periods. They estimated the bankfull depth to be a factor of 2 to 3 higher during the Late Middle Ages compared to the Early and Middle Holocene, and related this change to human and climate impact.

We conjecture that the change from laterally stable to meandering has occurred in some rivers for which increased Holocene fluvial activity was reported. The fact that such changes were not reported in the literature may either mean that critical conditions for channel pattern change were not reached or that evidence of such transitions is poorly preserved or left unnoticed. Both laterally stable and meandering rivers may display sinuous planforms, but the geomorphic processes in both rivers are different. Laterally stable channels are rivers without meandering processes, i.e. helicoidal flows causing bar formation and bank erosion at a significant rate (Kleinhans and Van den Berg, 2011; Nanson and Knighton, 1996; Seminara, 2006). In fact, the bends and channel cut-offs in laterally stable rivers may be the result of random and local perturbations (e.g. falling trees, beavers, bank collapse after heavy rainfall, etc.) leading to very limited and local displacement of the channel. Meandering and laterally stable rivers should therefore be distinguished by their different patterns of bar and floodplain formation, rather than merely by planform (Candel et al., 2017; Kleinhans and Van den Berg, 2011). We suggest that identifying channel pattern changes requires more detailed historic accounts or a much higher resolution of subsurface data than usually gathered because palaeochannels of laterally stable channels poorly preserve in the fluvial archive of meandering channel belts (Van de Lageweg et al., 2016), except when they have been cut off by random and local perturbations prior to the meandering phase. Using numeric (e.g. Oorschot et al., 2016) or scaled (e.g. Van Dijk et al., 2012) river simulation models is problematic for testing these ideas because these have not yet been capable of reproducing channel pattern changes. This reflects the lack of understanding of river processes and patterns (Kleinhans, 2010) and the need to gather such information from field studies.

This research entails a case study focussing on a river in which lateral activity during the past 500 to 600 years caused spectacular meandering: the Overijsselse Vecht in the Netherlands (Fig. 1). Previous work on this system has identified a transition from braiding to meandering during the Late Glacial (Huisink, 2000), while the meandering pattern remained throughout the Holocene until the river was channelized between 1896 and 1914 CE (Huisink, 2000; Neefjes et al., 2011). However, Quik and Wallinga (2018) found that the meanders were relatively young, with the oldest scroll bars dating from ca. 1400 to 1500 CE, by reconstructing meander formation using a combination of optically stimulated luminescence (OSL) dating of scroll bars and planform reconstruction based on historical maps. No fluvial deposits were found dating from before this period, except from a Holocene palaeochannel (here referred to as "Palaeochannel Q") in a ground-penetrating radar (GPR) profile recorded by Huisink (2000, p. 123) 13 km upstream near Hardenberg (Figs. 1b and 2). Palaeochannel Q is relatively small compared to the meandering channel, seems to lack scroll bars, and was already cut off on the historical map of 1720 CE. Therefore, it is questionable whether the Overijsselse Vecht meandered prior to ca. 1400 CE. Alternatively, the river changed from a laterally stable into a meandering river during the Late Middle Ages. Our aims are (1) to identify whether a channel pattern change has occurred from laterally stable to meandering by collecting and combining detailed subsurface and geochronological data of the river from the pronounced meandering phase and the preceding phase; (2) to develop a methodology to reconstruct bankfull discharge as a function of time using the scroll bar deposits and channel remnants as a geological archive of the former channel dimensions; (3) to test whether palaeohydrological changes may explain the potential channel pattern change; and (4) to elaborate on the potential causes for changes in the discharge and channel pattern.

2 Study area

The Overijsselse Vecht (Fig. 1) is a low energy, sand bed river flowing from Germany into the Netherlands, with an average annual discharge (Q_m) of 22.8 m^3 s^{-1} and a mean annual flood discharge (Q_{maf}) of 160 m^3 s^{-1} derived from the gauging station in Mariënberg for the period 1995 to 2015 (see location in Fig. 1b). The river has a length of 167 km, its catchment covers 3785 km^2 with the highest point +110 m above sea level (m a.s.l.), and a relatively uniform valley slope of 1.42×10^{-4} to 1.7×10^{-4} in the Dutch part of its trajectory (TAUW, 1992; Wolfert and Maas, 2007). The Overijsselse Vecht incised its current valley during the Late Glacial within fluvioperiglacial sands, locally covered by aeolian cover sands (Huisink, 2000; Ter Wee, 1966; Wolfert and Maas, 2007). During the Late Holocene, aeolian drift sands formed along the Overijsselse Vecht as a result of agricultural overexploitation (Van Beek and Groenewoudt, 2011). The Overijsselse Vecht was an actively meandering river until 1896, when weirs were constructed and parts of the river were channelized. The river was completely channelized after 1914 CE, with five weirs controlling the water levels. Recently, sinuous side channels bypassing the weirs have been created as part of river restoration aiming to restore past physical and ecological characteristics of the river.

At present the topography of the meandering phase is partly still intact in the floodplain (Maas, 1995). Wolfert and Maas (2007) reconstructed the pre-channelization planform from historical maps of 1720, 1850, and 1890 CE. Large differences in meander development and lateral migration rates were found between different river reaches. In particular in areas where non-cohesive aeolian sands formed the channel banks, large meanders formed and lateral migration reached rates up to 3 m yr^{-1}. In this research we will study two of the large meanders, named Prathoek and Junnerkoeland (Fig. 1), for which Quik and Wallinga (2018) reconstructed the scroll bar development using OSL dating in combination with historical maps.

Here we take advantage of the preservation of a palaeochannel (here referred to as "Palaeochannel X") with comparable dimensions as Palaeochannel Q (Huisink, 2000, p. 123) preserved in the Junnerkoeland as a sharp bend (Fig. 1c). Maas (1995) interpreted Palaeochannel X to be connected to the oldest swale of the Junnerkoeland scroll bar deposits (Fig. 1c). Palaeochannel X, however, was likely abandoned before the scroll bar formation because large differences in dimensions exist between Palaeochannel X and the meander bend, but the well-preserved nature suggests that Palaeochannel X is relatively young. The small dimensions of both Palaeochannels X and Q would suggest that the river had comparatively less energy and may have been relatively laterally stable prior to the meandering phase.

3 Methods

3.1 Lithological description

Cores were performed in a transect perpendicular to the scroll bars of both meander bends (Fig. 1c, d). An additional transect was cored perpendicular to Palaeochannel X (Fig. 1e). In the case that the deposit consisted of peat we used a gouge auger (Ø: 3 cm); in the case of unsaturated sand we used an Edelman auger, and in the case of saturated sand we used a Van der Staay suction corer (Van de Meene et al., 1979). In total, 68 cores were performed to a maximum depth of 7.3 m. The surface elevation of each coring site was either determined using a GPS combined with a DEM (Van Heerd and Van't Zand, 1999) or with a global navigation satellite system (GNSS) device. A standard method was used to describe the sediment cores in 10 cm thick intervals using the Dutch texture classification scheme, which approximately matches the USDA terminology (Berendsen and Stouthamer, 2001; De Bakker and Schelling, 1966). The median sediment grain size (D_{50}, m) of non-organic, sandy samples was visually checked in the field by comparison with a sand ruler. Grain size analysis was used to estimate a D_{50} for the entire scroll bar deposit (Sect. 3.3). In addition, the plant macroremains, any visible bedding, and colour were described. The percentage of gravel (> 2 mm) was estimated in the field using sieves. The lithogenesis was inferred from the lithological properties, facies geometries, and DEM topography, distinguishing fluvial, fluvioperiglacial, cover sand, drift sand, and residual channel-fill deposits (Huisink, 2000; Ter Wee, 1966).

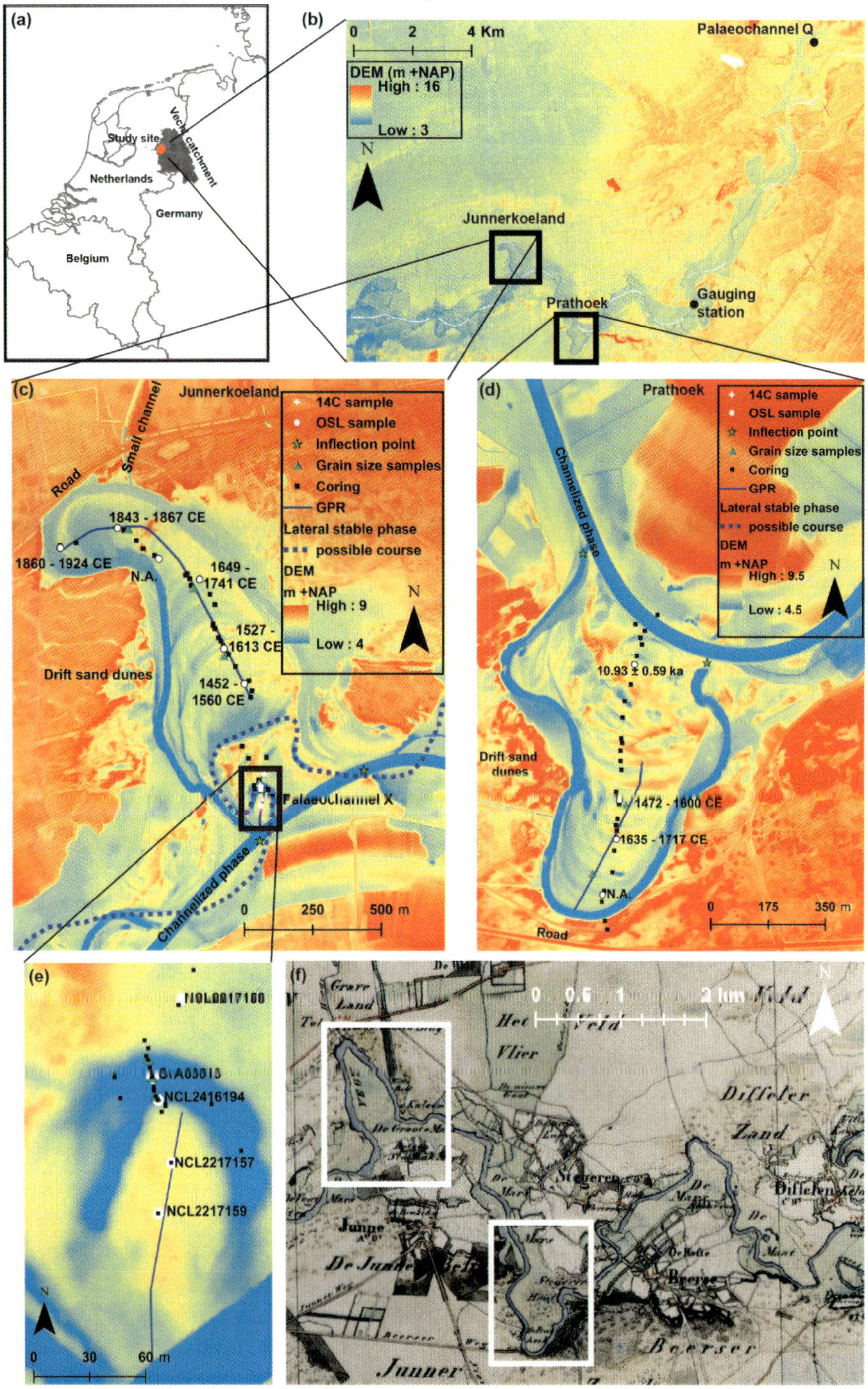

Figure 1. Maps of the Overijsselse Vecht. **(a)** Map showing the location of the Overijsselse Vecht catchment and the location of the study site. **(b)** Digital elevation map (DEM; Actueel Hoogtebestand Nederland, 0.5 × 0.5 m) (Van Heerd and Van't Zand, 1999) of the downstream section of the Overijsselse Vecht river, indicating both study sites: Junnerkoeland and Prathoek. DEM of the Junnerkoeland bend **(c)** and Prathoek bend **(d)**, including locations of cores, OSL samples by Quik and Wallinga (2018), the OSL and ^{14}C samples from this study, the GPR transects, the grain size samples, and inflection points. The possible historical course of Palaeochannel *X* according to Maas (1995) is indicated. **(e)** Zoomed-in figure of Palaeochannel *X*. **(f)** Topographical military map (TMK) dating from 1851 CE (CC-BY Kadaster, 2018; Van der Linden, 1973), showing the Overijsselse Vecht during its meandering phase.

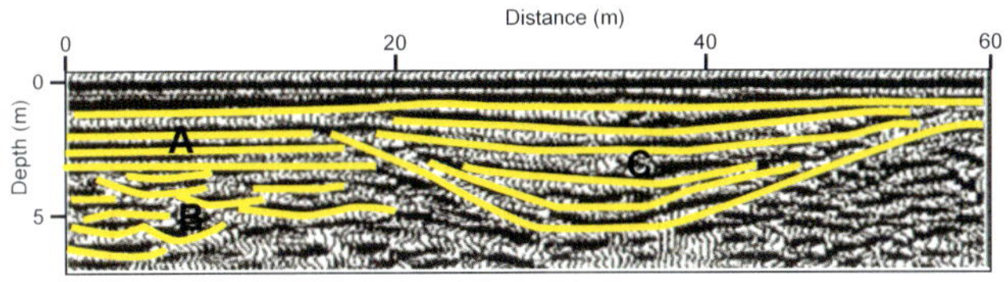

Figure 2. Interpretation by Huisink (2000) of subsurface strata from GPR data collected near Hardenberg 13 km upstream of Junnerkoeland (see location in Fig. 1b). Horizontal strata of cover sand deposits (A) on top of the channel deposits of an interpreted braiding system (B). A relatively small, symmetrical palaeochannel is present (C) within the Late Glacial deposits, hereafter referred to as "Palaeochannel Q". Figure adapted after Huisink (2000).

3.2 Ground-penetrating radar

Ground-penetrating radar (GPR) was used to reconstruct the channel dimensions of the scroll bars. GPR measurements were conducted with a pulseEKKO PRO 250 Hz with a SmartTow configuration. The GPR transects were placed along the centreline of the meander bends, perpendicular to the ridge and swale morphology (Fig. 1c, d). The electromagnetic-wave velocity was 0.060 m ns^{-1}, derived by using isolated reflector points (Neal, 2004; Van Heteren et al., 1998) and by comparing depths of recognizable layers with the coring data.

3.3 Grain size analysis

In total 33 samples for grain size analysis were taken from the scroll bar deposits and three samples were taken from Palaeochannel X. The samples of the scroll bar deposits were taken from each 0.5 m interval from the channel lag up to the swale surface at three locations in Junnerkoeland and two locations in Prathoek (Fig. 1c–e). The samples of Palaeochannel X were taken from three locations below the residual channel-fill from the former riverbed. Grain size samples were analysed in a laboratory with a Beckman Coulter LS230 laser particle sizer. This instrument has a measurement range of 0.1 to 2000 µm. Samples were sieved with a 2 mm sieve and prepared with HCl (1 M) and H_2O_2 (30 %). All data were processed using a Fraunhofer.rfd optical model because of the low clay–silt content (Agrawal et al., 1991). Finally, the average and standard deviation were calculated for both the scroll bar deposits and Palaeochannel X and used in the palaeohydrological calculations.

3.4 OSL dating

We used the modelled age–distance relationships determined by Quik and Wallinga (2018) in our calculations. Their obtained OSL ages from the scroll bar deposits were used as priors and combined with historical map data in a Bayesian sequence model using the OxCal software (Bronk Ramsey, 2009). For details on the method see Quik and Wallinga (2018). In this study, we took four additional samples for OSL dating on the inner and outer bank of Palaeochannel X. These samples were collected in an opaque PVC tube (Ø 4.5 cm) mounted on a hand auger, allowing for sampling without light exposure. The analysis in the laboratory followed the same procedure as in Quik and Wallinga (2018). The OSL age was determined at the Netherlands Centre for Luminescence dating, with equivalent doses measured on small aliquots of quartz using the SAR protocol (Murray and Wintle, 2003) and dose rates determined from activity concentrations measured using gamma-ray spectrometry. A bootstrapped version of the minimum age model (Cunningham and Wallinga, 2012) was used to derive the best estimate of the burial dose and deposition age. Given the limited amount of samples associated with Palaeochannel X and the absence of additional age constraints from historical maps, no Bayesian analysis was performed for these samples.

3.5 ^{14}C dating

A sample was taken in the deepest part of Palaeochannel X, at the sand–peat interface, using a piston corer (Ø: 6 cm). Macro-remains and leaf fragments from terrestrial species were selected from 1 cm intervals in the laboratory using a light microscope. Samples were stored in diluted HCl (4 %) at 5 °C. The sand content was measured for each interval to precisely determine the position of the sand–peat interface. Material with volumetric sand percentages lower than 10 % to 20 % was considered as peat (Bos et al., 2012). The macro-remains from the centimetre above this interface were selected for the ^{14}C analysis, providing a terminus ante quem date for the abandonment of the channel. The ^{14}C age was determined by accelerator mass spectrometry (AMS) at the Centre for Isotope Research (Groningen University). For calibration, the IntCal13 curve was used in the OxCal 4.2.4 software (Bronk Ramsey, 2009; Reimer et al., 2013).

3.6 Channel dimensions

The channel dimensions of Palaeochannel X were determined from the lithological cross section. The residual channel-fill was delineated along the sand–peat interface. Bankfull depth (H_{bf}) was defined from the bottom of the palaeochannel up to the first clear knick point on the bank, which was mapped with a GNSS device such that the width–depth ratio was minimal (Williams, 1986). Relative error of H_{bf} was assumed to be similar to the relative error of H_{bf} during the meandering phase (ca. 10 %) and used in the calculations (see details below) because both H_{bf} values were determined by using coring data. Additional dimensions were measured from the delineated channel, involving the bankfull width (W), cross-sectional area (A), and wetted perimeter (P). These channel dimensions were also measured for Palaeochannel Q from the GPR profile recorded by

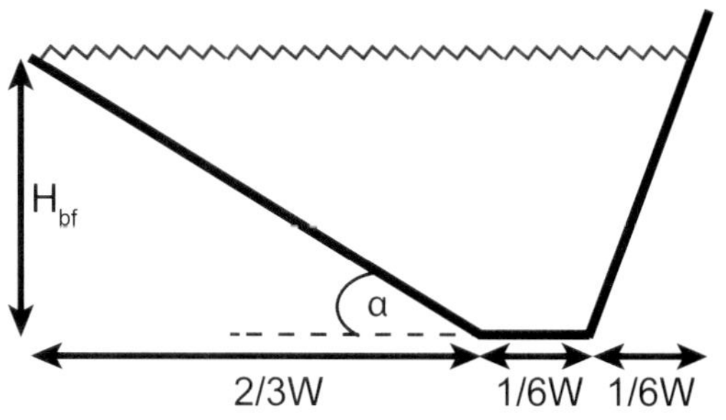

Figure 3. Sketch of the cross-sectional flow area of a meandering channel used for the bankfull palaeodischarge calculations (Allen, 1965; Hobo, 2015; Leeder, 1973).

Huisink (2000, p. 123) (Fig. 2). Here we assumed a similar relative error of W, A, and P as was taken for H_{bf}.

The river channel was assumed to have the channel dimensions as shown in Fig. 3 during the meandering phase. This sketch is based on Allen (1965), Leeder (1973), and Hobo (2015). The bankfull depth (H_{bf}) was estimated from the coring data taken from the bottom of the channel lag up to the surface elevation in the swales (Fig. 4). Small elevation differences were expected to result from local variation rather than real changes in H_{bf}, and therefore the average H_{bf} was calculated from the smoothed bottom and surface elevation. The standard deviation of H_{bf} was calculated from the actual variable bottom elevation over the length of the scroll bar. The transverse bed slope (α) of the inner bend was determined based on the GPR transects (Fig. 5), in which lateral accretion surfaces could be distinguished. The angle was measured on the steepest (middle) parts of the identified lateral accretion surfaces. The average and standard deviation of α were calculated and used in the calculations. The calculations of the channel dimensions follow from Fig. 3. The bankfull width (W, m) and cross-sectional area (A, m^2) were determined by Eqs. (1) and (2).

$$W = 1.5\frac{H_{\mathrm{bf}}}{\tan(\alpha)} \tag{1}$$

$$A = W H_{\mathrm{avg}} \tag{2}$$

H_{bf} is the bankfull depth (m), and $H_{\mathrm{avg}} = \frac{7H_{\mathrm{bf}}}{12}$ approximates the average water depth (m). The wetted perimeter (P, m) was calculated from the assumed channel geometry (Fig. 3) following Eq. (3).

$$P = \frac{H_{\mathrm{bf}}}{\sin(\alpha)} + \frac{W}{6} + \sqrt{H_{\mathrm{bf}}^2 + \left(\frac{W}{6}\right)^2} \tag{3}$$

The hydraulic radius (R, m) was calculated by Eq. (4).

$$R = \frac{A}{P} \tag{4}$$

For each swale visible on the DEM the sinuosity (s, -), radius of curvature (R_{curv}, m) and scroll bar surface area ($\mathrm{SB}_{\mathrm{surf}}$, m^2) were measured. The former channel sinuosity was estimated by the use of the DEM, measuring the distance along the swales relative to the distance along the valley between the inflection points (Fig. 1c, d). The sinuosity of Palaeochannel X was measured using the same approach (Fig. 1c). The channel slope (S_{c}, –) was calculated from the sinuosity and valley slope (S_{v}, –) determined by TAUW (1992) and Wolfert and Maas (2007) following Eq. (5).

$$S_{\mathrm{c}} = S_{\mathrm{v}}/s \tag{5}$$

The volumetric rate of scroll bar growth ($\mathrm{SB}_{\mathrm{vol}}$, $\mathrm{m}^3\,\mathrm{yr}^{-1}$) was determined from scroll bar surface area ($\mathrm{SB}_{\mathrm{surf}}$, $\mathrm{m}^2\,\mathrm{yr}^{-1}$) and the thickness between each swale and interpolated time interval following Eq. (6):

$$\mathrm{SB}_{\mathrm{vol}} = \frac{\mathrm{SB}_{\mathrm{surf}} H_{\mathrm{bf}}(1-\varphi)}{\Delta\mathrm{age}} \tag{6}$$

where φ is the porosity (here 0.3 to 0.35 volume fraction) (Nimmo, 2004), which was included to compare the $\mathrm{SB}_{\mathrm{vol}}$ with the sediment transport, and Δage is the age difference between the scroll bars (yr) based on the datings by Quik and Wallinga (2018). Although scroll bar deposits were absent, following Eq. (6) we also calculated the volumetric sediment transport for the fluvial deposits on the inside of Palaeochannel X.

3.7 Palaeodischarge

The channel dimensions were used to calculate the bankfull discharge (Q_{bf}, $\mathrm{m}^3\,\mathrm{s}^{-1}$). Bankfull discharge is commonly considered an approximation of the channel-forming discharge with a recurrence interval of 1 to 2 years (Dury, 1973; Wolman and Miller, 1960). We assumed that the bankfull discharge was similar for both Junnerkoeland and Prathoek, regarding the short distance between these river sections (Fig. 1b). Hence the bankfull discharge was presented by combining the bankfull discharges for both meander bends. The bankfull discharge was estimated by applying the Chézy equation, following Eq. (7):

$$Q_{\mathrm{bf}} = CA\sqrt{RS_{\mathrm{c}}}, \tag{7}$$

where Q_{bf} is the bankfull discharge ($\mathrm{m}^3\,\mathrm{s}^{-1}$), and C is the Chézy coefficient ($\mathrm{m}^{0.5}\,\mathrm{s}^{-1}$). The Chézy coefficient, i.e. flow resistance, was estimated by substituting Eq. (8) in Eq. (7). Equation (8) is an empirical relation (Brownlie, 1983):

$$Q_{\mathrm{bf}} = \left(\frac{R}{(0.3724 S_{\mathrm{c}}^{-0.2542} \sigma_{\mathrm{s}}^{0.105} D_{50})}\right)^{1.529} W g^{0.5} D_{50}^{1.5}, \tag{8}$$

where σ_{s} is the sorting of the bed material grain size (–) derived from the grain size analysis (Sect. 3.3) and approximated by $0.5(\frac{D_{50}}{D_{16}} + \frac{D_{84}}{D_{50}})$, D_{16} and D_{84} are the 16th and

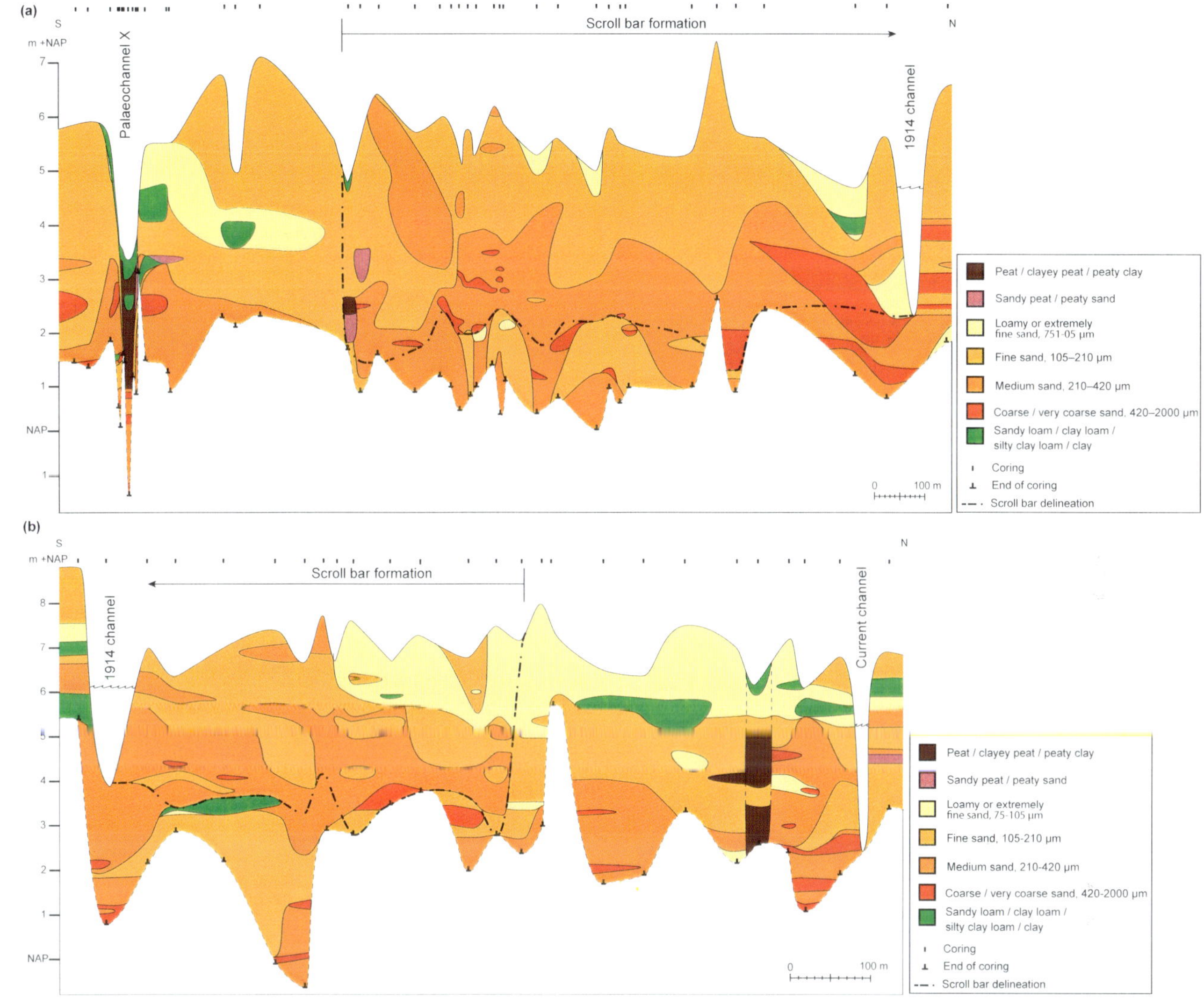

Figure 4.

84th percentile sediment grain size (m), respectively, and g is the gravitational acceleration ($m^2\,s^{-1}$). As a validation, the calculated Chézy coefficient was compared with average Chézy coefficients of 12 comparable low-energy, sand-bed rivers with scroll bars ($S_V < 0.001$, $90 < Q_{bf} < 320\,m^3\,s^{-1}$) calculated from a large river dataset (Kleinhans and Van den Berg, 2011; Van den Berg, 1995). The cross-sectional-averaged flow velocity (u_{bf}, $m\,s^{-1}$) was calculated by following Eq. (9).

$$u_{bf} = \frac{Q_{bf}}{A} \tag{9}$$

3.8 Sediment transport

The sediment transport was calculated to compare with the SB_{vol} (Eq. 6). Sediment transport was calculated in two different ways. The first method was the slightly modified Engelund and Hansen (1967) relation following Eq. (10):

$$Q_{s,bf} = \frac{0.05u^5 W t i}{(\frac{\rho_s}{\rho} - 1)^2 g^{0.5} D_{50} C^3 (1-\varphi)}, \tag{10}$$

where $Q_{s,bf}$ is the yearly sediment transport derived from the bankfull discharge ($m^3\,yr^{-1}$), t is the number of seconds in a year, i is the intermittency assumed to be 0.03 to 0.07 (Parker, 2008), ρ_s is the sediment density ($kg\,m^{-3}$), ρ is the water density ($kg\,m^{-3}$), and φ is the porosity assumed to be 0.3 to 0.35 (Nimmo, 2004). The relation of Engelund and Hansen was used because the relation is suitable for sand-bed rivers with relatively low flow velocities (Van den Berg and Van Gelder, 1993), and the input variables required were available.

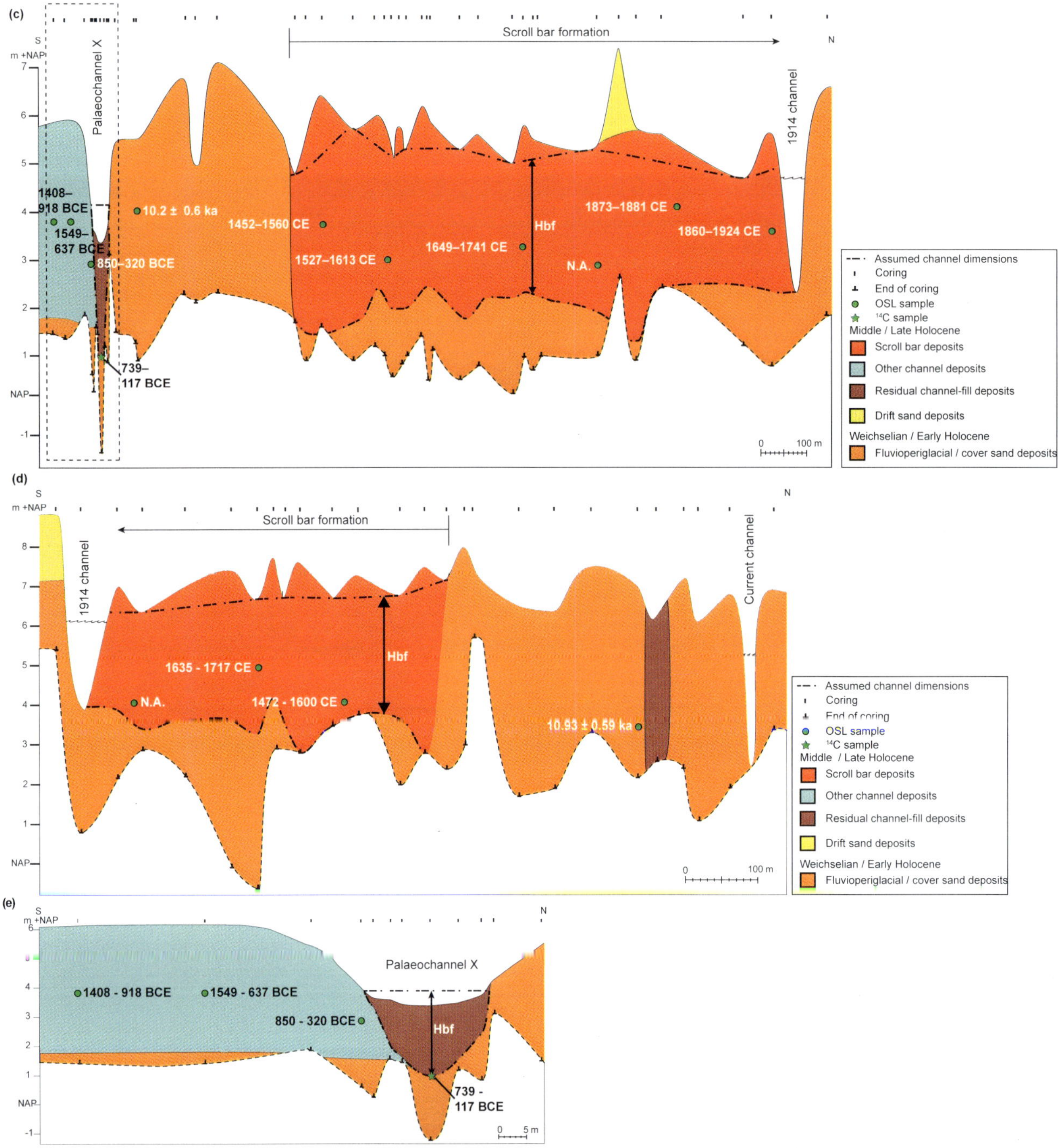

Figure 4. Stratigraphic cross sections of the study sites (for location see Fig. 1). Lithological cross sections of Junnerkoeland **(a)** and Prathoek **(b)**. Lithogenetic cross sections of Junnerkoeland **(c)** and Prathoek **(d)** including the OSL samples by Quik and Wallinga (2018) and OSL and ^{14}C dating results from this study. The surface and erosive base elevation are indicated with dashed lines, resulting in the inferred bankfull channel depth (H_{bf}). **(e)** Zoomed-in lithogenic cross section of Palaeochannel X. The thick dashed line indicates the bankfull level of the palaeochannel.

In the second method the sediment transport was determined for each discharge magnitude and related frequency ($Q_{s,freq}$) (Wolman and Miller, 1960) from present-day flow conditions by assuming that the current discharge frequency distribution also applied to the meandering phase. We used the hourly discharge data from 1995 to 2015 of the gauging station in Mariënberg (Fig. 1b). This gauging station is close to the study location and has the lowest amount of data gaps compared to the other stations. The flow duration was calculated for intervals of $10\,m^3\,s^{-1}$, and for each discharge interval the sediment transport was calculated using Eq. (10), excluding the intermittency factor. If the discharge was above bankfull, the flow would go across the floodplain. The Chézy coefficient for the floodplain was assumed to be half the Chézy coefficient in the channel because of the higher roughness of the floodplain compared to the channel. We assumed that the floodplain width was 350 m for the start of the meandering phase, which was estimated from the DEM (Fig. 1c), and that the width would increase proportionally with the lateral migration rate for each time step during the meandering phase.

3.9 Potential specific stream power

The potential specific stream power was calculated to plot into a stability diagram. Kleinhans and Van den Berg (2011) distinguished four different stability fields, further building on Van den Berg (1995) and Makaske et al. (2009): rivers with laterally stable channels, meandering rivers with scroll bars, meandering rivers with scroll and chute bars as well as moderately braided rivers, and braided rivers. In this research, only the first two stability fields are relevant. These stability fields are separated by a discriminator that represents the theoretical minimum energy needed for the channel pattern to occur (Kleinhans and Van den Berg, 2011). The potential specific stream power was calculated by applying the relationship presented by Kleinhans and Van den Berg (2011) following Eq. (11):

$$\omega_{\mathrm{pv}} = \frac{\rho g \sqrt{Q_{\mathrm{bf}}} S_{\mathrm{v}}}{\varepsilon}, \tag{11}$$

where ω_{pv} is the potential specific stream power ($\mathrm{W\,m^{-2}}$) and $\varepsilon = 4.7\sqrt{\mathrm{s}}\,\mathrm{m}^{-1}$ for sand-bed rivers (Van den Berg, 1995). The discriminator line was plotted by applying the relationships presented by Makaske et al. (2009) and Kleinhans and Van den Berg (2011) following Eq. (12):

$$\omega_{\mathrm{ia}} = 90 D_{50}^{0.42}, \tag{12}$$

where the subscript ia refers to the discrimination between laterally stable and meandering channels with scroll bars.

3.10 Bar regime

Bar regime was predicted by applying the relationships of Struiksma et al. (1985) and Kleinhans and Van den Berg (2011). River bends can be seen as an example of a perturbation to both the flow and bed sediment, which have different adaptation lengths over which they return to equilibrium. This difference in response is expressed by the interaction parameter (IP), which is the ratio between the adaptation length of bed perturbation and the adaptation length of flow. The adaptation length of flow was calculated following Eq. (13),

$$\lambda_w = \frac{C^2 H_{\mathrm{avg}}}{2g}, \tag{13}$$

and the adaptation length of a bed perturbation (m) is calculated following Eq. (14):

$$\lambda_{\mathrm{s}} = \frac{H_{\mathrm{avg}}}{\pi^2}\left(\frac{W}{H_{\mathrm{avg}}}\right)^2 f(\theta), \tag{14}$$

where $f(\theta)$ is the magnitude of the transverse slope effect (–) calculated following Eq. (15) (Talmon et al., 1995).

$$f(\theta) = 9\left(\frac{D_{50}}{H_{\mathrm{avg}}}\right)^{0.3}\sqrt{\theta} \tag{15}$$

Here, θ is the dimensionless shear stress (–) calculated following Eq. (16):

$$\theta = \frac{\tau}{(\rho_{\mathrm{s}} - \rho) g D_{50}}, \tag{16}$$

where τ is the shear stress (Pa) calculated following Eq. (17).

$$\tau = \rho g R S_{\mathrm{c}} \tag{17}$$

The interaction parameter (IP, –) was calculated following Eq. (18) to determine the bar regime for the historical and prehistorical Overijsselse Vecht and for comparison with the theoretical thresholds of bar regime (Crosato and Mosselman, 2009; Struiksma et al., 1985) by

$$\mathrm{IP} = \frac{\lambda_{\mathrm{s}}}{\lambda_{\mathrm{w}}}. \tag{18}$$

The IP is strongly related to the width–depth ratio and was therefore separately calculated for the meander bends Junnerkoeland and Prathoek. A low IP means that when a bar forms in response to a local perturbation, such as local curvature, the bar disappears within a short distance of the perturbation (Struiksma et al., 1985). This is called an overdamped regime and occurs in channels with a low width–depth ratio. The threshold between overdamped and underdamped can be calculated following Eq. (19):

$$\mathrm{IP} \le \frac{2}{n + 1 + 2\sqrt{2n - 2}}, \tag{19}$$

where n is the degree of non-linearity of sediment transport versus depth-averaged flow velocity (–). Following Crosato and Mosselman (2009) we chose $n = 4$, which corresponds to values for a sand-bed river. A higher IP, and hence a higher width–depth ratio, results in an underdamped regime associated with bars that also form further downstream of the perturbation. The thresholds can be calculated following Eq. (20).

$$\frac{2}{n + 1 + 2\sqrt{2n - 2}} < \mathrm{IP} < \frac{2}{n - 3} \tag{20}$$

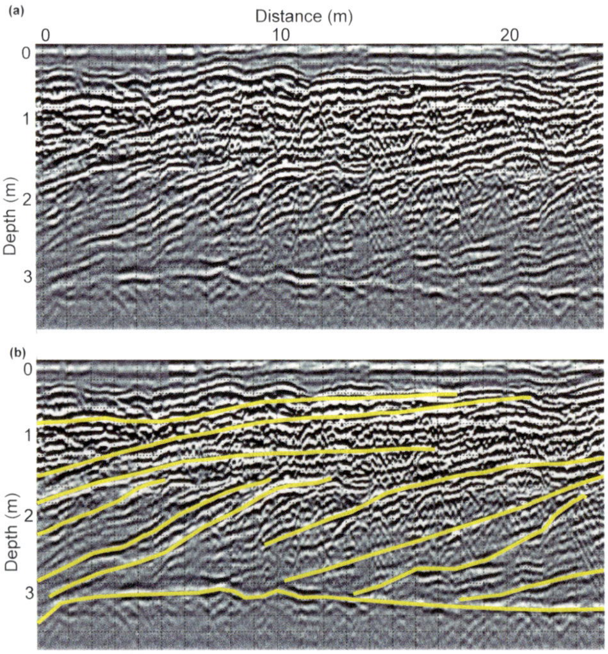

Figure 5. Example of a ground-penetrating radar (GPR) profile (250 Hz) in the Prathoek bend. **(a)** Original GPR profile and **(b)** interpreted GPR profile with lateral accretion surfaces and the channel lag, indicated by yellow lines.

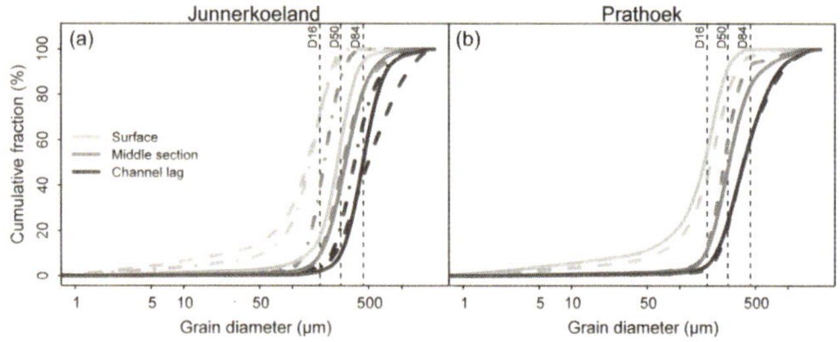

Figure 6. Cumulative grain size distributions of the scroll bar deposits in **(a)** Junnerkoeland and **(b)** Prathoek. Three series were made for Junnerkoeland and two for Prathoek, each indicated by a different line type. Each sample within a series is indicated by a different grey tone. The averages of D_{16}, D_{50}, and D_{84} are plotted. Figure 1c and d indicate the locations of the grain size samples.

3.11 Errors and uncertainty

The above described calculations (Eqs. 1 to 11 and 13 to 20) were run 10 000 times to take into account the random errors of the input parameters, following a stochastic approach by using Monte Carlo simulations. The uncertainty of these parameters was described above, relating to the transverse bed slope, bankfull depth of the meanders, valley slope, porosity, grain size, intermittency and the measured channel dimensions of Palaeochannels X and Q. Systematic errors were not taken into account because the palaeohydrological reconstruction was meant to distinguish relative differences between fluvial phases, rather than reconstructing absolute hydrological parameters. Results are plotted with average values from the Monte Carlo simulations when normally distributed, or median values when not-normally distributed, including the 16th and 84th quantile representing the uncertainty margin. All formulas and example data used are made available in the Supplement.

4 Results

4.1 Lithogenetic units

Several lithogenic units were distinguished (Fig. 4), following similar interpretations of the sedimentary units as Huisink (2000). The descriptions of the lithogenic units are summarized in Table 1. The cover sand deposits were sometimes difficult to distinguish in borehole descriptions from the fluvioperiglacial deposits when the latter had a relatively fine grain size. Because our interest is in the delineation of the scroll bar and residual channel-fill deposits, we combined both the fluvioperiglacial and cover sand deposits into one unit. The fining upward sequence within the scroll bar deposits (Table 1) can be recognized in the grain size analysis done for the scroll bar deposits at Junnerkoeland and Prathoek (Fig. 6). The depth-averaged grain size for both scroll bar complexes is 0.28 ± 0.05 mm. Commonly, at the base of the scroll bar deposits, a sharp transition occurs to the brightly coloured substratum of fluvioperiglacial deposits below, which lack organic material (Table 1). Cores that did not reach the fluvioperiglacial deposits below the scroll bar deposits indirectly indicate the boundary between these units because strongly consolidated layers are present in the fluvioperiglacial deposits that were difficult to core into. An example of a consolidated clay layer can be found directly below the southern part of the scroll bar deposits at Prathoek (Fig. 4b).

The GPR profiles clearly show the lateral accretion surfaces of the scroll bar deposits (see example in Fig. 5). The GPR results were poor only where the scroll bar deposits are relatively loamy or clayey on top (i.e. northern parts of Prathoek and Junnerkoeland). The bottom of the scroll bar deposits is mostly unrecognizable because of a low GPR reflection at this depth. In Fig. 5 the bottom of the scroll bars is visible because this part is located in the southern part of Prathoek where the above-mentioned clay layer was present (Fig. 4), which caused a strong reflection of the GPR signal. The well-preserved Palaeochannel X is a relatively symmetrical palaeochannel (Fig. 4e) similar to Palaeochannel Q of Huisink (2000) (Fig. 2). The outer bank consists of Weichselian–Early Holocene deposits (Fig. 4c). The average grain size of the Palaeochannel X bed sediments is 0.23 ± 0.12 mm. No lateral accretion surfaces can be observed in the GPR profile that was placed along the centreline of the Palaeochannel X bend (Fig. 1e and Supplement).

Table 1. Description of lithogenic units.

	Fluvioperiglacial deposits	Cover sand deposits	Other channel deposits	Residual channel-fill deposits	Scroll bar deposits	Drift sand deposits
Lithology	Mod. sort. 75–2000 µm Lenses of loam and loamy sand	Well sort. 75–21 µm Loamy sand	Mod. sort. 105–600 µm	Sandy peat or peaty sand Lenses of sand, silty clay loam or clay loam	Mod. sort. 75–600 µm Loamy sand near surface	Well sort. 75–210 µm
Colour	Light grey to brown	Light grey–brown or white	Light grey–brown	Dark brown or black	Light brown to dark grey	Greyish brown
Gravel (%)	0–20	< 1	< 1	< 1	< 40	< 1
Plant remains	Mostly absent	None	Sporadically near bottom	Abundant	Fragmented and abundant near bottom	Rare
Thickness (m)	> 2	< 2	4–5	4–5	4–5	1–5
Width (m)	> 1000	> 1000	< 100	20–40	> 100	10–100
Beds	cm to dm thick	None	None	None	cm to dm thick	None
Additional		Palaeo-podzol in top	Slightly coarser near bottom	May be poorly preserved	Fining upward, lateral accretion surfaces (GPR)	Micro-podzol in top

4.2 Dating results

The channel deposits on the inside of Palaeochannel X date from 850–320 and 1408–918 BCE. Palaeochannel X was cut off at 739–117 BCE (Figs. 1e, 4c, e and Table 2).

4.3 Meander and channel geometry

The reconstructed transverse bed slopes do not show a spatial trend (Fig. 7a, b), and hence the mean and standard deviations were used in the palaeohydrological calculations. The transverse bed slope at Prathoek is higher ($4.5 \pm 1.0°$) than at Junnerkoeland ($3.3 \pm 1.3°$), but much lower than the transverse bed slope of Palaeochannel X ($16.9 \pm 1.9°$) and of Palaeochannel Q ($28.8 \pm 3.8°$). The age as a function of distance of lateral accretion follows from Fig. 7c, d. This relation was used for the meander and channel geometry calculations (Fig. 8). The bankfull depths of Palaeochannels X and Q are comparable to the bankfull depths of the meanders Prathoek and Junnerkoeland ca. 1500 CE (Fig. 8a, b) (3 to 4 m). The bankfull depths at Junnerkoeland decreased relatively fast ca. 1800 CE because the erosive base elevation rises towards the cut-off channel (Fig. 4c). At Prathoek, the bankfull depth decreased more gradually over time. The reconstructed bankfull width of Palaeochannels X and Q is much lower compared to the meandering phase (Fig. 8c, d), resulting in a relatively small cross-sectional area of Palaeochannels X and Q (Fig. 8e, f).

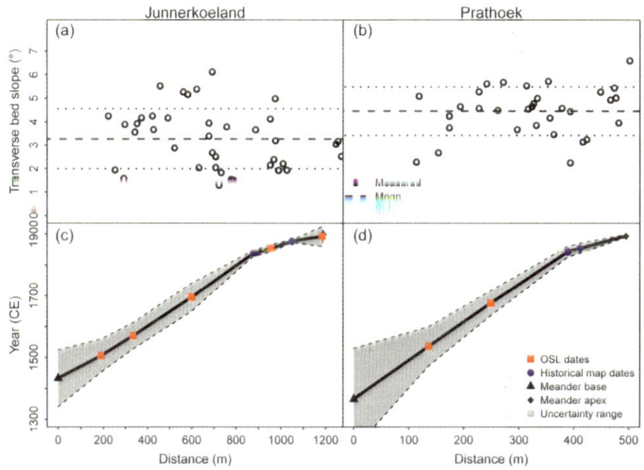

Figure 7. Transverse bed slope derived from GPR cross sections from the inner point bar to the outer bend for Junnerkoeland **(a, c)** and Prathoek **(b, d)** as well as lateral migration distance plotted against age for both bends. Panels **(a)** and **(b)** show the transverse bed slope of lateral accretion surfaces measured in the GPR profile (example in Fig. 4), including the mean and standard deviation of all measurements. Panels **(c)** and **(d)** show the relation between age and migration distance of the bends. Shading indicates standard deviation of the Bayesian deposition model determined by Quik and Wallinga (2018) for the OSL and historical map dates.

4.4 Palaeohydrology

The reconstructed Q_{bf} is 3 to 9 times higher at the start of the meandering phase ($85–194\ m^3\ s^{-1}$) compared to the preceding phase represented by Palaeochannels X and Q ($19–32\ m^3\ s^{-1}$) (Fig. 9). The difference in Q_{bf} between 400 BCE and 1500 CE is significant despite the relatively large uncertainty. A similar discharge in 400 BCE compared to 1500 CE would require a cross-sectional area 5 times larger than currently estimated (Fig. 8e) or a 50 times higher valley slope, which falls outside the uncertainty ranges of these parameters. The Q_{bf} eventually declines over time and drops to $32–70\ m^3\ s^{-1}$ ca. 1850 CE. The calculated Chézy coefficients for the meandering phase ($47.5 \pm 0.9\ m^{0.5}\ s^{-1}$; Eqs. 7 and 8) were comparable to average Chézy coefficients derived from 12 low-energy rivers ($44.8 \pm 13\ m^{0.5}\ s^{-1}$) from the river dataset by Kleinhans and Van den Berg (2011).

Table 2. OSL and ^{14}C dating results from Palaeochannel X. Locations are indicated in Fig. 1c, d and Fig. 4c.

Sample code	Material	Elevation (m+NAP)	^{14}C age (a BP)	Palaeo-dose (Gy)	Dose rate (Gy / ka)	Age (ka)	Age (BCE)	Lat, Long (RD)
NCL2416194	Fluvial sand	1.10		2.1 ± 0.2	0.81 ± 0.03	2.6 ± 0.3	850–320	229242, 505286
NCL2217157	Fluvial sand	3.65		3.3 ± 0.2	1.06 ± 0.05	3.1 ± 0.5	1549–637	229249, 505254
NCL2217158	Aeolian sand	3.99		12.5 ± 0.5	1.23 ± 0.05	10.2 ± 0.6	8761–7641	229254, 505338
NCL2217159	Fluvial sand	3.55		3.6 ± 0.2	1.14 ± 0.05	3.2 ± 0.2	1408–918	229242, 505228
GrA69519	Selected macrofossils	1.14	2300 ± 100			2.4 ± 0.3	739–117	229239, 505298

Combining the frequency of each discharge interval with the sediment transport rate (Fig. 10a) results in a histogram of the sediment transport contribution as a function of discharge ($Q_{s,freq}$, Fig. 10b). The highest measured discharge at the gauging station Mariënberg between 1995 and 2015 is 185.5 $m^3\,s^{-1}$. The most frequent discharge occurring in the channelized Overijsselse Vecht is 0 to 10 $m^3\,s^{-1}$, with a frequency of 8.2 % (Fig. 10a). When discharge is still below bankfull, sediment transport increases relatively fast with an increasing discharge. Above bankfull, additional discharge largely flows across the more flow-resistant floodplain, and hence the sediment transport rates increase less. The effective discharge (Q_{eff}) is 29 $m^3\,s^{-1}$, represented by the highest sediment transport contribution (Fig. 10a, b).

Calculated sediment transport rates were higher than the inner bank growth or scroll bar growth, suggesting the channel deposition can be explained entirely by the reconstructed sediment transport (Fig. 10c). The $Q_{s,bf}$ of the laterally stable phase was much lower than for the meandering channels, explaining the large difference between the growth rate of the channel deposits on the inner bank at Palaeochannel X (7.0 $m^3\,yr^{-1}$) and the scroll bars of Junnerkoeland and Prathoek at the start of the meandering phase ($1.8 \times 10^3\,m^3\,yr^{-1}$). Both the sediment transport and average scroll bar growth decreased during the meandering phase.

Figure 11a shows that the river theoretically had insufficient stream power for meandering ca. 400 BCE, and the bar regime was overdamped (Fig. 11b). The stream power seemed sufficient for meandering ca. 1500 CE, and the bar regime was underdamped. The potential for meandering gradually decreased during the meandering phase and became again insufficient when the potential specific stream power drops relatively fast ca. 1850 CE. The damping regime also gradually decreased, but remained underdamped ca. 1850 CE.

5 Discussion

5.1 Laterally stable phase

A relatively laterally stable phase existed prior to the meandering phase, which is corroborated by the geochronological and palaeohydrological reconstruction. Palaeochannel X formed by extremely slow channel displacement of ca. 6 $cm\,yr^{-1}$, assuming a constant channel displacement rate, shown by the OSL dates taken from the channel deposits on the inside of Palaeochannel X (Figs. 1e, 4c and Table 2). The lateral migration rate of the Junnerkoeland meander bend was ca. 40 times higher (Quik and Wallinga, 2018; Wolfert and Maas, 2007). The outer bank of Palaeochannel X consists of Weichselian and Early Holocene deposits. No Middle Holocene deposits were found in the corings (Fig. 4c, d), reflecting the stable character of the Overijsselse Vecht during this period.

The preservation potential of deposits associated with the laterally stable phase is likely very small. Deposits and dimensions of active channel reaches are not preserved during the stable to meandering transition because channel belt dimensions increase. Hence, channel reaches are only preserved when they were cut off prior to the stable–meandering transition, e.g. due to local perturbations. A channel cut-off probably caused Palaeochannels X and Q of the laterally stable phase to become disconnected from the main river before the meandering phase started. In this way these reaches escaped from later lateral erosion during the meandering phase. Consequently, the lateral stability of the river is not immediately evident from these preserved channel reaches because the perturbations led to very slow channel displacement, as was found for Palaeochannel X. However, scroll bar deposits did not form and lateral accretion surfaces were lacking (Figs. 2, 4e and the Supplement), showing that the displacement was not related to meandering in which helicoidal flows cause bar formation and bank erosion at a significant rate and all along the channel (Seminara, 2006). The laterally stable phase lacked the potential to meander given its low position in Fig. 11a and is characterized by an overdamped regime (Fig. 11b, c) and low sediment transport (Fig. 10c). Consequently, the formation of bars was suppressed and the inner bank deposition was small (Fig. 10c).

The bend curvature is also an indication for the channel stability. Palaeochannel X comprises a very sharp bend ($\frac{R_{curv}}{W} = 1.4 \pm 0.2$) compared to the meandering phase ($\frac{R_{curv}}{W} = 2.1 \pm 0.4$), which is often found in low-energy rivers in which lateral migration is limited (Candel et al., 2018; Hickin and Nanson, 1984). Large similarities exist between the laterally stable phase reported here and the laterally stable channels in highly cohesive sediment on the intertidal mud-

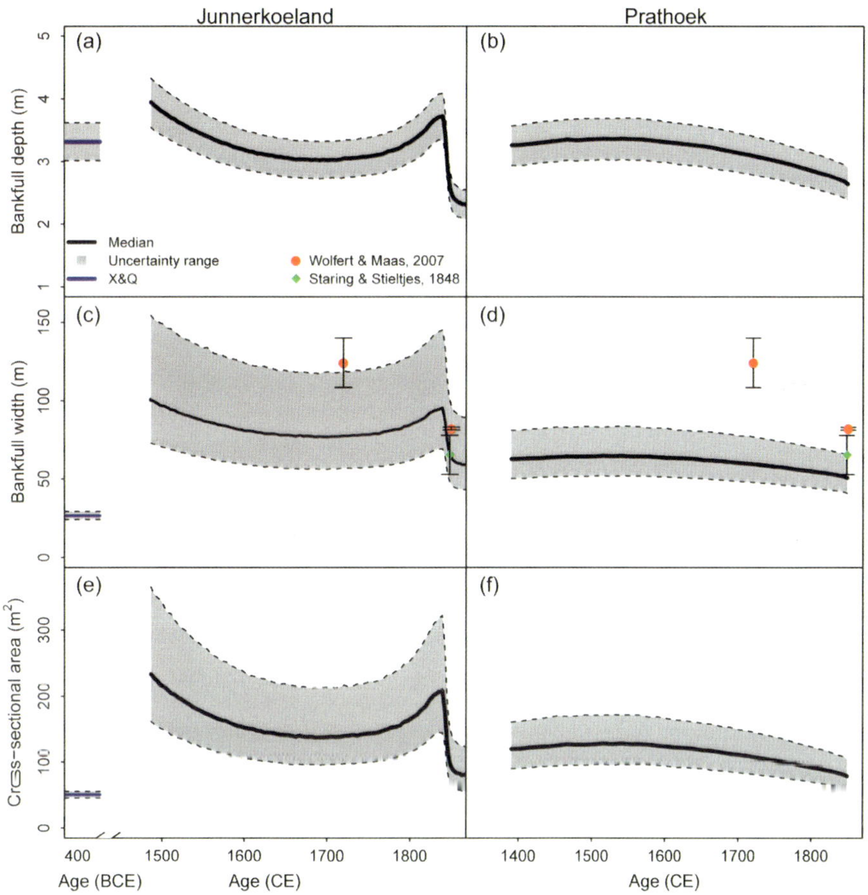

Figure 8. Reconstructed meander and channel geometry over time, assuming the date–distance relations (see Fig. 7c, d) over the scroll bars. Panels **(a)** and **(b)** show the bankfull depth (H_{bf}) derived from the coring data taken from the bottom of the channel lag to the inferred bankfull water surface (Fig. 4c, d), for both the Junnerkoeland **(a, c, e)** and Prathoek **(b, d, f)**. Panels **(c)** and **(d)** show the bankfull width derived from the bankfull depth and reconstructed transverse bed slope (Eq. 1). The river width data from Wolfert and Maas (2007) observed on historical maps and the bankfull river width data from Staring and Stieltjes (1848) were included for comparison. Panels **(e)** and **(f)** show the cross-sectional area derived from the bankfull width and water depth (Eq. 2). Shading indicates the 16th and 84th quantile. *X*&*Q* refers to Palaeochannels *X* and *Q*.

flat, which are mostly laterally stable except for some sharp bends at which bank failure and flow separation result in very limited and local channel migration (Kleinhans et al., 2009).

5.2 Channel pattern change

The Overijsselse Vecht river changed from a laterally stable into a meandering river. Differences in palaeohydrological conditions between the two phases were large enough to distinguish despite the large uncertainties in the palaeohydrological reconstruction. Bar regime changed from an overdamped regime into an underdamped regime (Fig. 11b, c), leading to overdeepening of the outer bend pool and enhancement of the point bars in the inner bend (Struiksma et al., 1985; Crosato and Mosselman, 2009; Kleinhans and Van den Berg, 2011). The significantly higher bankfull discharge (factor of 3 to 9; Fig. 9) explains the potential to meander (Fig. 11a), the high sediment transport, and the high scroll bar growth (Fig. 10c) at the start of the meandering phase.

The exact moment of the channel pattern change is between the cut-off of Palaeochannel *X* (400 ± 300 BCE) and the reconstructed initiation of scroll bar formation (1504 ± 52 CE). Most likely, the transition occurred shortly before the latter because both point bars had a relatively similar meander start age (Figs. 2 and 4), the surrounding floodplain is formed by Late Glacial or Early Holocene deposits (Fig. 4), and there is no evidence of older scroll bar deposits in the vicinity of the studied meander bends. Mature meandering river systems would always leave traces of older scroll bar deposits, channel cut-offs, or meander scars because these are never completely being removed by the river (Toonen et al., 2012; Van de Lageweg et al., 2016).

The palaeohydrological reconstruction shows that the increasing bankfull discharge likely explains the channel pat-

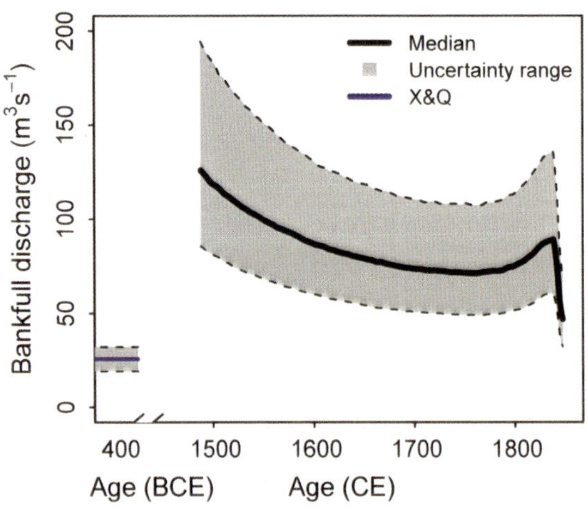

Figure 9. Bankfull discharge over time, combined for Junnerkoeland and Prathoek. Shading indicates the 16th and 84th quantile. *X&Q* refers to Palaeochannels *X* and *Q*.

tern change. The increasing bankfull discharge may reflect an increase in annual discharge, but could also be related to a more irregular discharge regime because the bankfull discharge largely represents the higher discharges in a river (Dury, 1973; Wolman and Miller, 1960). Consequently, the discharge may have been constant over a year with low peak discharges and a relatively high base flow during the laterally stable phase, changing into a more peaked discharge regime with a relatively low base flow at the start of the meandering phase.

A potential cause of the discharge regime and channel pattern change may be the climate change at the start of the Little Ice Age (14th to 19th century) (Grove, 1988) given the overlap in time with the meandering phase (Fig. 2c, d). Although geomorphological responses differ for each river during the Little Ice Age, enhanced lateral migration or incision was generally observed for most rivers in north-western Europe (Rumsby and Macklin, 1996). The increased bankfull discharge in the Overijsselse Vecht may have been caused by higher run-off relative to precipitation due to reduced evapotranspiration rates and frozen soils (Rumsby and Macklin, 1996; Van Engelen et al., 2001) and/or a higher snowfall to rainfall ratio due to lower winter temperatures in the Netherlands and Germany (Behringer, 1999; Lenke, 1968). Higher snowfall rates were also recorded for the United Kingdom (Manley, 1969), where they led to more flooding during the snowmelt period (Archer, 1992). Studies on historical observations of rivers nearby the Overijsselse Vecht (IJssel, Elbe, Lower Rhine, and Meuse) suggested a significantly higher flooding rate during the Little Ice Age compared to more recent flooding rates (Glaser et al., 2010; Glaser and Stangl, 2003; Mudelsee et al., 2003, 2004).

An additional cause for an increasing bankfull discharge may have been land use change in the catchment (Kondolf et al., 2002), which affects the discharge regime due to the direct relation with evapotranspiration (Fohrer et al., 2001). For the Overijsselse Vecht catchment, peat reclamation started in the 12th and 13th century (Gerding, 1995; Van Beek et al., 2015b) and intensified from the 14th century onwards (Borger, 1992; Van Beek et al., 2015a). Reclamation of peatlands partly comprised digging canals to drain the land, and although the reclamation was mainly limited to the margins of peatlands, the hydrological consequences were large. The margins are a natural seal of the peat bog, with a low hydraulic conductivity compared to the remainder of the bog, ensuring peat dome growth. Destruction of these margins will result in drainage of the entire peat bog (Baird et al., 2008; Van der Schaaf, 1999). After several centuries, focus shifted from peat reclamation to exploitation by excavating large peatland areas for fuel during the 17th and 18th century (Gerding, 1995). The largest part of the peat has currently disappeared. Yearly average discharges in peatlands can increase by 40 % in the Dutch climatological setting due to evapotranspiration differences for reclaimed peat areas compared to undisturbed peat areas (Baden and Eggelsmann, 1964; Streefkerk and Casparie, 1987; Uhden, 1967). This increase cannot fully explain the large increase in bankfull discharge in the Overijsselse Vecht (factor of 3 to 9) because peat covered just ca. 27 % of the Overijsselse Vecht catchment area during the 14th century (Casparie and Streefkerk, 1992; Vos et al., 2011), and hence the yearly average discharge of the catchment increased by ca. 11 % due to evapotranspiration differences.

However, several studies have also shown that an increased drainage network in peatlands resulted in higher discharge peaks with a fast discharge response to precipitation (Conway and Millar, 1960; Holden et al., 2004, 2006; Streefkerk and Casparie, 1987). For example, the run-off to rainfall ratio was a factor of 3 higher in a drained Irish peatland compared to an undrained Irish peatland (Burke, 1975), which is comparable to the observed bankfull discharge increase in the Overijsselse Vecht. Finally, canals were not only dug for peat reclamation, but also for shipping and effective generation of water power starting in the 11th and 12th century (Driessen et al., 2000), which may have promoted the higher peak flows even more. New canals resulted in a faster run-off, but also changed the watershed delineation (Driessen et al., 2000). We conclude that both climatic and land use changes were likely responsible for an increase in both total discharge and peak flows, resulting in the transition of a relatively stable river to a highly dynamic meandering system.

5.3 Meandering phase

Interestingly, the bankfull discharge declined during the meandering phase (Fig. 9), leading to decreasing sediment transport relative to the scroll bar growth (Fig. 10c) and insufficient potential specific stream power for meandering after ca. 1850 CE (Fig. 11a). This decline was corroborated by observations of river width from previous studies, which can be

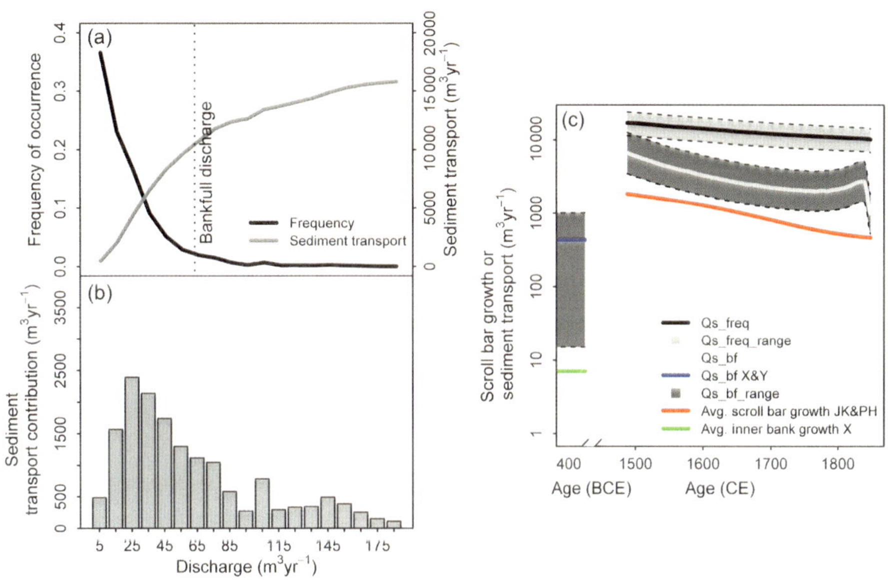

Figure 10. Sediment transport budgets calculated from present-day flow conditions and from meander migration. **(a)** Discharge and sediment transport characteristics of the Overijsselse Vecht derived from hourly discharge data from 1995 to 2015 of the gauging station Mariënberg, including the frequency of each discharge class over a year on a frequency scale from 0 to 1 and the sediment transport as a function of discharge for a randomly selected year (1546 CE) in the Junnerkoeland meander bend. **(b)** Histogram of the sediment transport contribution as a function of discharge. **(c)** The sediment transport and average scroll bar growth over time (JK: Junnerkoeland, PH: Prathoek, X & Q: Palaeochannels X and Q). The abbreviations $Q_{s,freq}$ and $Q_{s,bf}$ are explained in Sect. 3.8. The inner bank growth X refers to the growth rate of the channel deposits on the inner bank at Palaeochannel X, assuming a constant lateral migration rate. Shading indicates the 16th and 84th quantile.

compared to the reconstructed widths (Fig. 8c, d). These observations included measurements from historical maps by Wolfert and Maas (2007) and measurements of the bankfull river width over a large river section in 1848 CE by Staring and Stieltjes (1848). The river width data from Wolfert and Maas (2007) largely fall in the range of reconstructed bankfull widths at Junnerkoeland and show a similar decreasing trend (Fig. 8c). However, the historical maps used by them may result in large uncertainties because the water stage that these maps represent is unknown (Quik and Wallinga, 2018). The measured widths by Staring and Stieltjes (1848) are in line with the predicted width at Junnerkoeland, falling within the uncertainty range. The predicted width at Prathoek is underestimated compared to the measured widths by Wolfert and Maas (2007) and Staring and Stieltjes (1848). This underestimation may explain the lower cross-sectional area compared to Junnerkoeland (Fig. 8f) and hence an underestimated bankfull discharge (Fig. 9) and potential specific stream power (Fig. 11a).

The observed decline of bankfull discharge would suggest that the hydrological forcing disappeared or diminished and had a temporary character, which would fit the Little Ice Age that ended in the 19th century as a potential cause. Consequently, it would be expected that the channel pattern reorganized and became laterally stable again. However, the river was still laterally migrating until channelization between 1896 and 1914 CE (Wolfert and Maas, 2007), which may be related to the presence of an underdamped regime enhancing point bar formation in the inner bend (Fig. 11b, c). Additionally, historical bank stability changes may have promoted the river meandering during this period. For example, floodplains were intensively used for cattle grazing, which may have weakened the banks, enhancing meandering after 1850 CE (Beschta and Ripple, 2012; Trimble and Mendel, 1995; Wolfert et al., 1996). Drift sand activity was also initiated by intensive land use since the Late Middle Ages (Fig. 1c, d) (Koster et al., 1993), which may have affected the bank stability.

6 Conclusions

We show that bankfull discharge and the associated river parameters can be reconstructed by following a stochastic approach and through detailed geochronological and lithological analysis of scroll bar deposits and palaeochannels. For the Overijsselse Vecht river we demonstrate that an increase in bankfull discharge ca. 1400 to 1500 CE resulted in a river channel pattern change from laterally stable to meandering. Geochronological data confirmed our hypothesis on the lateral stability of the river prior to the meandering phase, in contrast to previous assumptions that were made of continuous meandering during the Holocene. We show that the reconstructed river parameters are consistent with

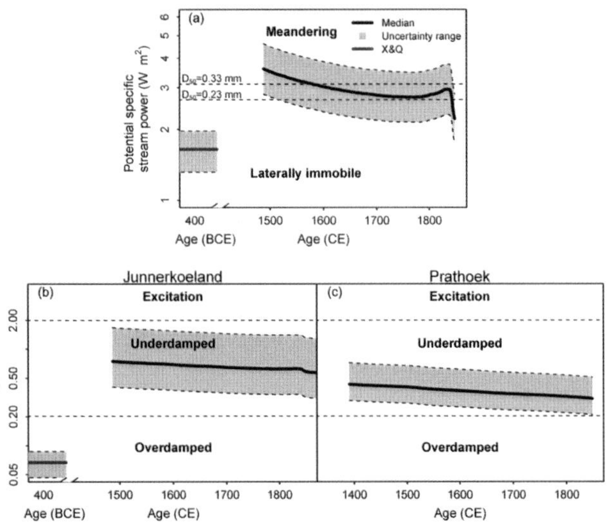

Figure 11. The potential for meandering with time. **(a)** The potential specific stream power in a stability diagram (Eq. 11). Two discriminators were plotted for a range of median particle sizes of the bed sediment, which is the range of particle sizes found in the scroll bars and Palaeochannels *X* and *Q* (Fig. 6). Panels **(b)** and **(c)** show the bar regime for both Junnerkoeland and Prathoek, determined with the interaction parameter (IP) (Eq. 18) and compared to the thresholds (Eqs. 19 and 20). Shading indicates the 16th and 84th quantile. *X*&*Q* refers to Palaeochannels *X* and *Q*.

both the laterally stable and meandering channel pattern by applying empirical channel and bar pattern models. Potential causes for the discharge regime changes include climate change (Little Ice Age) and land use changes (peat reclamation, peat exploitation, digging of canals). We conjecture that the change from laterally stable to meandering has occurred in other rivers for which increased Holocene fluvial activity was reported.

Author contributions. The authors contributed in the following proportions to concept and design, data collection, analysis, and conclusions and paper preparation: JHJC (40 %, 40 %, 80 %, 60 %), MGK (20 %, 0 %, 10 %, 5 %), BM (20 %, 0 %, 0 %, 20 %), WZH (0 %, 40 %, 5 %, 0 %), CQ (0 %, 20 %, 5 %, 0 %), JW (20 %, 0 %, 0 %, 15 %).

Competing interests. The authors declare that they have no conflict of interest.

Acknowledgements. This research is part of the research programme RiverCare, supported by the Netherlands Organization for Scientific Research (NWO) and the Dutch Foundation of Applied Water Research (STOWA), and is partly funded by the Ministry of Economic Affairs under grant number P12-14 (perspective programme). Maarten G. Kleinhans was also supported by the NWO (grant Vici 016.140.316/13710). The paper has benefited greatly from reviews by Peter Houben and two anonymous reviewers. The authors would like to thank the following persons for their help with the different methods used in this research: Joep Storms, Gerard Heuvelink, Marijn van der Meij, Wobbe Schuurmans, Alice Versendaal, Erna Voskuilen, Aldo Bergsma, Marjolein Gouw-Bouman, and UU students Karianne van der Werf, Sjoukje de Lange, Jip Zinsmeister, and Pascal Born. We would also like to thank the De Roos family, Staatsbosbeheer, and Water Board Vechtstromen for access to and inside knowledge of the field sites.

References

Agrawal, Y., McCave, I., and Riley, J.: Laser diffraction size analysis, Principles, methods, and application of particle size analysis, Cambridge University Press, New York, 119–128, 1991.

Allen, J. R. L.: The sedimentation and palaeogeography of the Old Red Sandstone of Anglesey, north Wales, P. Yorks. Geol. Soc., 35, 139–185, 1965.

Archer, D.: Land of Singing Waters-rivers and great floods of Northumbria, Spredden Press, 1992.

Baden, W. and Eggelsmann, R.: Der Wasserkreislauf eines nordwestdeutschen Hochmoores, The water balance of a high moor in northwest Germany, Schriftenreihe des Kuratoriums fur Kulturbauwesen, 12, 156 pp., 1964.

Baird, A. J., Eades, P. A., and Surridge, B. W.: The hydraulic structure of a raised bog and its implications for ecohydrological modelling of bog development, Ecohydrology, 1, 289–298, 2008.

Behringer, W.: Climatic change and witch hunting: the impact of the Little Ice Age on mentalities, Climate Change, 43, 335–351, 1999.

Berendsen, H. J. and Stouthamer, E.: Palaeogeographic development of the Rhine-Meuse delta, the Netherlands, Koninklijke van Gorcum, 2001.

Beschta, R. L. and Ripple, W. J.: The role of large predators in maintaining riparian plant communities and river morphology, Geomorphology, 157, 88–98, 2012.

Borger, G. J.: Draining – digging – dredging; the creation of a new landscape in the peat areas of the low countries, in: Fens and bogs in the Netherlands, edited by: Verhoeven, J. T. A., Springer, Geobotany, 1992.

Bos, I. J., Busschers, F. S., and Hoek, W. Z.: Organic-facies determination: a key for understanding facies distribution in the basal peat layer of the Holocene Rhine-Meuse delta, The Netherlands, Sedimentology, 59, 676–703, 2012.

Brewer, P. and Lewin, J.: Planform cyclicity in an unstable reach: complex fluvial response to environmental change, Earth Surf. Proc. Land., 23, 989–1008, 1998.

Bronk Ramsey, C.: OxCal Program v4. 2, available at: http://www.rlaha.ox.ac.uk/orau/oxcal.html (last access: 13 March 2017), 2009.

Brownlie, W. R.: Flow depth in sand-bed channels, J. Hydraul. Eng., 109, 959–990, 1983.

Burke, W.: Effect of drainage on the hydrology of blanket bog, Irish J. Agr. Res., 14, 145–162, 1975.

Candel, J. H. J., Makaske, B., Storms, J. E. A., and Wallinga, J.: Oblique aggradation: a novel explanation for sinuosity of low-energy streams in peat-filled valley systems, Earth Surf. Proc. Land., 42, 2679–2696, 2017.

Candel, J. H. J., Makaske, B., Kijm, N., Storms, J. E. A., and Wallinga, J.: Decreasing lateral migration and increasing planform complexity of the Dommel River during the Holocene, NCR Days 2018 Proceedings, 42-2018, 32–33, 2018.

Casparie, W. A. and Streefkerk, J.: Climatological, stratigraphic and palaeo-ecological aspects of mire development, in: Fens and Bogs in the Netherlands, Springer, Kluwer Academic Publishers, Dordrecht, 1992.

Conway, V. and Millar, A.: The hydrology of some small peat-covered catchments in the northern Pennines, J. Inst. Wat. Eng., 14, 415–424, 1960.

Crosato, A. and Mosselman, E.: Simple physics-based predictor for the number of river bars and the transition between meandering and braiding, Water Resour. Res., 45, W03424, https://doi.org/10.1029/2008WR007242, 2009.

Cunningham, A. C. and Wallinga, J.: Realizing the potential of fluvial archives using robust OSL chronologies, Quat. Geochronol., 12, 98–106, 2012.

De Bakker, H. and Schelling, J.: Systeem van bodemclassificatie voor Nederland, De hogere niveaus, Grondboor & Hamer, 20, 229–229, 1966.

De Moor, J., Kasse, C., Van Balen, R., Vandenberghe, J., and Wallinga, J.: Human and climate impact on catchment development during the Holocene–Geul River, the Netherlands, Geomorphology, 98, 316–339, 2008.

Driessen, A. M. A. J., Van de Ven, G. P., and Wasser, H. J.: Gij beken eeuwig vloeijend, Water in de streek van Rijn en IJssel, Matrijs, Utrecht, 2000.

Dury, G.: Magnitude-frequency analysis and channel morphology, Fluvial Geomorphology: New York, State University of New York, 91–121, 1973.

Eekhout, J., Fraaije, R., and Hoitink, A.: Morphodynamic regime change in a reconstructed lowland stream, Earth Surf. Dynam., 2, 279–293, 2014.

Engelund, F. and Hansen, E.: A monograph on sediment transport in alluvial streams, Tekniskforlag Skelbrekgade 4 Copenhagen V, Denmark, 1967.

Ferguson, R.: Hydraulic and sedimentary controls of channel pattern, River channels: environment and process, in: River channels: environment and process, edited by: Richards, K. S., Blackwell, Oxford, UK, 129–158, 1987.

Fohrer, N., Haverkamp, S., Eckhardt, K., and Frede, H.-G.: Hydrologic response to land use changes on the catchment scale, Phys. Chem. Earth. Pt. B, 26, 577–582, 2001.

Friedkin, J. F.: Laboratory study of the meandering of alluvial rivers, Vicksburg, Mississippi, 1945.

Gerding, M.: Vier eeuwen turfwinning, De verveningen in Groningen, Friesland, Drenthe en Overijssel tussen, 1550, 1995.

Gibling, M. R. and Davies, N. S.: Palaeozoic landscapes shaped by plant evolution, Nat. Geosci., 5, 99–105, 2012.

Glaser, R. and Stangl, H.: Historical floods in the Dutch Rhine Delta, Nat. Hazards Earth Syst. Sci., 3, 605–613, https://doi.org/10.5194/nhess-3-605-2003, 2003.

Glaser, R., Riemann, D., Schönbein, J., Barriendos, M., Brázdil, R., Bertolin, C., Camuffo, D., Deutsch, M., Dobrovolný, P., and van Engelen, A.: The variability of European floods since AD 1500, Climatic Change, 101, 235–256, 2010.

Grove, J.: The Little Ice Age, 498 pp., Methuen, London, 1988.

Gurnell, A.: Plants as river system engineers, Earth Surf. Proc. Land., 39, 4–25, 2014.

Hesselink, A. W., Weerts, H. J., and Berendsen, H. J.: Alluvial architecture of the human-influenced river Rhine, The Netherlands, Sediment. Geol., 161, 229–248, 2003.

Hickin, E. J. and Nanson, G. C.: Lateral migration rates of river bends, J. Hydraul. Eng., 110, 1557–1567, 1984.

Hobo, N.: The sedimentary dynamics in natural and human-influenced delta channel belts, Utrecht Stud Earth Sci., 97–200, 2015.

Hobo, N., Makaske, B., Wallinga, J., and Middelkoop, H.: Reconstruction of eroded and deposited sediment volumes of the embanked River Waal, the Netherlands, for the period ad 1631–present, Earth Surf. Proc. Land., 39, 1301–1318, 2014.

Hoffmann, T., Lang, A., and Dikau, R.: Holocene river activity: analysing 14C-dated fluvial and colluvial sediments from Germany, Quaternary Sci. Rev., 27, 2031–2040, 2008.

Holden, J., Chapman, P., and Labadz, J.: Artificial drainage of peatlands: hydrological and hydrochemical process and wetland restoration, Prog. Phys. Geog., 28, 95–123, 2004.

Holden, J., Evans, M., Burt, T., and Horton, M.: Impact of land drainage on peatland hydrology, J. Environ. Qual., 35, 1764–1778, 2006.

Huisink, M.: Changing river styles in response to Weichselian climate changes in the Vecht valley, eastern Netherlands, Sediment. Geol., 133, 115–134, 2000.

Janssens, M. M., Kasse, C., Bohncke, S. J. P., Greaves, H., K. M. Cohen, N., Wallinga, J., and Hoek, W. Z.: Climate-driven fluvial development and valley abandonment at the last glacial-interglacial transition (Oude IJssel-Rhine, Germany), Neth. J. Geosci., 91, 37–62, 2012.

Kadaster: Topographische en Militaire Kaart van het Koninkrijk der Nederlanden, 1 : 50.000, map sheet 22 (publication date 1859), Kadaster, Apeldoorn, Digital file, 2018.

Kasse, C., Hoek, W. Z., Bohncke, S. J. P., Konert, M., Weijers, J. W. H., Cassee, M. L., and Van Der Zee, R. M.: Late Glacial fluvial response of the Niers-Rhine (western Germany) to climate and vegetation change, J. Quaternary Sci., 20, 377–394, 2005.

Kasse, C., Van Balen, R., Bohncke, S., Wallinga, J., and Vreugdenhil, M.: Climate and base-level controlled fluvial system change and incision during the last glacial–interglacial transition, Roer river, The Netherlands–western Germany, Neth. J. Geosci., 96, 71–92, 2016.

Kleinhans, M. G.: Sorting out river channel patterns, Prog. Phys. Geog., 34, 287–326, 2010.

Kleinhans, M. G. and Van den Berg, J. H.: River channel and bar patterns explained and predicted by an empirical and a physics-based method, Earth Surf. Proc. Land., 36, 721–738, 2011.

Kleinhans, M. G., Schuurman, F., Bakx, W., and Markies, H.: Meandering channel dynamics in highly cohesive sediment on an intertidal mud flat in the Westerschelde estuary, the Netherlands, Geomorphology, 105, 261–276, 2009.

Kondolf, G. M., Piégay, H., and Landon, N.: Channel response to increased and decreased bedload supply from land use change: contrasts between two catchments, Geomorphology, 45, 35–51, 2002.

Koster, E. A., Castel, I. I., and Nap, R. L.: Genesis and sedimentary structures of late Holocene aeolian drift sands in northwest Europe, Geol. Soc. Spec. Publ., 72, 247–267, 1993.

Leeder, M.: Fluviatile fining-upwards cycles and the magnitude of palaeochannels, Geol. Mag., 110, 265–276, 1973.

Lenke, W.: Das Klima Ende des 16. und Anfang des 17. Jahrhunderts nach Beobachtungen von Tycho de Brahe auf Hven im Sund DK, Leopold III, Treutwein in Fiirstenfeld Oberbayern und David Fabricius in Ostfriesland, Berichte des Dt. Wetterdienstes, 15, 1968.

Leopold, L. B. and Wolman, M. G.: River channel patterns: braided, meandering, and straight, USGS Professional Paper, 1957, 39–86, 1957.

Lespez, L., Viel, V., Rollet, A., and Delahaye, D.: The anthropogenic nature of present-day low energy rivers in western France and implications for current restoration projects, Geomorphology, 251, 64–76, 2015.

Lewin, J. and Macklin, M. G.: Floodplain catastrophes in the UK Holocene: messages for managing climate change, Hydrol. Process., 24, 2900–2911, 2010.

Lewin, J., Davies, B., and Wolfenden, P.: Interactions between channel change and historic mining sediments, Wiley, New York, 1977.

Liébault, F. and Piégay, H.: Assessment of channel changes due to long-term bedload supply decrease, Roubion River, France, Geomorphology, 36, 167–186, 2001.

Maas, G.: De Overijsselse Vecht – Geomorfogenetische gesteldheid anno 1890, DLO – Staring Centrum Wageningen, Wageningen, 1995.

Macklin, M. G., Jones, A. F., and Lewin, J.: River response to rapid Holocene environmental change: evidence and explanation in British catchments, Quaternary Sci. Rev., 29, 1555–1576, 2010.

Makaske, B., Smith, D. G., Berendsen, H. J. A., de Boer, A. G., van Nielen-Kiezebrink, M. F., and Locking, T.: Hydraulic and sedimentary processes causing anastomosing morphology of the upper Columbia River, British Columbia, Canada, Geomorphology, 111, 194–205, 2009.

Manley, G.: Snowfall in Britain over the past 300 years, Weather, 24, 428–437, 1969.

Millar, R. G.: Influence of bank vegetation on alluvial channel patterns, Water Resour. Res., 36, 1109–1118, 2000.

Mudelsee, M., Börngen, M., Tetzlaff, G., and Grünewald, U.: No upward trends in the occurrence of extreme floods in central Europe, Nature, 425, 166–169, 2003.

Mudelsee, M., Börngen, M., Tetzlaff, G., and Grünewald, U.: Extreme floods in central Europe over the past 500 years: Role of cyclone pathway "Zugstrasse Vb", J. Geophys. Res.-Atmos., 109, D23101, https://doi.org/10.1029/2004JD005034, 2004.

Murray, A. S. and Wintle, A. G.: The single aliquot regenerative dose protocol: potential for improvements in reliability, Radiat. Meas., 37, 377–381, 2003.

Nanson, G. C. and Croke, J. C.: A genetic classification of floodplains, Geomorphology, 4, 459–486, 1992.

Nanson, G. C. and Knighton, A. D.: Anabranching rivers: their cause, character and classification, Earth Surf. Proc. Land., 21, 217–239, 1996.

Neal, A.: Ground-penetrating radar and its use in sedimentology: principles, problems and progress, Earth-Sci. Rev., 66, 261–330, 2004.

Neefjes, J., Brinkkemper, O., Jehee, L., and Van de Griendt, W.: Cultuur-historische atlas van de Vecht, WBooks, Zwolle, 2011.

Nimmo, J.: Porosity and pore size distribution, in: Encyclopedia of Soils in the Environment, edited by: Hillel, D., Columbia University, New York, NY, USA, Elsevier Academic Press, Amsterdam, 3, 295–303, 2004.

Notebaert, B. and Verstraeten, G.: Sensitivity of West and Central European river systems to environmental changes during the Holocene: A review, Earth-Sci. Rev., 103, 163–182, 2010.

Notebaert, B., Broothaerts, N., and Verstraeten, G.: Evidences of anthropogenic tipping points in fluvial dynamics in Europe, Global Planet. Change, 164, 27–38, 2018.

Oorschot, M. v., Kleinhans, M., Geerling, G., and Middelkoop, H.: Distinct patterns of interaction between vegetation and morphodynamics, Earth Surf. Proc. Land., 41, 791–808, 2016.

Parker, G.: Transport of gravel and sediment mixtures, in: Sedimentation engineering: Processes, measurements, modeling, and practice, edited by: Garcia, M. H., Am. Soc. of Civ. Eng., New York, 110, 165–252, 2008.

Passmore, D. G., Macklin, M. G., Brewer, P. A., Lewin, J., Rumsby, B. T., and Newson, M. D.: Variability of late Holocene braiding in Britain, Geol. Soc. Spec. Publ., 75, 205–229, 1993.

Peakall, J., Ashworth, P. J., and Best, J. L.: Meander-bend evolution, alluvial architecture, and the role of cohesion in sinuous river channels: a flume study, J. Sediment. Res., 77, 197–212, 2007.

Quik, C. and Wallinga, J.: Reconstructing lateral migration rates in meandering systems; a novel Bayesian approach combining OSL dating and historical maps, Earth Surf. Dynam. Discuss., https://doi.org/10.5194/esurf-2018-30, in review, 2018.

Reimer, P. J., Bard, E., Bayliss, A., Beck, J. W., Blackwell, P. G., Bronk Ramsey, C., Buck, C. E., Cheng, H., Edwards, R. L., and Friedrich, M.: IntCal13 and Marine13 radiocarbon age calibration curves 0–50 000 years cal BP, Radiocarbon, 51, 1111–1150, 2013.

Rumsby, B. T. and Macklin, M. G.: River response to the last neoglacial (the "Little Ice Age") in northern, western and central Europe, Geol. Soc. Spec. Publ., 115, 217–233, 1996.

Seminara, G.: Meanders, J. Fluid Mech., 554, 271–297, 2006.

Słowik, M.: Transformation of a lowland river from a meandering and multi-channel pattern into an artificial canal: retracing a path of river channel changes (the Middle Obra River, W Poland), Reg. Environ. Change, 13, 1287–1299, 2013.

Słowik, M.: Is history of rivers important in restoration projects? The example of human impact on a lowland river valley (the Obra River, Poland), Geomorphology, 251, 50–63, 2015.

Staring, W. C. A. and Stieltjes, T. J.: De Overijsselsche wateren, s.n., Zwolle, 1848.

Streefkerk, J. and Casparie, W.: De hydrologie van hoogveen systemen, Staatsbosbeheer-rapport, 19, 1–119, 1987.

Struiksma, N., Olesen, K., Flokstra, C., and de Vriend, H.: Bed deformation in curved alluvial channels, J. Hydraul. Res., 23, 57–79, 1985.

Surian, N. and Rinaldi, M.: Morphological response to river engineering and management in alluvial channels in Italy, Geomorphology, 50, 307–326, 2003.

Talmon, A., Struiksma, N., and Van Mierlo, M.: Laboratory measurements of the direction of sediment transport on transverse alluvial-bed slopes, J. Hydraul. Res., 33, 495–517, 1995.

TAUW: Bouwstenen voor een natuurontwikkelingsvisie voor de Overijsselse Vecht, Directie Natuur, Bos, Landschap en Fauna, nr. 3187381., Deventer, 1992.

Toonen, W. H. J., Kleinhans, M. G., and Cohen, K. M.: Sedimentary architecture of abandoned channel fills, Earth Surf. Proc. Land., 37, 459–472, 2012.

Trimble, S. W. and Mendel, A. C.: The cow as a geomorphic agent – a critical review, Geomorphology, 13, 233–253, 1995.

Turowski, J. M., Hovius, N., Meng-Long, H., Lague, D., and Men-Chiang, C.: Distribution of erosion across bedrock channels, Earth Surf. Proc. Land., 33, 353–363, 2008.

Uhden, O.: Niederschlags-und Abflußbeobachtungen auf unberührten, vorentwässerten und kultivierten Teilen eines nordwestdeutschen Hochmoores, der Esterweger Dose am Küstenkanal bei Papenburg, Verlag Wasser und Boden, Hamburg, 1967.

Van Beek, R. and Groenewoudt, B.: An Odyssey along the River Vecht in the Dutch-German border area: A Regional Analysis of Roman-period Sites in Germania Magna, Germania, 89, 157–190, 2011.

Van Beek, R., Gouw-Bouman, M., and Bos, J.: Mapping regional vegetation developments in Twente (the Netherlands) since the Late Glacial and evaluating contemporary settlement patterns, Neth. J. Geosci., 94, 229–255, 2015a.

Van Beek, R., Maas, G. J., and van den Berg, E.: Home Turf: an interdisciplinary exploration of the long-term development, use and reclamation of raised bogs in the Netherlands, Landscape History, 36, 5–34, 2015b.

Van de Lageweg, W. I., van Dijk, W. M., Box, D., and Kleinhans, M. G.: Archimetrics: a quantitative tool to predict three-dimensional meander belt sandbody heterogeneity, The Depositional Record, 2, 22–46, 2016.

Van de Meene, E., Van der Staay, J., and Teoh, L. H.: The Van der Staay suction-corer: a simple apparatus for drilling in sand below groundwater table, Rijks Geologische Dienst, 1979.

Van den Berg, J. H. and Gelder, A.: Prediction of suspended bed material transport in flows over silt and very fine sand, Water Resour. Res., 29, 1393–1404, 1993.

Van den Berg, J. H.: Prediction of alluvial channel pattern of perennial rivers, Geomorphology, 12, 259–279, 1995.

Van der Linden, J. A.: Topographische en Militaire kaart van het Koningrijk der Nederlanden, Fibula-Van Dischoeck, 1973.

Van der Schaaf, S.: Analysis of the hydrology of raised bogs in the Irish Midlands: a case study of Raheenmore Bog and Clara Bog, PhD, Wageningen Agricultural University, Wageningen, 1999.

Ter Wee, M.: Toelichtingen bij de geologische kaart van Nederland 1:50 000: blad Steenwijk oost (16 O), Geologische Stichting, 1966.

Van Dijk, W., Lageweg, W., and Kleinhans, M.: Experimental meandering river with chute cutoffs, J. Geophys. Res.-Earth, 117, F03023, https://doi.org/10.1029/2011JF002314, 2012.

Van Engelen, A. F., Buisman, J., and Ijnsen, F.: A millennium of weather, winds and water in the low countries, In: History and Climate, Springer, 2001.

Van Heerd, R. and Van't Zand, R.: Productspecificatie Actueel Hoogtebestand Nederland, Rijkswaterstaat Meetkundige Dienst, Delft, 1999.

Van Heteren, S., Fitzgerald, D. M., Mckinlay, P. A., and Buynevich, I. V.: Radar facies of paraglacial barrier systems: coastal New England, USA, Sedimentology, 45, 181–200, 1998.

Vandenberghe, J.: Timescales, climate and river development, Quaternary Sci. Rev., 14, 631–638, 1995.

Vandenberghe, J.: The relation between climate and river processes, landforms and deposits during the Quaternary, Quaternary Int., 91, 17–23, 2002.

Vandenberghe, J., Kasse, C., Bohncke, S., and Kozarski, S.: Climate-related river activity at the Weichselian-Holocene transition: a comparative study of the Warta and Maas rivers, Terra Nova, 6, 476–485, 1994.

Vargas-Luna, A., Crosato, A., Hoitink, A., Groot, J., and Uijttewaal, W.: Effects of riparian vegetation development in a restored lowland stream, in: River Flow 2016, CRC Press, 2016.

Vos, P., Bazelmans, J., Weerts, H., and Van der Meulen, M.: Atlas van Nederland in het Holoceen, Bakker, Amsterdam, 2011.

Williams, G. P.: River meanders and channel size, J. Hydrol., 88, 147–164, 1986.

Wolfert, H. and Maas, G.: Downstream changes of meandering styles in the lower reaches of the River Vecht, the Netherlands, Neth. J. Geosci., 86, 257–271, 2007.

Wolfert, H. P., Maas, G., and Dirkx, G.: Het meandergedrag van de Overijsselse Vecht: historische morfodynamiek en kansrijkdom voor natuurontwikkeling, DLO-Staring Centrum, 1996.

Wolman, M. G. and Miller, J. P.: Magnitude and frequency of forces in geomorphic processes, J. Geol., 68, 54–74, 1960.

2

Geomorphometric delineation of floodplains and terraces from objectively defined topographic thresholds

Fiona J. Clubb[1], Simon M. Mudd[1], David T. Milodowski[2], Declan A. Valters[3], Louise J. Slater[4], Martin D. Hurst[5], and Ajay B. Limaye[6]

[1]School of GeoSciences, University of Edinburgh, Drummond Street, Edinburgh, EH8 9XP, UK
[2]School of GeoSciences, University of Edinburgh, Crew Building, King's Buildings, Edinburgh, EH9 3JN, UK
[3]School of Earth, Atmospheric, and Environmental Science, University of Manchester, Oxford Road, Manchester, M13 9PL, UK
[4]Department of Geography, Loughborough University, Loughborough, LE11 3TU, UK
[5]School of Geographical and Earth Sciences, East Quadrangle, University of Glasgow, Glasgow, G12 8QQ, UK
[6]Department of Earth Sciences and St. Anthony Falls Laboratory, University of Minnesota, Minneapolis, Minnesota, USA

Correspondence to: Fiona J. Clubb (f.clubb@ed.ac.uk)

Abstract. Floodplain and terrace features can provide information about current and past fluvial processes, including channel response to varying discharge and sediment flux, sediment storage, and the climatic or tectonic history of a catchment. Previous methods of identifying floodplain and terraces from digital elevation models (DEMs) tend to be semi-automated, requiring the input of independent datasets or manual editing by the user. In this study we present a new method of identifying floodplain and terrace features based on two thresholds: local gradient, and elevation compared to the nearest channel. These thresholds are calculated statistically from the DEM using quantile–quantile plots and do not need to be set manually for each landscape in question. We test our method against field-mapped floodplain initiation points, published flood hazard maps, and digitised terrace surfaces from seven field sites from the US and one field site from the UK. For each site, we use high-resolution DEMs derived from light detection and ranging (lidar) where available, as well as coarser resolution national datasets to test the sensitivity of our method to grid resolution. We find that our method is successful in extracting floodplain and terrace features compared to the field-mapped data from the range of landscapes and grid resolutions tested. The method is most accurate in areas where there is a contrast in slope and elevation between the feature of interest and the surrounding landscape, such as confined valley settings. Our method provides a new tool for rapidly and objectively identifying floodplain and terrace features on a landscape scale, with applications including flood risk mapping, reconstruction of landscape evolution, and quantification of sediment storage and routing.

1 Introduction

Identifying the location of floodplains and fluvial terrace features can provide important insights into geomorphic and hydrological processes. Understanding the controls on floodplain inundation carries increasing societal importance, as the frequency of flood events is predicted to increase with the rise in global temperatures and varying patterns of precipitation caused by climate change (Schreider et al., 2000; Booij, 2005; Hartmann et al., 2013). Although there are still large uncertainties regarding the impacts of climate change on flood frequency (Booij, 2005), identifying floodplains is crucial for forecasting and planning purposes. On longer timescales, the morphology and structure of fluvial terraces can provide important information on channel response to climatic, tectonic, and base-level variations (Bull, 1991; Merritts et al., 1994; Pazzaglia et al., 1998); the relative importance of lateral and vertical channel incision (Finnegan and Dietrich, 2011); and sediment storage and dynamics (Pazzaglia, 2013; Gran et al., 2013).

Attempts to identify floodplains can be classified into two broad families of methods: (i) flood risk mapping and hydrological modelling, and (ii) geometric terrain classification. Traditionally, identification of floodplains has relied upon the creation of flood hazard maps, produced through detailed hydraulic modelling studies (e.g. Noman et al., 2001; Grimaldi et al., 2013). These studies tend to incorporate historical flood event information, hydrological analyses, and hydraulic flow propagation models (Degiorgis et al., 2012). These mature techniques can lead to accurate flood inundation predictions down to the level of a single building (e.g. Horritt and Bates, 2002; Cobby et al., 2003; Guzzetti et al., 2005; Hunter et al., 2007; Kim et al., 2012). However, these models can be computationally expensive and time-consuming to run, even in one dimension, requiring the calibration of large numbers of parameters, all with their own uncertainties (e.g. Beven, 1993; Horritt and Bates, 2002; Liu and Gupta, 2007). This means that hydraulic simulations are usually performed at cross sections across the channel and interpolated to cover the rest of the stream network (Noman et al., 2001; Dodov and Foufoula-Georgiou, 2006). For example, floodplain mapping tools have been developed that incorporate either field-based or modelled stage-duration information at multiple cross sections along the channel and interpolate a three-dimensional water surface between these sections (e.g Yang et al., 2006; Belmont, 2011).

The introduction of high-resolution digital elevation models (DEMs) has provided the opportunity to map floodplain features much more rapidly and over larger spatial scales than previously possible (Noman et al., 2001). This had led to the development of many different methods that rely on extracting a variety of topographic indices from DEMs, such as local slope, contributing area, and curvature (Manfreda et al., 2014). One common metric used to predict floodplains is the topographic index ($\phi = \ln(A/(\tan\beta))$), where A is the contributing area to each cell (m^2) and β is the local slope in degrees (e.g. Kirkby, 1975; Beven and Kirkby, 1979; Beven et al., 1995; Quinn et al., 1995; Beven, 1997). The contributing area term reflects the tendency of water to accumulate at certain regions of the basin, whereas the slope term represents the tendency for gravity to transport water downhill. Therefore, high values of the topographic index represent areas which are likely to saturate first, as they have a large contributing area compared to local slope (Beven, 1997). Manfreda et al. (2011) suggested a modified version of the topographic index, changing the weighting on the area term by raising it to an exponent n. This modification allows the relative importance of slope or contributing area to be changed by varying the n parameter. They proposed that floodplains can be identified as cells with a modified topographic index (ϕ_m) greater than a threshold value, τ. However, this method requires calibration of the parameters τ and n through comparing the output floodplain map with a pre-existing hazard map and noting the occurrence of true and false positives and negatives (Manfreda et al., 2011).

Another geometric method that has been developed to identify floodplains uses a series of linear binary classifiers for a number of topographic metrics (Degiorgis et al., 2012). Five different parameters are sampled from the DEM (slope, contributing area, elevation from nearest channel, distance from nearest channel, and curvature), and each cell is classified as either 1 (floodplain) or 0 (non-floodplain) depending on whether these parameters are above or below threshold values. Each of these five metrics can be considered in isolation or in pairs. The thresholds are calibrated using flood hazard maps, where the number of true and false positives and negatives are noted, similar to the approach of Manfreda et al. (2011). For each parameter and threshold value the receiver operating characteristics (ROC) curve (e.g. Fawcett, 2006) is calculated, which is defined by the number of true and false positives. The maximum area under the curve is determined to allow the threshold value for each parameter to be calibrated, as well as comparisons between each parameter to be found. The pair of best-performing features was identified as the distance (D) and elevation (H) from the nearest channel (m). This method is also semi-automated, as it requires the existence of flood hazard maps for at least some part of the catchment in order to select the correct binary classifiers for floodplain identification.

Dodov and Foufoula-Georgiou (2006) present an algorithm for identifying floodplains over large scales based on information on bankfull channel depths. They suggest that the morphology of the floodplain is defined by the lateral channel migration rate through time and controlled by the transport of water and sediment by the channel. Therefore, they assume that the geometry of the floodplain is related to that of the channel, and demonstrate a relationship between bankfull channel depths and floodplain inundation depths which is linear over a range of scales (Dodov and Foufoula-Georgiou, 2006). Floodplain delineation is carried out by locally filling the DEM up to the depth of inundation, which is determined based on bankfull channel depths, calibrated using data from United States Geological Survey (USGS) gauging stations across Oklahoma and Kansas, along with field measurements. The depth of inundation at points along the channel network is then used to find the lateral extent of the floodplain by using the planform curvature of the channel. This method also requires significant user input, as the channel bankfull depths are required in order to estimate the inundation depth.

The extraction of fluvial terraces (the remnants of previous floodplains) represents a closely related problem to the delineation of presently active floodplain surfaces. Previous studies have also used a geometric approach to identify terrace features from DEMs. For example, Demoulin et al. (2007) identified terrace surfaces based on local slope and height of each pixel compared to the channel. They used these attributes in order to reconstruct palaeo-channel profiles from terrace surfaces, but their methodology was not designed to produce a map of terrace extents on a wider landscape scale. Therefore, following on from their approach, Stout and Belmont (2014) presented the TerEx toolbox, a semi-automated tool to identify potential terrace surfaces based on thresholds of local relief, minimum area, and maximum distance from the channel. After potential terrace surfaces are identified, their area and height above the local channel are measured. The tool then allows the user to edit the terrace surfaces based on comparison with field data. Hopkins and Snyder (2016) evaluated the TerEx toolbox, along with two other semi-automated methods for identifying terrace surfaces (Wood, 1996; Walter et al., 2007) at the Sheepscot River, Maine. They found that all of the methods over-predicted terrace areas compared to the field-mapped terraces, and the accuracy of the methods decreased in lower-relief landscapes. These semi-automated methods allow the user to manually clip over-predicted terrace surfaces based on field data and DEM observations, and remove selected surfaces that do not represent terraces, such as roads, alluvial fans, or water bodies (Stout and Belmont, 2014).

The geomorphic methods of mapping both terraces and floodplains outlined above are all semi-automated, requiring independent datasets and significant user input. For example, the method proposed by Manfreda et al. (2011) requires the parameters to be optimised using flood inundation maps from hydraulic simulations. The linear binary classifiers outlined by Degiorgis et al. (2012) and tested by Manfreda et al. (2014) use flood hazard maps to select the correct threshold for floodplain prediction from the geomorphic indices. The TerEx toolbox, developed by Stout and Belmont (2014), requires significant user input in order to manually edit the predicted terrace surfaces. No existing approach to mapping either floodplains or terraces from topographic data includes objective criteria for setting the thresholds that identify floodplains and terraces. As a result, the different thresholds that a user might select can result in varying floodplain and terrace maps for the same input DEM, complicating efforts to consistently map geomorphic features between different landscapes.

Here we introduce a new method of identifying floodplain and terrace surfaces from topographic data. This method uses two geometric thresholds that can be readily extracted from DEMs: the gradient of each pixel, and the elevation of each pixel relative to the nearest channel. Importantly, this method does not require calibration using any independent datasets, as the thresholds are statistically calculated from the DEM using quantile–quantile plots. We test our method against field-mapped floodplain initiation points, published flood hazard maps, and digitised terrace surfaces from seven field sites throughout the US and one site in the UK (Fig. 1).

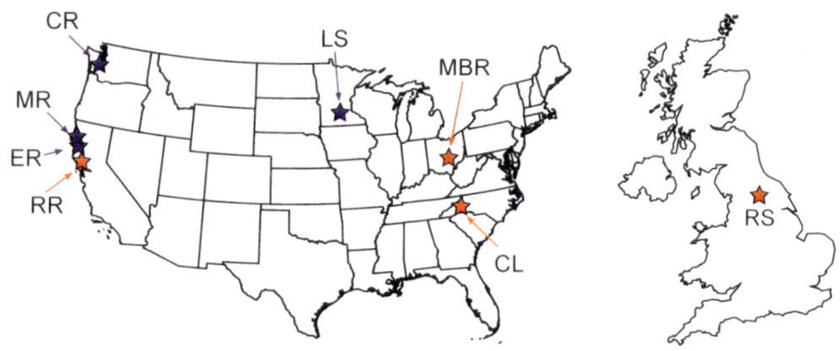

Figure 1. Maps of the US and UK showing the location of the eight field sites in the study. Red stars represent floodplain sites; blue stars represent terrace sites. RR: Russian River, CA; ER: South Fork Eel River, CA; MR: Mattole River, CA; CR: Clearwater River, WA; LS: Le Sueur River, MN; MBR: Mid Bailey Run, OH; CL: Coweeta Hydrologic Laboratory, NC; RS: River Swale, Yorkshire, UK.

For each site, where available, we use high-resolution lidar-derived DEMs, as well as the corresponding national elevation datasets (10 m resolution for the US and 5 m for the UK) in order to test the sensitivity of our method to grid resolution.

2 Methodology

Floodplain and terrace surfaces can be defined as low relief, quasi-planar areas capped by alluvium and found proximal to the modern river channel. Therefore, field mapping campaigns typically identify these surfaces as spatially continuous areas with low gradients that occur next to the channel. We present a new geometric method which replicates this field approach as closely as possible by using two metrics which can be readily extracted from the DEM: elevation compared to the nearest channel, and local gradient. Our method is efficient to run and is based on the statistical selection of topographic thresholds, requiring no input of independent datasets or field mapping. We outline below the DEM pre-processing steps followed by the methodology for identifying floodplain and terrace features.

2.1 DEM pre-processing

The first step of the algorithm is to smooth the DEM in order to remove micro-topographic noise. Gaussian filters are often used to smooth DEMs, where the smoothing can be described by linear diffusion. A Gaussian filter results in the DEM being smoothed uniformly at all locations and in all directions (e.g. Lashermes et al., 2007). However, one consequence of the Gaussian filtering is the loss of information where there are sharp boundaries between features due to the uniform smoothing. Therefore, we filter the input DEM using a non-linear filter proposed by Perona and Malik (1990) and applied to channel extraction from high-resolution topography by Passalacqua et al. (2010a). The Perona–Malik filter is an adaptive filter in which the degree of smoothing decreases as topographic gradient increases (Perona and Malik, 1990; Passalacqua et al., 2010a). This non-linear diffusion equation can be described as

$$\partial_t h(x, y, t) = \nabla \cdot [p(|\nabla h|)\nabla h], \quad (1)$$

where h is the elevation at location (x, y) and time t, ∇ is the gradient operator, and $p(|\nabla h|)$ is an edge-stopping function that specifies where to stop diffusion across feature boundaries, where

$$p(|\nabla h|) = \frac{1}{1 + (|\nabla h|/\lambda)^2}, \quad (2)$$

where λ is a constant. Importantly for the identification of low-gradient surfaces, the Perona–Malik filtering enhances the transitions between features, such as the low-gradient valley floor and the surrounding hillslopes, while preferentially smoothing low-gradient reaches of the DEM. Following the methodology of Passalacqua et al. (2010a), we set the number of iterations (t) to 50 and the calculation of λ as the 90 % quantile. We keep these parameters constant across each site tested in the study. A full explanation of these parameters and derivation of the Perona–Malik filter is described by Passalacqua et al. (2010a).

After the DEM is smoothed, we then extract the channel network. Many studies have proposed different methods for identifying channel networks from high-resolution topography (e.g. Lashermes et al., 2007; Tarolli and Dalla Fontana, 2009; Passalacqua et al., 2010b, 2012; Pelletier, 2013; Clubb et al., 2014). Grieve et al. (2016c) tested the validity of channel extraction algorithms at coarsening DEM resolution and found that a geometric method of channel extraction was consistent up to DEM resolutions of 30 m. This method, described by Grieve et al. (2016b), uses an optimal Wiener filter to remove micro-topographic noise from the DEM (Wiener, 1949; Pelletier, 2013). The optimal Wiener filter is only used to extract the channel network: we use the Perona–Malik filtering to extract the floodplains and terraces. Channelised portions of the landscape are selected using a tangential curvature threshold (Pelletier, 2013), which is defined using quantile–quantile plots as described by Lashermes et al. (2007) and Passalacqua et al. (2010a). These channelised portions of the landscape are combined into a channel network using a connected components algorithm outlined by He et al. (2008) and thinned using the algorithm of Zhang and Suen (1984). We chose this algorithm for channel extraction to allow consistency when running our method on DEMs of varying grid resolutions.

2.2 Floodplain and terrace identification

After smoothing the DEM, the user can choose to run the terrace and floodplain mapping algorithm across the whole DEM or to extract the floodplains and terraces relative to a specific channel of interest. If the algorithm is run on the whole DEM, the local gradient, S, and relief relative to the nearest channel, R_c, are calculated for each pixel. These two parameters were chosen on the basis that floodplains and terraces tend to form low-gradient regions that are close to the elevation of the modern channel. Local gradient has been used in previous geometric methods of floodplain and terrace identification, both in the calculation of the topographic index (Kirkby, 1975; Manfreda et al., 2011) and in combination with other topographic metrics (e.g. Degiorgis et al., 2012; Stout and Belmont, 2014; Limaye and Lamb, 2016). Local gradient was calculated by fitting a polynomial surface to the DEM with a circular window (e.g. Lashermes et al., 2007; Roering et al., 2010; Hurst et al., 2012; Grieve et al., 2016a). The radius of the window is calculated by identifying breaks in the standard deviation and interquartile range of curvature with increasing window size, following Grieve et al. (2016a). This allows the window size to be calculated for each DEM to ensure that the slope values are representative at the hillslope scale, rather than being influenced by smaller-scale variations from vegetation (e.g. Roering et al., 2010; Hurst et al., 2012). R_c has also been used in previous geometric methods (e.g. Degiorgis et al., 2012; Manfreda et al., 2014; Limaye and Lamb, 2016) and is calculated as the difference in elevation between the starting pixel and the nearest channel pixel, identified using a steepest descent flow routing algorithm (O'Callaghan and Mark, 1984; Braun and Willett, 2013). A threshold Strahler stream order is set by the user such that the nearest channel must have a stream order greater than the threshold. This is necessary so that each pixel is mapped to the main channel along which floodplains or terraces have formed, rather than narrow tributary valleys. We found that a threshold of third-order channels was appropriate for each of our field sites, based on a visual inspection of the DEM. One of the outputs of our software package is a raster of the channel network labelled by the Strahler stream order. The user can identify an appropriate threshold stream order based on visual inspection of floodplain and terrace surfaces compared to this network.

As well as running the algorithm on the whole landscape, the user can also choose to extract floodplains or terraces relative to a specific channel of interest. The user must provide the latitude and longitude of two points defining the upstream and downstream end of the channel. The algorithm then defines a channel network between these points using a steepest descent flow routing algorithm (O'Callaghan and Mark, 1984; Braun and Willett, 2013). After the identification of the channel, a swath profile is created along it following the method outlined in Hergarten et al. (2014) and applied by Dingle et al. (2016). The user must specify the width of the swath, which can be estimated by a visual inspection of the DEM, to provide a sufficiently wide swath compared to the valleys in the landscape. The same two parameters (S and R_c) are used for feature classification for each pixel in the swath profile, except that R_c is calculated compared to the nearest point on the reference channel.

After the calculation of slope and R_c, we identify thresholds for each metric in order to provide a binary classification of each pixel as either floodplain/terrace (1) or hillslope (0). A key feature of our new method is that the thresholds for R_c and local gradient do not need to be set by the user based on independent validation but are calculated statistically from the DEM. Many methods of channel extraction employ statistical selection of topographic thresholds (e.g. Lashermes et al., 2007; Thommeret et al., 2010; Passalacqua et al., 2010a; Pelletier, 2013; Clubb et al., 2014), but this has yet to be developed for the identification of floodplains or terraces. We identify thresholds for R_c and S using quantile–quantile plots, which have previously been used in the detection of hillslope–valley transitions (e.g. Lashermes et al., 2007; Passalacqua et al., 2010a). Quantile–quantile plots are used to determine if a probability density function of real data can be described by a Gaussian distribution. The transition between process domains can be determined by the value at which the probability density function of the real data deviates from the Gaussian function (Lashermes et al., 2007). The real data are plotted against the corresponding standard normal variate, which indicates how many standard deviations an element is from the mean. For example, if a value has a standard normal variate (or z score) of 1, then it is 1 standard deviation above the mean, which has a z score of 0. A Gaussian distribution plots as a straight line on a quantile–quantile plot, and is modelled for each DEM based on a lower and upper percentile of the real data. The percentiles chosen to represent the reference Gaussian distribution can be set by the user based on the landscape in question, but are generally set as the 25th and 75th percentile (Passalacqua et al., 2010a). For each value of the real data, we calculate the difference between the real data and the Gaussian distribution as a fraction of the range of the real data (Fig. 2). The threshold values for R_c and slope are then identified as the lowest value at which there is less than 1 % difference between the two distributions. Figure 3 shows an example of the channel relief and slope maps for the Russian River field site, with the calculated thresholds for each field site presented in Table 1. If the user wishes to extract only the terraces, then a threshold height above the modern river channel must be set: any pixels below this height will be identified as floodplain, and any pixels above this height will be identified as terraces. This threshold height can also be determined based on a visual inspection of the DEM. Our method allows the analysis of spatial extent of floodplain and terrace features (if run across the whole DEM) as well as the distribution along a specific channel of interest (if run with the swath mode). For example, in swath mode, the elevation and slope of the terraces can be mapped as a function of distance upstream along the channel network. This provides numerous potential applications of the method for understanding controls on terrace formation and morphology.

Table 1. Channel relief and slope threshold for each field site

Field site	Channel relief threshold	Slope threshold
Mid Bailey Run, OH	23.69	0.15
Coweeta, NC	32.80	0.11
Russian River, CA	43.51	0.81
River Swale, UK	39.40	0.05
South Fork Eel River, CA	42.96	0.05
Le Sueur River, MN	9.42	0.05
Mattole River, CA	50.25	0.17
Clearwater River, WA	12.67	0.06

2.3 Comparison with published data

In order to test the results of our method we compare the predicted floodplain and terrace locations to field-mapped floodplain initiation points, published flood hazard maps, and digitised terrace surfaces. In order to quantify the performance of our methods compared to these datasets, we assess the rates of true positives (TP), false positives (FP), true negatives (TN), and false negatives (FN) (e.g. Heipke et al., 1997; Molloy and Stepinski, 2007; Tarolli et al., 2010; Orlandini et al., 2011; Manfreda et al., 2014; Clubb et al., 2014). Each pixel is assigned to one of the four categories:

1. True positive, TP: the pixel is identified as floodplain/terrace by both the geomorphic method and the independent dataset.

2. False positive, FP: the pixel is identified as floodplain/terrace by the geomorphic method, but not by the independent dataset.

3. True negative, TN: the pixel is not identified as floodplain/terrace by either dataset.

4. False negative, FN: the pixel is identified as floodplain/terrace by the independent dataset but not by the geomorphic method.

We report the reliability (r), sensitivity (s), and overall quality (Q) for each field site:

$$r = \frac{\sum \mathrm{TP}}{\sum \mathrm{TP} + \sum \mathrm{FP}}, \tag{3a}$$

$$s = \frac{\sum \mathrm{TP}}{\sum \mathrm{TP} + \sum \mathrm{FN}}, \tag{3b}$$

$$Q = \frac{\sum \mathrm{TP}}{\sum \mathrm{TP} + \sum \mathrm{FP} + \sum \mathrm{FN}}. \tag{3c}$$

The reliability, r, is a measure of the ability of the method to not generate false positives. The r value can vary between

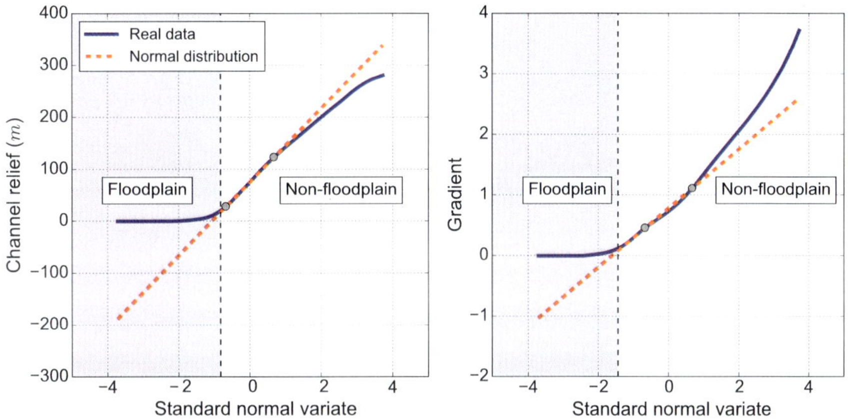

Figure 2. Example quantile–quantile plots for Mid Bailey Run, Ohio, showing probability density function of relief relative to the channel and slope. The probability density function of each is shown in blue, with the reference normal distribution shown by the red dashed line. The threshold (black dashed line) is selected where there is less than 1 % difference between the real and reference distributions. The blue box highlights the portion of the distribution identified as floodplain.

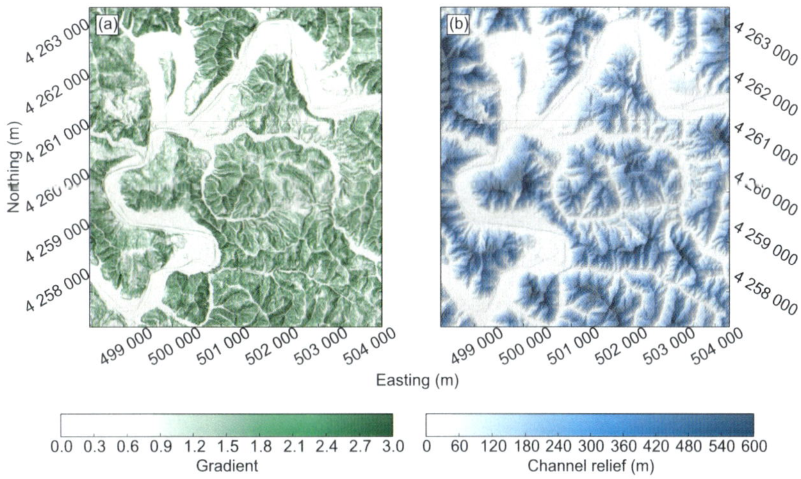

Figure 3. Maps showing **(a)** gradient and **(b)** relief relative to the nearest channel, R_C, for the Russian River field site. The areas of the landscape identified as below the threshold are shown in white, with values above the threshold then grading to darker colours. In order to be selected as floodplain, each pixel must be below the threshold for both gradient and R_C. The coordinate system is UTM Zone 10° N.

0 and 1: if the r value is low, then the method is predicting a large amount of pixels as floodplain or terrace which are not identified by the independent dataset, whereas high r value indicates that the majority of pixels mapped as floodplain or terrace are also identified by the independent map. The sensitivity, s, is a measure of the ability of the method to not generate false negatives: a low s value indicates that the method is not identifying many of the floodplain or terrace pixels selected by the published maps. The overall quality, Q, combines both the number of false positives and false negatives to give an overall "goodness" of the feature classification. It also varies between 0 and 1, where 0 represents no correlation between the predicted and observed features, and 1 represents a perfect match (Heipke et al., 1997).

3 Study areas

We ran our new method on a total of eight field sites, located in Fig. 1. Four of these field sites (the Russian River, CA; Mid Bailey Run, OH; Coweeta NC; and the River Swale, UK) were selected to test the ability of the algorithm to identify floodplains, using published flood maps for the regions. The remaining four sites were selected to validate the algorithm against digitised terrace maps (South Fork Eel River, CA; Le Sueur River, MN; Clearwater River, WA; and Mattole River, CA). Table 2 summarises the mean annual precipitation and mean annual temperature of each site, based on data from the PRISM Climate Group (http://prism.oregonstate.edu) for the US sites and the Met Office (http://www.metoffice.gov.uk/public/weather/climate/) for the UK site. It also summarises

Table 2. Details of climate and lithology for each field site.

Field site	UTM Zone	MAP (mm)	MAT (°C)	Lithology	Comparison datasets	Grid res. (m)
Russian River, CA	10° N	1396	14.1	Sandstones and shales, Quaternary alluvial deposits	FEMA flood hazard maps	1
Mid Bailey Run, OH	17° N	1005	10.9	Sandstones, siltstones, shales	FEMA flood hazard maps Field-mapped FIPs	1
Coweeta, NC	17° N	1792	12.3	Meta-sedimentary units	FEMA flood hazard maps Field-mapped FIPs	1
River Swale, UK	30° N	898	8.4	Limestones and sandstones	EA flood hazard maps	5
South Fork Eel River, CA	10° N	2009	12.7	Greywackes and shales	Digitised terraces (Seidl and Dietrich, 1992)	1
Le Sueur River, MN	15° N	793	7.5	Pleistocene tills and Ordovician dolostones	Digitised terraces (Gran et al., 2009)	1
Mattole River, CA	10° N	2593	12.8	Sandstones and shales, Quaternary alluvial deposits	Digitised terraces (Dibblee and Minch, 2008; Limaye and Lamb, 2016)	10
Clearwater River, WA	10° N	3126	9.9	Sandstones with interbedded shales	Digitised terraces (Wegmann and Pazzaglia, 2002; Limaye and Lamb, 2016)	10

the underlying lithology, the source of the data used for validation, and the grid resolution. The algorithm was run based on topographic data derived from 1 m lidar data for the sites where these were available (the Russian River, CA; Mid Bailey Run, OH; Coweeta, NC; the South Fork Eel River, CA; and the Le Sueur River, MN). For the remaining field sites the topographic data were generated from the United States Geological Survey National Elevation Dataset 1/3 arcsec DEM, sampled at 10 m resolution for the US sites, and from the Ordnance Survey Terrain 5 dataset for the UK site, sampled at 5 m resolution. All DEMs were converted to the Universal Transverse Mercator (UTM) coordinate system using the WGS84 datum.

4 Results

4.1 Comparison with mapped floodplains

We compare the floodplain extent predicted by our method to field-mapped floodplain initiation points (FIPs) from two of the four study areas: Mid Bailey Run, OH, and Coweeta, NC. An FIP was defined as the upstream limit of low-gradient surfaces at the same elevation as the channel banks. As the valley opens out from its more confined upper reaches, these surfaces transition from discontinuous depositional pockets to more continuous floodplain surfaces (Jain et al., 2008). In this study we consider the FIP to start at the onset of alluviation outside the channel banks: therefore, we mapped the start of the discontinuous floodplain pockets at the FIPs in each channel. The onset of alluviation often occurred at multiple locations along the same channel: in these cases we took the location of each FIP downstream along the channel.

A total of 19 FIPs were mapped in Mid Bailey Run, OH, during May–June 2011, and eight FIPs were mapped in the Coweeta catchment, NC, in May 2014. FIPs in the Mid Bailey Run catchment were mapped using a Trimble GeoXM GeoExplorer 2008 series GPS with a mean horizontal accuracy of 6 m. Point locations in the Coweeta catchment were mapped using a Trimble GeoXR GeoExplorer 6000 series GPS with a mean horizontal accuracy of 1.01 m and a mean precision of 1.3 m. Figure 4 shows the relationship between the field-mapped initiation points and predicted floodplain extent. In order to compare these field-mapped FIPs to our predicted floodplain extents, we measured the flow distance between the field-mapped point and the furthest upstream point of the nearest predicted floodplain patch. The distances for each FIP are reported in Table 3, where negative values indicate that the predicted floodplain initiation was upstream of the mapped, and vice versa for positive values. We also report the r, s, and Q values for the predicted floodplain initiation points. Following the methodology of Orlandini et al. (2011), we classify a point as a TP if the predicted FIP is within a 30 m radius of the mapped FIP. The comparison with the mapped FIPs resulted in $r = 0.83$, $s = 0.67$, and $Q = 0.59$ for Mid Bailey Run, and $r = 0.78$, $s = 1$, and $Q = 0.78$ for Coweeta.

Along with these field-mapped floodplain initiation points, we also compare our predicted floodplain extent to published flood risk maps for three out of the four study areas. For the sites in the US, flood risk maps were obtained from the Fed-

Table 3. Flow distances between the field-mapped FIPs and predicted floodplain extents

Field site	Mapped FIP	Easting (m)	Northing (m)	Flow distance*
Mid Bailey Run, OH	T2FPI1	401 513	4 364 940	59
	T3FPI1	401 622	4 364 773	85
	T3FPI2	401 661	4 364 732	−49
	WBT1FPI	400 090	4 363 977	−23
	WBT2FPI1	399 865	4 364 215	−1
	T4FPI	401 342	4 365 472	28
	T5FPI2	401 072	4 365 675	0
	T7FPI2	400 670	4 366 152	2
	T5FPI1	401 208	4 365 807	0
	T1FPI1	401 443	4 365 150	0
	TX3D3-FPI0	400 718	4 366 277	−42
	TX3FPI1	400 644	4 366 126	−5
	MBFPI	400 449	4 366 130	−34
	T7FPI1	400 600	4 366 074	−19
	T4FPI2	401 391	4 365 514	92
	T6FPI1	400 900	4 365 921	−20
Coweeta, NC	SF5	277 212.380	3 882 554.000	−51
	BC1	276 326.800	3 880 661.200	−3
	HCW	277 641.5	3 881 694.2	2
	BC3	277 584.633	3 881 138.653	−3
	HW1	278 252.652	3 881 715.719	13
	CB1	278 089.041	3 882 301.638	12
	HB1	277 444.900	3 882 919.685	−16
	CC2	277 098.745	3 882 348.108	−2

* The distance between the mapped FIP and the upstream extent of the nearest floodplain patch predicted by our geomorphic method.

Figure 4. Shaded relief maps of Mid Bailey Run and Coweeta field sites showing the relationship between the predicted floodplain (blue) and the mapped floodplain initiation points (red). The UTM zone is 17° N.

eral Emergency Management Agency's (FEMA) National Flood Hazard Layer (https://msc.fema.gov/portal/). The National Flood Hazard Layer is a compilation of GIS data consisting of a US-wide flood insurance rate map. It contains information on the flood zone, base flood elevation, and floodway status for a location. Floodplain extents are calculated using a hydraulic model, such as HEC-RAS (Hydrologic Engineering Center-River Analysis System), incorporating discharge data, cross sectional survey data, and stream characteristics. These studies can be expensive, with a detailed survey on a mile-long reach typically costing between USD 10 000 and 25 000 (Committee on FEMA Flood Maps, 2009). The original data were in the geographic projection NAD1983, and were converted to the projected UTM

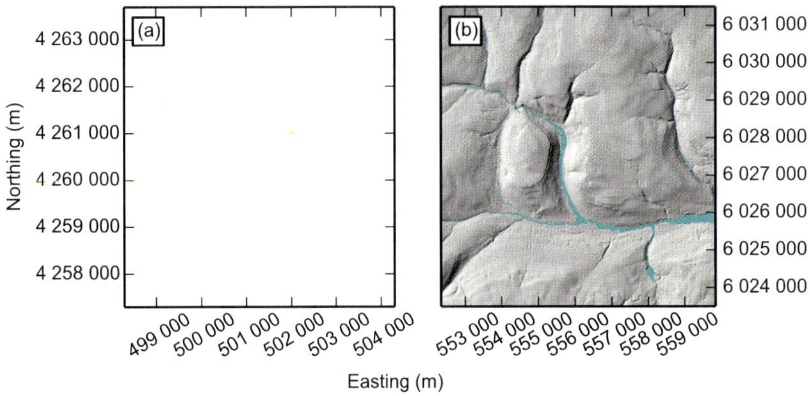

Figure 5. Shaded relief maps showing **(a)** FEMA flood risk map for the Russian River, CA, UTM Zone 10° N, and **(b)** EA flood risk map for the River Swale, UK, UTM Zone 30° N. In some parts of the landscape the published flood maps do not extend all the way up the catchments.

WGS84 coordinate system (Ohio and NC Zone 17° N, Russian River Zone 10° N). We separate the flood zones into two categories: areas within the 100-year flood (blue), with a 1 % annual chance of flooding, and areas with a greater than 100-year flood risk (less than 1 % annual risk of flooding). In order to compare these maps to our method, we gridded the FEMA flood risk maps with a resolution of 1 m. The Coweeta field site in North Carolina did not have a complete flood risk map for the catchment and therefore could not be included in this analysis.

For the River Swale field site in the UK, flood risk maps were obtained from the Environment Agency's (EA) Risk of Flooding from Rivers and Sea dataset, which divides the landscape into 50 by 50 m cells (https://data.gov.uk/dataset/risk-of-flooding-from-rivers-and-sea1). Each cell is categorised into one of four flood risk likelihood categories: high (3.3 % annual chance of flooding), medium (between 3.3 and 1 %), low (between 1 and 0.1 %), or very low (< 0.1 %). The dataset is created by hydraulic modelling, including information about the state of flood defenses and local stage heights as inputs to the model. The data were re-projected from the British National Grid coordinate system to the UTM WGS84 datum, Zone 30° N. In order to keep the comparison consistent with the sites from the US, each pixel was classified into the same two categories as for the FEMA maps, with areas of flood risk identified as having greater than 1 % annual chance of flooding. The dataset is provided as vector data: to compare with the floodplain identified by the our method, we gridded the vector dataset at 5 m resolution (the same as the input DEM). Figure 5 shows examples of the FEMA and EA flood maps for each study area.

The r, s, and Q values for each site are reported in Table 4, with a visual comparison between the method and the published flood maps shown in Fig. 6. We also report the quality values for floodplains extracted from the United States Geological Survey's 1/3 arcsec National Elevation Dataset (NED), gridded at 10 m, in order to test the sensitivity of our method to grid resolution. The USGS NED is a seamless dataset created for the conterminous US, using a variety of elevation products which is updated on a 2-month cycle. The method was most similar to the flood risk maps for the Russian River, CA, with the highest overall quality value ($Q = 0.67$ for the 1 m DEM and 0.68 for the 10 m DEM). The method has a higher sensitivity than reliability for both DEM datasets, with $s = 0.97$ and $r = 0.74$ for the 1 m DEM, compared to $s = 0.96$ and $r = 0.70$ for the 10 m DEM. For both the Mid Bailey Run and Russian River field sites, the sensitivity is higher than the reliability for all of the DEM resolutions tested (Table 4). However for the River Swale site, the reliability is higher than the sensitivity ($r = 0.84$, $s = 0.65$).

Table 4. Results of the reliability (r), sensitivity (s), and overall quality (Q) analysis for each site

Field site	Grid resolution (m)	r	s	Q
Mid Bailey Run, OH	1	0.73	0.76	0.59
	10	0.77	0.80	0.65
Russian River, CA	1	0.74	0.97	0.67
	10	0.70	0.96	0.68
River Swale, UK	5	0.84	0.65	0.58
South Fork Eel River, CA	1	0.65	0.72	0.52
Le Sueur River, MN	1	0.58	0.54	0.39
Mattole River, CA	10	0.58	0.65	0.44
Clearwater River, WA	10	0.56	0.55	0.39

4.2 Comparison with mapped terraces

We also compare the features extracted by our method to field-mapped terraces from four field sites throughout the US: the South Fork Eel River, CA (Seidl and Dietrich, 1992);

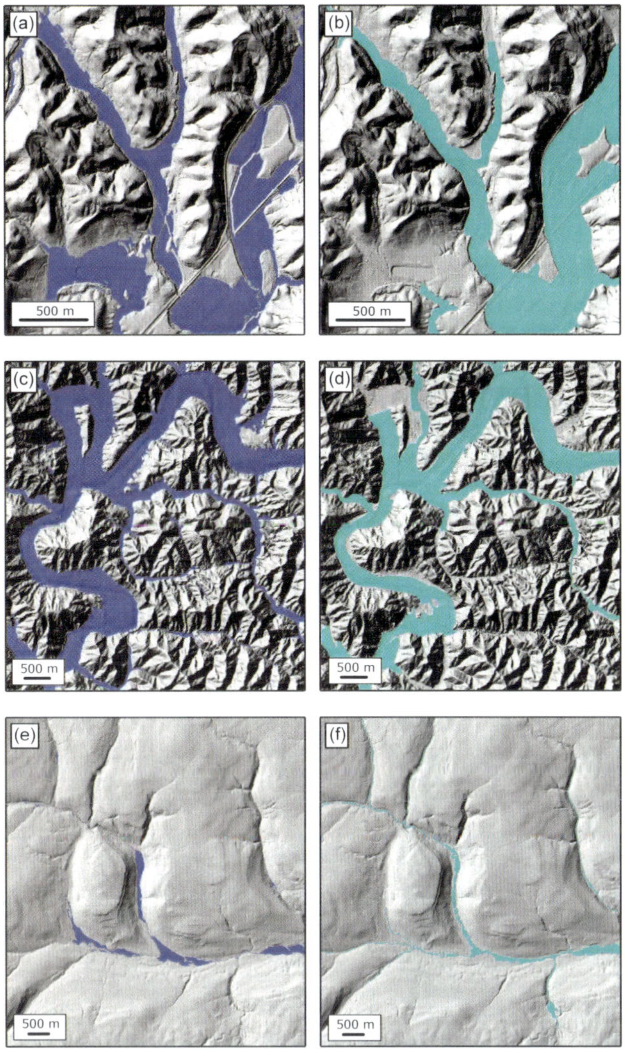

Figure 6. Shaded relief maps for each field site showing a comparison between the predicted floodplains **(a, c, e)** and the published FEMA/EA maps **(b, d, f)**. **(a–b)** Mid Bailey Run, OH. **(c–d)** Russian River, CA. **(e–f)** River Swale, UK.

the Le Sueur River, MN (Gran et al., 2009); the Mattole River, CA (Dibblee and Minch, 2008); and the Clearwater River, WA (Wegmann and Pazzaglia, 2002). Two of these sites had 1 m lidar-derived DEMs (the South Fork Eel and Le Sueur rivers). For the remaining two sites, 10 m DEMs were derived from the USGS 1/3 arcsec NED, following Limaye and Lamb (2016). Terraces in the South Fork Eel River and the Le Sueur River were digitised from field mapping carried out in previous studies (Seidl and Dietrich, 1992; Gran et al., 2009), constrained by the hillshaded DEMs. Terraces from the Mattole River and the Clearwater River were digitised by Limaye and Lamb (2016) from geological maps, with the terraces mapped by Dibblee and Minch (2008) for the Mattole River, and Wegmann and Pazzaglia (2002) for the Clearwater River. We ran our method in the swath setting for each of these sites, so that the terraces were mapped compared to the main stem channel of interest in each site. The thresholds for terrace identification (R_c and S) were set statistically for each site using the quantile–quantile plots. In order to quantify the difference between our method and the digitised terraces, we calculated the r and s values following the same methodology as for the floodplain comparison (Table 4).

Figure 7 shows a visual comparison of the predicted and digitised terraces from the two sites with 1 m lidar-derived DEMs. In general there was good spatial correlation between the two terrace datasets for each field site, although in some cases the automated method did not identify all terraces at high elevations compared to the modern channel. The South Fork Eel River had the highest values of both r (0.65) and s (0.72). The comparison between the two terrace datasets for the field sites with 10 m DEMs is shown in Fig. 8. These sites had lower r and s values than that of the South Fork Eel River, but were comparable to the values for the Le Sueur River (e.g. Table 4).

5 Discussion

5.1 Floodplains

The results outlined above compare our method of automatic feature extraction to various datasets of both floodplains and terraces. In order to test the ability of our method in identifying floodplains, we compared the delineated geomorphic floodplain to both field-mapped floodplain initiation points and hydrological modelling predictions. We found that our method predicts the location of the field-mapped FIPs to within tens of metres for both field sites (Mid Bailey Run, OH, and Coweeta, NC). The reliability and sensitivity values were highest for the Coweeta field sites, with a value of $r = 0.78$ and $s = 1$, which indicates that there were no false negatives in this field site. Table 3 shows that in many cases the error between the mapped and predicted FIPs is within the same order of magnitude as the error on the field-mapped coordinates (≈ 1 m for Coweeta and ≈ 6 m for Mid Bailey Run). In isolated cases in the Mid Bailey Run site, the error was higher between the mapped and predicted FIPs (around 90 m for two of the points), where the mapped FIP was located in narrow headwater valleys (Fig. 4). Furthermore, the predicted floodplain in the majority of cases was located downstream of the mapped FIPs in Mid Bailey Run (Table 3). This is not surprising, as our method is based on identifying areas of low gradient, which is calculated based on polynomial surface fitting with a specified window radius (Sect. 2.2). Small pockets of alluviation in narrow valleys may therefore be missed by the method if the width of the floodplain is less than that of the window radius or the DEM resolution.

We also validated our method against published flood maps for three of our field sites (Mid Bailey Run, OH; Russian River, CA; and River Swale, UK). The quality analy-

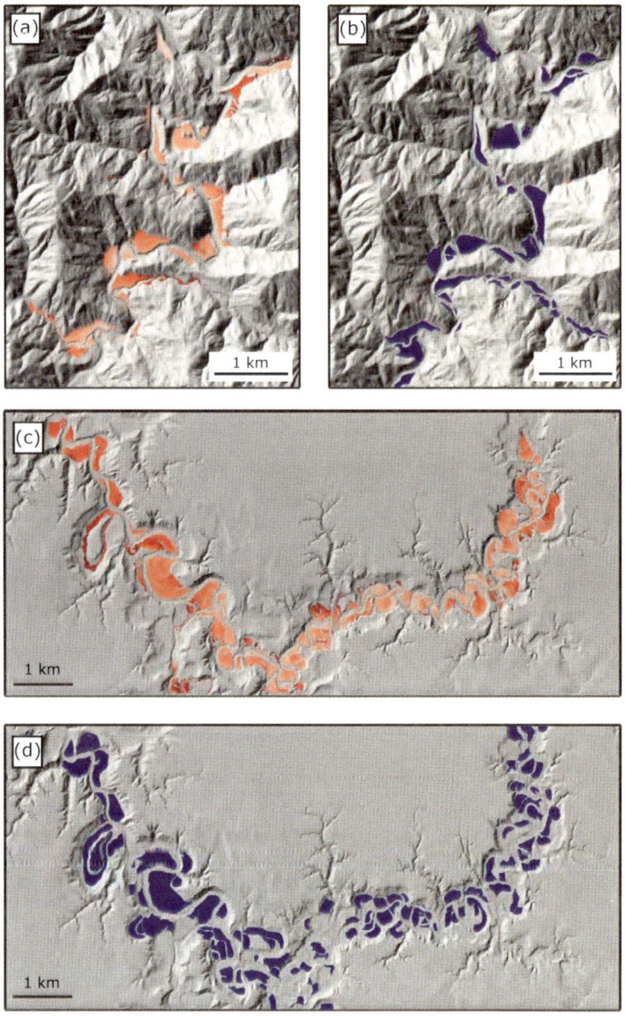

Figure 7. Shaded relief maps for the two field sites with lidar-derived DEMs showing a comparison between the predicted terraces (red) and the digitised terraces (blue). The predicted terraces are coloured by elevation compared to the channel, where darker red indicates higher elevation. **(a–b)** South Fork Eel River, CA. Maximum terrace height is 43 m. **(c–d)** Le Sueur River, MN. Maximum terrace height is 9.5 m.

sis for this comparison (Table 4 and Fig. 6) suggests that there is in general a good correlation between our method and the published flood maps, with high values for reliability ($r \geq 0.7$), sensitivity ($s \geq 0.65$), and overall quality ($Q \geq 0.58$) for each field site. The results for both the Russian River and Mid Bailey Run showed higher sensitivity values than reliability, suggesting that the our method predicted more false positives than false negatives. In each field site, the published flood maps were classified to define the 1 % annual chance of flooding, or the 100-year return period flood event. It may therefore be expected that our geomorphic-based method would delineate a larger floodplain than is flooded in a 100-year return period event. The results for the River Swale, however, show a higher reliability than sensitivity, suggesting that more false negatives were predicted than false positives. This may be due to methodological differences in the production of this flood map by the Environment Agency (UK) compared to the US sites. Figure 6f shows the published flood map for the River Swale site, which, in comparison to the FEMA flood maps (Fig. 6b and d), extends into the headwaters of the channel network. As these areas do not have low-gradient surfaces next to the channel, they may not be selected by our method. This may account for the higher number of false negatives predicted at this site.

Published flood maps are useful in providing an independent estimate of likely floodplains in each field site. However, there are potential limitations to these maps which must be carefully considered and may result in some of the differences compared to geomorphic floodplain prediction techniques. Hydrodynamic models have a large number of parameters, which require careful calibration with field and hydraulic data, such as channel roughness and discharge data from gauging stations. Furthermore, due to the time-consuming and expensive nature of these studies, flood maps are often not produced for small catchment sizes and may therefore be incomplete on a landscape scale (e.g. Fig. 5). There may also be differences in the methodology used in producing these maps for each site, depending on the input topographic data and modelling software used. However, despite these discrepancies between the flood maps we find a good spatial correlation between these and the predictions from our method (Fig. 6).

In order to test the sensitivity of our method to grid resolution, we also ran the floodplain extraction using 10 m DEMs derived from the USGS NED for two of the field sites (Russian River, CA, and Mid Bailey Run, OH), as well as testing it on the River Swale in the UK (5 m resolution DEM). We found there was little difference in the reliability and sensitivity results when compared to the 1 m DEMs (Table 4). This suggests that our method is relatively insensitive to grid resolution, allowing the identification of floodplain features on coarser-resolution DEMs. Furthermore, in the Mid Bailey Run field site, the method performed better on the 10 m data compared to the 1 m DEM. High-resolution topographic data may contain both small-wavelength topographic noise caused by tree throw and biotic activity (Roering et al., 2010; Marshall and Roering, 2014), as well as synthetic noise from point cloud processing (Liu, 2008; Meng et al., 2010). This noise may affect the calculation of topographic metrics (Grieve et al., 2016c), potentially leading to differences in the location of extracted floodplains or terraces compared to the lower-resolution data.

5.2 Terraces

We also tested the ability of our method to identify fluvial terraces in four field sites (South Fork Eel River, CA; Le Sueur River, MN; Mattole River, CA; and Clearwater River, WA) by comparing to digitised terrace maps. Two of these field sites had 1 m lidar-derived DEMs (Fig. 7), whereas two had 10 m DEMs from the USGS NED (Fig. 8). The quality analysis for the 1 m DEMs showed the higher reliability and sensitivity values for the South Fork Eel River site ($r = 0.65$ and $s = 0.72$), with comparable values for the remaining three field sites. This may be due to the influence of topographic structure on terrace identification. The portion of the Eel River DEM analysed here has higher relief, with a maximum elevation of 290 m above the nearest channel, compared to the lower-relief landscape covered by the DEM for the Le Sueur River, with a maximum elevation of 40 m above the nearest channel. As our method relies on the distribution of relief relative to the channel in order to select the threshold for terrace identification, it will work best in areas where there is a greater contrast between the slope and relief of the terrace surfaces compared to the surrounding topography, such as steep mountainous areas. This is similar to other semi-automated terrace extraction methods (e.g. Stout and Belmont, 2014; Hopkins and Snyder, 2016). The Le Sueur River is currently incising through Pleistocene tills, forming a low-gradient surface or plateau (Fisher, 2003; Gran et al., 2009; Belmont et al., 2011a). High-altitude, low-gradient surfaces, such as relict plateaus, may result in error in the method due to the difficulty in distinguishing the distribution of terrace elevations from these low-relief surfaces. The Le Sueur River basin is also heavily influenced by human land use, which makes feature extraction challenging (Passalacqua et al., 2012). The results of the quality analysis for the eight field sites (Table 4) showed that the method performed better in the floodplain identification compared to the terrace identification. This may be due to the fact that, with the exception of the South Fork Eel River, the sites used for terrace extraction are lower relief than those used to test the floodplain extraction (e.g. Figs. 6–8).

Another potential cause of error between the predicted and digitised terrace locations may be problems in distinguishing whether features represent the modern floodplain or terraces. In our method a minimum height above the modern channel is set, where pixels above this height are classified as terrace and below this height as floodplain. In some cases, particularly where the terraces are at a similar elevation to that of the modern channel, our method may mistakenly identify terraces as being part of the modern floodplain, or vice versa. An example of this may be the Clearwater River site, where our method had lower indices of r and Q (Fig. 8c and d and Table 4). In this site, the digitised terraces are close in elevation to the modern channel, with a maximum terrace height of 13 m. Furthermore, in some cases our method did not select all of the terraces identified by the field mapping, particularly at the highest elevations compared to the modern channel (e.g. Fig. 7c and d). This may be the case if the threshold for elevation compared to the channel selected by the quantile–quantile plot is lower than that of the highest terrace elevations. This can be examined for the landscape in question by a visual inspection of the quantile–quantile plots

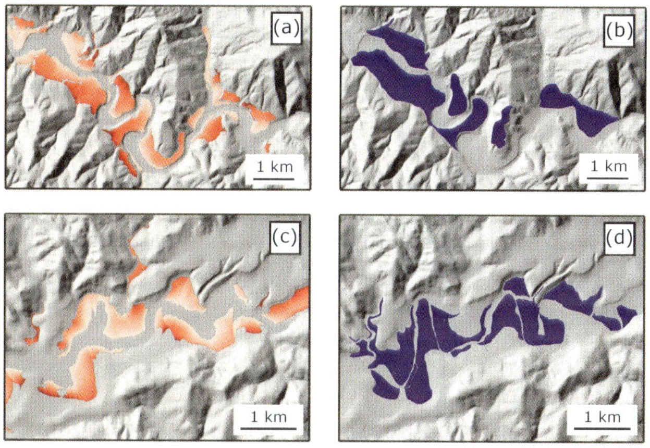

Figure 8. Shaded relief maps for the two field sites with 10 m resolution DEMs from the USGS NED showing a comparison between the predicted terraces (red) and the digitised terraces (blue). The predicted terraces are coloured by elevation compared to the channel, where darker red indicates higher elevation. **(a–b)** Mattole River, CA. Maximum terrace height is 50 m. **(c–d)** Clearwater River, WA. Maximum terrace height is 13 m.

and the location of the threshold compared to the distribution of channel relief (e.g. Fig. 2). Our method fits a Gaussian distribution to the quantile–quantile plots, and selects the thresholds as the deviation of the real data from this distribution, as a simple general model of elevation distributions that can be applied across multiple landscapes. However, in some landscapes, the distribution of elevations may not be accurately represented by a Gaussian distribution. A future avenue for development of this method may be to include multiple models for elevation distributions from which to select the thresholds of elevation and gradient.

However, despite these limitations, the selection of the threshold from quantile–quantile plots means that our method does not require the input of any independent datasets or field mapping. Semi-automated methods of terrace identification, where the terrace polygons are manually edited by the user, are particularly useful in areas where independent datasets of terrace locations are available for calibration, and may be more appropriate than our method on site-specific scales (e.g. Stout and Belmont, 2014). However, the selection of thresholds based on a statistical approach means that our method can be applied in areas where these data do not exist, on a broader landscape scale, or as a rapid first-order predictor of terrace locations.

In addition to the field sites with lidar-derived DEMs, we also tested our method against digitised terraces from two sites with 10 m DEMs gridded from the USGS NED, to examine the performance of the method at lower grid resolution. Figure 8 shows the results of the terrace identification on the 10 m resolution data. The reliability and sensitivity of the method for these two sites (Table 4) was lower than that of the South Fork Eel River, but comparable to that of the

Le Sueur River. This suggests that the method is able to successfully select terraces at lower grid resolutions. Although there are some differences between the terraces predicted by the method and those digitised in the field, the majority of the terrace features evident from a visual inspection of the hillshaded DEMs are correctly identified by the algorithm (Fig. 8). In some cases, some terrace-like features that can be seen on the hillshaded DEMs are not identified in the digitised terrace maps (e.g. Fig. 8b). This may be due to error in the mapping of terrace surfaces in the field, or discrepancies resulting from the digitisation process.

An objective, landscape-scale method of identifying floodplain and terrace features has numerous applications in the geomorphological and hydrological communities. For example, terrace surfaces have been used to examine the response of fluvial systems to tectonic and climatic perturbations (e.g. Merritts et al., 1994), and to investigate the relative importance of lateral and vertical channel incision (e.g. Finnegan and Dietrich, 2011). Analysis of terrace areas can be used to quantify sediment budgets and estimate storage volumes over millennial timescales (e.g. Trimble, 1999; Brown et al., 2009; Belmont et al., 2011b; Blöthe and Korup, 2013). Our new method facilitates the rapid extraction of terrace surfaces either across the whole landscape or compared to a representative channel of interest. It allows the user to investigate how various metrics, such as elevation compared to the channel, slope, and curvature, vary both within and between individual terrace surfaces (e.g. Fig. 7). These metrics could be used in order to examine how terrace heights vary with distance along channel profiles, for example, or to identify signatures of deformation corresponding to tectonic processes (Avouac and Peltzer, 1993; Lavé and Avouac, 2000; Pazzaglia and Brandon, 2001; Viveen et al., 2014).

5.3 Research needs: fully automated feature extraction

A key goal for the Earth surface research community is to develop fully automated methods of feature extraction from DEMs in order to avoid expensive and time-consuming field mapping, and to investigate the controls on geomorphic processes at a landscape scale. Our new method of floodplain and terrace delineation attempts to meet some of these research needs by allowing the statistical determination of the thresholds for feature extraction. However, our method still requires the input of some user-defined parameters. If the method is run across the whole landscape, the user must set a threshold stream order for the calculation of elevation compared to the nearest channel. This is necessary so that each pixel is mapped to the main channel along which floodplains or terraces have formed, rather than narrow tributary valleys. This threshold can be determined by the user based on a visual inspection of the DEM compared to the channel network. If the user runs the method based on the swath mode, the width of the swath profile must be set. This can also be done based on a visual inspection of the DEM to provide a sufficiently wide swath compared to the valleys in the landscape. Furthermore, if the method is run in the swath mode, then a minimum terrace height must be set in order to delineate between floodplains and fluvial terraces.

However, future development of new algorithms, such as extraction of valley widths, would allow these parameters to be set based on the topographic data alone. Our method represents a first step towards this goal of fully automated geomorphic feature identification, which can be improved upon with future research. The combination of different algorithms for terrain analysis, such as hillslope flow routing, channel network extraction, floodplains, and fluvial terraces, would allow an objective landscape-scale investigation of the controls on geomorphic processes.

6 Conclusions

We have presented a novel method for the geomorphometric delineation of floodplain and fluvial terrace features from topographic data. Unlike previous methods, which tend to require calibration with additional datasets, our method selects floodplain and terrace features using thresholds of local gradient and elevation compared to the nearest channel, which are calculated statistically from the DEM. Furthermore, the floodplain or terrace surfaces do not need to be manually edited by the user at any point during the process. Our method can be run either across the whole landscape or from a topographic swath profile where features can be compared to a specific channel of interest.

In order to test the performance of our method we have compared it to field-mapped floodplains and terraces from eight field sites with a range of topographies and grid resolutions. We find that our method performs well when compared to field-mapped floodplain initiation points, published flood risk maps, and digitised terrace surfaces. Our method works particularly well in higher-relief areas, such as the Russian and South Fork Eel rivers (CA), where the floodplain and terrace features are constrained within valleys. It is relatively insensitive to grid resolution, allowing the successful extraction of floodplain and terrace features at resolutions of 1–10 m.

Our new method has numerous applications in both the hydrological and geomorphological communities. It can allow the rapid extraction of floodplain features in areas where the data required for detailed hydrological modelling studies are unavailable, facilitating investigation of flood response, sediment transport, and alluviation. Furthermore, the automated extraction of terrace locations, heights, and other metrics could be used to examine the response of fluvial systems to climatic and tectonic perturbations, as well as the relative importance of lateral and vertical channel incision.

Author contributions. FJC, SMM, DTM, and DAV wrote the software for the feature extraction. MDH, LJS, and FJC collected the field data for floodplain validation; ABL collected the field data for terrace validation. FJC performed the analyses, created the figures, and wrote the manuscript with contributions from the other authors.

Competing interests. The authors declare that they have no conflict of interest.

Acknowledgements. Fiona J. Clubb is funded by the Carnegie Trust for the Universities of Scotland and NERC grant NE/P012922/1. Simon M. Mudd is funded by NERC grant NE/P015905/1 and U.S. Army Research Office contract number W911NF-13-1-0478. David T. Milodowski is funded by NERC grant NE/K01627X/1 and Declan A. Valters is funded by NERC grant NE/L501591/1. Louise J. Slater was supported by a NERC PhD studentship. Ajay B. Limaye acknowledges support from the National Center for Earth-Surface Dynamics 2 Synthesis Postdoctoral Program. We are also grateful for additional financial support from the British Society for Geomorphology and the Royal Geographical Society with IBG. We would like to thank Greg Hancock, Andrew Wickert, and two other reviewers for their helpful comments and suggestions, along with Stuart Grieve and Elizabeth Dingle for their help with fieldwork.

References

Avouac, J. P. and Peltzer, G.: Active tectonics in southern Xinjiang, China: Analysis of terrace riser and normal fault scarp degradation along the Hotan-Qira Fault System, J. Geophys. Res.-Sol. Ea., 98, 21773–21807, https://doi.org/10.1029/93JB02172, 1993.

Belmont, P.: Floodplain width adjustments in response to rapid base level fall and knickpoint migration, Geomorphology, 128, 92–102, https://doi.org/10.1016/j.geomorph.2010.12.026, 2011.

Belmont, P., Gran, K., Jennings, C. E., Wittkop, C., and Day, S. S.: Holocene landscape evolution and erosional processes in the Le Sueur River, central Minnesota, Field Guides, 24, 439–455, https://doi.org/10.1130/2011.0024(21), 2011a.

Belmont, P., Gran, K. B., Schottler, S. P., Wilcock, P. R., Day, S. S., Jennings, C., Lauer, J. W., Viparelli, E., Willenbring, J. K., Engstrom, D. R., and Parker, G.: Large Shift in Source of Fine Sediment in the Upper Mississippi River, Environ. Sci. Technol., 45, 8804–8810, https://doi.org/10.1021/es2019109, 2011b.

Beven, K.: Prophecy, reality and uncertainty in distributed hydrological modelling, Adv. Water Resour., 16, 41–51, https://doi.org/10.1016/0309-1708(93)90028-E, 1993.

Beven, K.: TOPMODEL: A critique, Hydrol. Process., 11, 1069–1085, https://doi.org/10.1002/(SICI)1099-1085(199707)11:9<1069::AID-HYP545>3.0.CO;2-O, 1997.

Beven, K. J. and Kirkby, M. J.: A physically based, variable contributing area model of basin hydrology/Un modèle á base physique de zone d'appel variable de l'hydrologie du bassin versant, Hydrological Sciences Bulletin, 24, 43–69, https://doi.org/10.1080/02626667909491834, 1979.

Beven, K. J., Lamb, R., Quinn, P., Romanowicz, R., and Freer, J.: TOPMODEL, in: Computer Models of Watershed Hydrology, edited by: Singh, V. P., 627–668, Water Resource Publications, Colorado, 1995.

Blöthe, J. H. and Korup, O.: Millennial lag times in the Himalayan sediment routing system, Earth Planet. Sc. Lett., 382, 38–46, https://doi.org/10.1016/j.epsl.2013.08.044, 2013.

Booij, M. J.: Impact of climate change on river flooding assessed with different spatial model resolutions, J. Hydrol., 303, 176–198, https://doi.org/10.1016/j.jhydrol.2004.07.013, 2005.

Braun, J. and Willett, S. D.: A very efficient O(n), implicit and parallel method to solve the stream power equation governing fluvial incision and landscape evolution, Geomorphology, 180–181, 170–179, https://doi.org/10.1016/j.geomorph.2012.10.008, 2013.

Brown, A. G., Carey, C., Erkens, G., Fuchs, M., Hoffmann, T., Macaire, J.-J., Moldenhauer, K.-M., and Walling, D. E.: From sedimentary records to sediment budgets: Multiple approaches to catchment sediment flux, Geomorphology, 108, 35–47, https://doi.org/10.1016/j.geomorph.2008.01.021, 2009.

Bull, W. B.: Geomorphic responses to climatic change, Oxford University Press, New York, available at: https://www.osti.gov/scitech/biblio/5603696 (last access: 9 March 2017), 1991.

Clubb, F. J., Mudd, S. M., Milodowski, D. T., Hurst, M. D., and Slater, L. J.: Objective extraction of channel heads from high-resolution topographic data, Water Resour. Res., 50, 4283–4304, https://doi.org/10.1002/2013WR015167, 2014.

Clubb, F. J., Mudd, S. M., Grieve, S. W. D., Milodowski, D. T., Valters, D. A., and Hurst, M. D.: LSDTerraceModel v1.0, Zenodo, https://doi.org/10.5281/zenodo.824205, 2017.

Cobby, D. M., Mason, D. C., Horritt, M. S., and Bates, P. D.: Two-dimensional hydraulic flood modelling using a finite-element mesh decomposed according to vegetation and topographic features derived from airborne scanning laser altimetry, Hydrol. Process., 17, 1979–2000, https://doi.org/10.1002/hyp.1201, 2003.

Committee on FEMA Flood Maps: Mapping the Zone: Improving Flood Map Accuracy, The National Academies Press, 500 Fifth Street, N.W. Washington, DC 20001, available at: http://www.nap.edu/catalog/12573/mapping-the-zone-improving-flood-map-accuracy (last access: 23 June 2016), 2009.

Degiorgis, M., Gnecco, G., Gorni, S., Roth, G., Sanguineti, M., and Taramasso, A. C.: Classifiers for the detection of flood-prone areas using remote sensed elevation data, J. Hydrol., 470–471, 302–315, https://doi.org/10.1016/j.jhydrol.2012.09.006, 2012.

Demoulin, A., Bovy, B., Rixhon, G., and Cornet, Y.: An automated method to extract fluvial terraces from digital elevation models: The Vesdre valley, a case study in eastern Belgium, Geomorphology, 91, 51–64, https://doi.org/10.1016/j.geomorph.2007.01.020, 2007.

Dibblee, T. W. and Minch, J. A.: Geologic Map of the Point Delgada and Garberville 15 minute Quadrangles, Humboldt and Mendocino Counties, California, 2008.

Dingle, E. H., Sinclair, H. D., Attal, M., Milodowski, D. T., and Singh, V.: Subsidence control on river morphology and grain size in the Ganga Plain, Am. J. Sci., 316, 778–812, https://doi.org/10.2475/08.2016.03, 2016.

Dodov, B. and Foufoula-Georgiou, E.: Floodplain morphometry extraction from a high-resolution digital elevation model: a simple algorithm for regional analysis studies, IEEE Geosci. Remote Sens. Lett., 3, 410–413, https://doi.org/10.1109/LGRS.2006.874161, 2006.

Fawcett, T.: An introduction to ROC analysis, Pattern Recognition Letters, 27, 861–874, https://doi.org/10.1016/j.patrec.2005.10.010, 2006.

Finnegan, N. J. and Dietrich, W. E.: Episodic bedrock strath terrace formation due to meander migration and cutoff, Geology, 39, 143–146, https://doi.org/10.1130/G31716.1, 2011.

Fisher, T. G.: Chronology of glacial Lake Agassiz meltwater routed to the Gulf of Mexico, Quaternary Res., 59, 271–276, https://doi.org/10.1016/S0033-5894(03)00011-5, 2003.

Gran, K. B., Belmont, P., Day, S. S., Jennings, C., Johnson, A., Perg, L., and Wilcock, P. R.: Geomorphic evolution of the Le Sueur River, Minnesota, USA, and implications for current sediment loading, Geol. S. Am. S., 451, 119–130, https://doi.org/10.1130/2009.2451(08), 2009.

Gran, K. B., Finnegan, N., Johnson, A. L., Belmont, P., Wittkop, C., and Rittenour, T.: Landscape evolution, valley excavation, and terrace development following abrupt postglacial base-level fall, Geol. Soc. Am. Bull., 125, 1851–1864, https://doi.org/10.1130/B30772.1, 2013.

Grieve, S. W., Mudd, S. M., and Hurst, M. D.: How long is a hillslope?, Earth Surf. Proc.d Land., 41, 1039–1054, https://doi.org/10.1002/esp.3884, 2016a.

Grieve, S. W. D., Mudd, S. M., Hurst, M. D., and Milodowski, D. T.: A nondimensional framework for exploring the relief structure of landscapes, Earth Surf. Dynam., 4, 309–325, https://doi.org/10.5194/esurf-4-309-2016, 2016b.

Grieve, S. W. D., Mudd, S. M., Milodowski, D. T., Clubb, F. J., and Furbish, D. J.: How does grid-resolution modulate the topographic expression of geomorphic processes?, Earth Surf. Dynam., 4, 627–653, https://doi.org/10.5194/esurf-4-627-2016, 2016c.

Grimaldi, S., Petroselli, A., Arcangeletti, E., and Nardi, F.: Flood mapping in ungauged basins using fully continuous hydrologic–hydraulic modeling, J. Hydrol., 487, 39–47, https://doi.org/10.1016/j.jhydrol.2013.02.023, 2013.

Guzzetti, F., Stark, C. P., and Salvati, P.: Evaluation of Flood and Landslide Risk to the Population of Italy, Environ. Manage., 36, 15–36, https://doi.org/10.1007/s00267-003-0257-1, 2005.

Hartmann, D., Klein Tank, A., Rusticucci, M., Alexander, L. V., Bronnimann, S., Charabi, Y., Dentener, F., Dlugokencky, E. J., Easterling, D. R., Kaplan, A., Soden, B. J., Thorne, P. W., Wild, M., and Zhai, P. M.: Observations: Atmosphere and Surface, in: Climate Change 2013: The Physical Science Basis. Contribution of Working Group I to the Fifth Assessment Report of the Intergovernmental Panel on Climate Change, edited by: Stocker, T. F., Qin, D., Plattner, G., Tignor, M., Allen, S., Boschung, J., Nauels, A., Xia, Y., Bex, V., and Midgley, P., Cambridge University Press, Cambridge, UK and New York, NY, USA, 2013.

He, L., Chao, Y., and Suzuki, K.: A Run-Based Two-Scan Labeling Algorithm, IEEE T. Image Process., 17, 749–756, https://doi.org/10.1109/TIP.2008.919369, 2008.

Heipke, C., Mayer, H., Wiedemann, C., and Jamet, O.: Automated reconstruction of topographic objects from aerial images using vectorized map information, Int. Arch. Photogramm. Remote Sens. Special Inf. Sci., 23, 47–56, 1997.

Hergarten, S., Robl, J., and Stüwe, K.: Extracting topographic swath profiles across curved geomorphic features, Earth Surf. Dynam., 2, 97–104, https://doi.org/10.5194/esurf-2-97-2014, 2014.

Hopkins, A. J. and Snyder, N. P.: Performance evaluation of three DEM-based fluvial terrace mapping methods, Earth Surf. Proc. Land., 41, 1144–1152, https://doi.org/10.1002/esp.3922, 2016.

Horritt, M. S. and Bates, P. D.: Evaluation of 1D and 2D numerical models for predicting river flood inundation, J. Hydrol., 268, 87–99, https://doi.org/10.1016/S0022-1694(02)00121-X, 2002.

Hunter, N. M., Bates, P. D., Horritt, M. S., and Wilson, M. D.: Simple spatially-distributed models for predicting flood inundation: A review, Geomorphology, 90, 208–225, https://doi.org/10.1016/j.geomorph.2006.10.021, 2007.

Hurst, M. D., Mudd, S. M., Walcott, R., Attal, M., and Yoo, K.: Using hilltop curvature to derive the spatial distribution of erosion rates, J. Geophys. Res.-Earth, 117, F02017, https://doi.org/10.1029/2011JF002057, 2012.

Jain, V., Fryirs, K., and Brierley, G.: Where do floodplains begin? The role of total stream power and longitudinal profile form on floodplain initiation processes, Geol. Soc. Am. Bull., 120, 127–141, https://doi.org/10.1130/B26092.1, 2008.

Kim, J., Warnock, A., Ivanov, V. Y., and Katopodes, N. D.: Coupled modeling of hydrologic and hydrodynamic processes including overland and channel flow, Adv. Water Resour., 37, 104–126, https://doi.org/10.1016/j.advwatres.2011.11.009, 2012.

Kirkby, M. J.: Hydrograph modelling strategies, in: Process in Physical and Human Geography, edited by: Peel, R., Chisholm, M., and Haggett, P., 69–90, Heinemann, London, 1975.

Lashermes, B., Foufoula-Georgiou, E., and Dietrich, W. E.: Channel network extraction from high resolution topography using wavelets, Geophys. Res. Lett., 34, L23S04, https://doi.org/10.1029/2007GL031140, 2007.

Lavé, J. and Avouac, J. P.: Active folding of fluvial terraces across the Siwaliks Hills, Himalayas of central Nepal, J. Geophys. Res.-Sol. Ea., 105, 5735–5770, https://doi.org/10.1029/1999JB900292, 2000.

Limaye, A. B. S. and Lamb, M. P.: Numerical model predictions of autogenic fluvial terraces and comparison to climate change expectations, J. Geophys. Res.-Earth, 121, 512–544, https://doi.org/10.1002/2014JF003392, 2016.

Liu, X.: Airborne LiDAR for DEM generation: some critical issues, Prog. Phys. Geogr., 32, 31–49, https://doi.org/10.1177/0309133308089496, 2008.

Liu, Y. and Gupta, H. V.: Uncertainty in hydrologic modeling: Toward an integrated data assimilation framework, Water Resour. Res., 43, W07401, https://doi.org/10.1029/2006WR005756, 2007.

Manfreda, S., Di Leo, M., and Sole, A.: Detection of Flood-Prone Areas Using Digital Elevation Models, J. Hydrol. Eng., 16, 781–790, https://doi.org/10.1061/(ASCE)HE.1943-5584.0000367, 2011.

Manfreda, S., Nardi, F., Samela, C., Grimaldi, S., Taramasso, A. C., Roth, G., and Sole, A.: Investigation on the use of geomorphic approaches for the delineation of flood prone areas, J. Hydrol., 517, 863–876, https://doi.org/10.1016/j.jhydrol.2014.06.009, 2014.

Marshall, J. A. and Roering, J. J.: Diagenetic variation in the Oregon Coast Range: Implications for rock strength, soil production, hillslope form, and landscape evolution, J. Geophys. Res.-Earth, 119, 1395–1417, https://doi.org/10.1002/2013JF003004, 2014.

Meng, X., Currit, N., and Zhao, K.: Ground Filtering Algorithms for Airborne LiDAR Data: A Review of Critical Issues, Remote Sens., 2, 833–860, https://doi.org/10.3390/rs2030833, 2010.

Merritts, D. J., Vincent, K. R., and Wohl, E. E.: Long river profiles, tectonism, and eustasy: A guide to interpreting fluvial terraces, J. Geophys. Res.-Sol. Ea., 99, 14031–14050, https://doi.org/10.1029/94JB00857, 1994.

Molloy, I. and Stepinski, T. F.: Automatic mapping of valley networks on Mars, Comput. Geosci., 33, 728–738, https://doi.org/10.1016/j.cageo.2006.09.009, 2007.

Noman, N. S., Nelson, E. J., and Zundel, A. K.: Review of Automated Floodplain Delineation from Digital Terrain Models, J. Water Res. Pl.-ASCE, 127, 394–402, https://doi.org/10.1061/(ASCE)0733-9496(2001)127:6(394), 2001.

O'Callaghan, J. F. and Mark, D. M.: The extraction of drainage networks from digital elevation data, Computer Vision, Graphics, and Image Processing, 28, 323–344, https://doi.org/10.1016/S0734-189X(84)80011-0, 1984.

Orlandini, S., Tarolli, P., Moretti, G., and Dalla Fontana, G.: On the prediction of channel heads in a complex alpine terrain using gridded elevation data, Water Resour. Res., 47, W02538, https://doi.org/10.1029/2010WR009648, 2011.

Passalacqua, P., Do Trung, T., Foufoula-Georgiou, E., Sapiro, G., and Dietrich, W. E.: A geometric framework for channel network extraction from lidar: Nonlinear diffusion and geodesic paths, J. Geophys. Res.-Earth, 115, F01002, https://doi.org/10.1029/2009JF001254, 2010a.

Passalacqua, P., Tarolli, P., and Foufoula-Georgiou, E.: Testing space-scale methodologies for automatic geomorphic feature extraction from lidar in a complex mountainous landscape, Water Resour. Res., 46, W11535, https://doi.org/10.1029/2009WR008812, 2010b.

Passalacqua, P., Belmont, P., and Foufoula-Georgiou, E.: Automatic geomorphic feature extraction from lidar in flat and engineered landscapes, Water Resour. Res., 48, W03528, https://doi.org/10.1029/2011WR010958, 2012.

Pazzaglia, F. J.: 9.22 Fluvial Terraces, in: Treatise on Geomorphology, edited by: Shroder, J. F., 379–412, Academic Press, San Diego, https://doi.org/10.1016/B978-0-12-374739-6.00248-7, 2013.

Pazzaglia, F. J. and Brandon, M. T.: A Fluvial Record of Long-term Steady-state Uplift and Erosion Across the Cascadia Forearc High, Western Washington State, Am. J. Sci., 301, 385–431, https://doi.org/10.2475/ajs.301.4-5.385, 2001.

Pazzaglia, F. J., Gardner, T. W., and Merritts, D. J.: Bedrock Fluvial Incision and Longitudinal Profile Development Over Geologic Time Scales Determined by Fluvial Terraces, in: Rivers Over Rock: Fluvial Processes in Bedrock Channels, edited by: Tinkler, K. J. and Wohl, E. E., 207–235, American Geophysical Union, https://doi.org/10.1029/GM107p0207, 1998.

Pelletier, J. D.: A robust, two-parameter method for the extraction of drainage networks from high-resolution digital elevation models (DEMs): Evaluation using synthetic and real-world DEMs, Water Resour. Res., 49, 75–89, https://doi.org/10.1029/2012WR012452, 2013.

Perona, P. and Malik, J.: Scale-space and edge detection using anisotropic diffusion, IEEE T. Pattern Anal., 12, 629–639, https://doi.org/10.1109/34.56205, 1990.

Quinn, P. F., Beven, K. J., and Lamb, R.: The $\ln(a/\tan\beta)$ index: How to calculate it and how to use it within the topmodel framework, Hydrol. Process., 9, 161–182, https://doi.org/10.1002/hyp.3360090204, 1995.

Roering, J. J., Marshall, J., Booth, A. M., Mort, M., and Jin, Q.: Evidence for biotic controls on topography and soil production, Earth Planet. Sc. Lett., 298, 183–190, https://doi.org/10.1016/j.epsl.2010.07.040, 2010.

Schreider, S. Y., Smith, D. I., and Jakeman, A. J.: Climate Change Impacts on Urban Flooding, Climatic Change, 47, 91–115, https://doi.org/10.1023/A:1005621523177, 2000.

Seidl, M. and Dietrich, W. E.: The problem of channel erosion into bedrock, Catena, 23, 101–124, 1992.

Stout, J. C. and Belmont, P.: TerEx Toolbox for semi-automated selection of fluvial terrace and floodplain features from lidar, Earth Surf. Proc. Land., 39, 569–580, https://doi.org/10.1002/esp.3464, 2014.

Tarolli, P. and Dalla Fontana, G.: Hillslope-to-valley transition morphology: New opportunities from high resolution DTMs, Geomorphology, 113, 47–56, https://doi.org/10.1016/j.geomorph.2009.02.006, 2009.

Tarolli, P., Sofia, G., and Fontana, G. D.: Geomorphic features extraction from high-resolution topography: landslide crowns and bank erosion, Natural Hazards, 61, 65–83, https://doi.org/10.1007/s11069-010-9695-2, 2010.

Thommeret, N., Bailly, J. S., and Puech, C.: Extraction of thalweg networks from DTMs: application to badlands, Hydrol. Earth Syst. Sci., 14, 1527–1536, https://doi.org/10.5194/hess-14-1527-2010, 2010.

Trimble, S. W.: Decreased Rates of Alluvial Sediment Storage in the Coon Creek Basin, Wisconsin, 1975–93, Science, 285, 1244–1246, https://doi.org/10.1126/science.285.5431.1244, 1999.

Viveen, W., Schoorl, J. M., Veldkamp, A., and van Balen, R. T.: Modelling the impact of regional uplift and local tectonics on fluvial terrace preservation, Geomorphology, 210, 119–135, https://doi.org/10.1016/j.geomorph.2013.12.026, 2014.

Walter, R., Merritts, D. J., and Rahnis, M.: Estimating volume, nutrient conent, and rates of stream bank erosion of legacy sediment in the Piedmont and Valley and Ridge physiographic provinces, Southeastern and Central PA, Tech. rep., Pennsylvania Department of Environmental Protection, Harrisburg, PA, 2007.

Wegmann, K. W. and Pazzaglia, F. J.: Holocene strath terraces, climate change, and active tectonics: The Clearwater River basin, Olympic Peninsula, Washington State, Geol. Soc. Ame. Bull., 114, 731–744, https://doi.org/10.1130/0016-7606(2002)114<0731:HSTCCA>2.0.CO;2, 2002.

Wiener, N.: Extrapolation, interpolation, and smoothing of stationary time series: with engineering applications, Technology Press of the Massachusetts Institute of Technology, 1949.

Wood, J.: The geomorphological characterisation of Digital Elevation Models, PhD thesis, University of Leicester, available at: https://lra.le.ac.uk/handle/2381/34503 (last access: 9 March 2017), 1996.

Yang, J., Townsend, R. D., and Daneshfar, B.: Applying the HEC-RAS model and GIS techniques in river network floodplain delineation, Can. J. Civil Eng., 33, 19–28, https://doi.org/10.1139/l05-102, 2006.

Zhang, T. Y. and Suen, C. Y.: A Fast Parallel Algorithm for Thinning Digital Patterns, Commun. ACM, 27, 236–239, https://doi.org/10.1145/357994.358023, 1984.

3

Identification of stable areas in unreferenced laser scans for automated geomorphometric monitoring

Daniel Wujanz[1], Michael Avian[2], Daniel Krueger[3], and Frank Neitzel[1]

[1]Institute of Geodesy and Geoinformation Science, Technische Universität Berlin, Berlin, Germany
[2]Austrian Institute of Technology GmbH, Tulln, Austria
[3]GFaI – Society for the Advancement of Applied Computer Science, Berlin, Germany

Correspondence: Daniel Wujanz (daniel.wujanz@tu-berlin.de)

Abstract. Current research questions in the field of geomorphology focus on the impact of climate change on several processes subsequently causing natural hazards. Geodetic deformation measurements are a suitable tool to document such geomorphic mechanisms, e.g. by capturing a region of interest with terrestrial laser scanners which results in a so-called 3-D point cloud. The main problem in deformation monitoring is the transformation of 3-D point clouds captured at different points in time (epochs) into a stable reference coordinate system. In this contribution, a surface-based registration methodology is applied, termed the iterative closest proximity algorithm (ICProx), that solely uses point cloud data as input, similar to the iterative closest point algorithm (ICP). The aim of this study is to automatically classify deformations that occurred at a rock glacier and an ice glacier, as well as in a rockfall area. For every case study, two epochs were processed, while the datasets notably differ in terms of geometric characteristics, distribution and magnitude of deformation. In summary, the ICProx algorithm's classification accuracy is 70 % on average in comparison to reference data.

1 Introduction

Monitoring surface changes in hazardous areas has been an important task of the geodetic community in the last decades. Due to the predicted increase of natural disasters (Anderson and Bausch, 2006), this problem domain will hence gain importance in the future. Potential forms of geomorphic natural hazards are rockfalls or slides, landslides, as well as slow surface movements which may be triggered or enforced by heavy rainfall events. Apart from geomorphic changes the characteristics and geomorphic consequences of glacial retreat are a crucial part of scientific research (Avian et al., 2018). This interest can be justified by, e.g. the function of glaciers as water reservoirs (Bradley et al., 2006) that act as water supply for agriculture and communities downstream (Tirado et al., 2010). A valuable tool for the quantitative analysis of the Earth's surface is represented by geomorphometry. The most common forms of geomorphometric monitoring aim to quantify kinematic (Schwalbe et al., 2008), areal (Lambrecht and Kuhn, 2007) and/or volumetric changes over time. A particularly appropriate measurement technique for this task is terrestrial laser scanning (TLS) due to its combination of high precision and information density (Avian et al., 2009). In order to process the acquired TLS data, an appropriate deformation model has to be chosen (Heunecke and Welsch, 2000). Deformation measurements require the repeated acquisition of a region of interest (ROI) at different points in time that are referred to as epochs. Based on these multi-temporal measurements geometric changes that occurred in between, epochs can be identified. Therefore, a stable reference frame is required that is represented by temporally stable points and their 3-D coordinates, respectively. In order to provide this vital condition, deformed areas within the ROI have to be identified. Stable points or areas within the ROI can then be used to determine transformation parameters. These parameters allow the transformation of a successively captured epoch into the coordinate system of the reference epoch.

1.1 The processing chain of deformation monitoring

For the determination of geomorphometric measures, TLS can be seen as an established acquisition technique (Oppikofer et al., 2008), whose measurements yield so-called point clouds, i.e. 3-D Cartesian coordinates of the captured object in a local coordinate system with its origin in the centre of the laser scanner. In order to quantify geometric differences in different epochs, it is recommended to carry out all steps of the following processing chain:

- viewpoint planning (Bechtold and Höfle, 2016; Wujanz et al., 2016a),
- data acquisition at different points in time (van Veen et al., 2017),
- transformation of individual epochs into a stable reference frame and
- quantification of deformation (Jaboyedoff et al., 2012; Lindenbergh and Pietrzyk, 2015).

The most delicate step in this processing chain is the transformation of single epochs into a common reference coordinate system, which is also referred to as registration or matching of point clouds. Erroneous effects that occur in this step have an immediate and systematic impact on the quantification of deformation. Thus, all conclusions that are drawn based on the generated results are falsified. The detectability of geometric changes itself is influenced by the precision of the applied laser scanner (Böhler et al., 2003; Wujanz et al., 2017) as well as the chosen algorithm for the actual quantification of deformation; see, e.g. Cignoni et al. (1998), Girardeau-Montaut et al. (2005) or Lague et al. (2013). While random influences onto the results are inevitable, systematic effects, such as usage of uncalibrated sensors (Neitzel, 2006; Salo et al., 2008; Al-Manasir and Lichti, 2015) or the aforementioned problem when computing transformation parameters, have to be avoided at all costs that otherwise yield bogus deformation.

1.2 State of research

Up to now, the transformation into a reference coordinate system has been preferably carried out by usage of artificial targets and/or monumental reference points. Several disadvantages can be associated with this strategy, such as (i) the effort of distributing the targets in the area of interest, (ii) the required survey of the targets in order to determine their 3-D coordinates in a reference system and (iii) the limited extent of the placed targets within the region of interest. In addition, the size of the targets must increase with growing range between scanner and ROI – for ultra-long-range scanners with ranges of up to several kilometres, the targets would have to measure several metres. For built-in systems that monitor a ROI over longer periods of time, rotational deviations of the sensor's home position (Lichti and Licht, 2006) that may occur in between epochs are usually refined by using artificial targets; see, e.g. Adams et al. (2013). For this, the assumption has to be made that the majority of targets remain geometrically stable over the course of the survey campaign – which might not be the case. A geodetic principle in the context of computing transformation parameters is that corresponding points should surround the ROI. If this is not the case, extrapolative effects occur, which quite likely falsify the outcome. The reason why targets are still commonly used in practice despite their numerous drawbacks can be found in the fact that all well-established operational procedures for total stations are directly transferable to TLS. Numerous alternatives for referencing point clouds exist which can be classified as follows:

- registration based on radiometric features,
- direct georeferencing of terrestrial laser scanners and
- use of geometric information.

Point clouds captured by terrestrial scanners also contain radiometric information (Höfle and Pfeifer, 2007), which is also referred to as intensity, in addition to the geometric content. Intensity values are assigned to individual points and are based on the strength of a reflected signal. If the topology of points within a dataset is known, this information can be used to convert a point cloud into an image where intensity values represent the brightness of individual pixels. By doing this, well-established techniques from the field of image matching can be applied to 3-D datasets. As a first step, so-called "keypoints" have to be extracted, which are distinct features within a local neighbourhood in terms of their grey scale values. After keypoint extraction, descriptors are used to establish correspondences between keypoints from different datasets (Lowe, 1999). Based on this information, transformation parameters can be computed. Since the long-term stability of radiometric information captured by TLS has not yet been studied, this strategy is not considered in the following. The successful application over short periods of time has been demonstrated by Böhm and Becker (2007).

Instead of computing transformation parameters based on established correspondences, the relation between different local TLS coordinate systems can also be achieved by observing their location and orientation within a superior coordinate frame. For this, additional sensors are required, while the general concept is well known, for instance, in aerial photogrammetry (Cramer et al., 2000). Methods for direct georeferencing of TLS were initially just of scientific interest (e.g. Paffenholz et al., 2010), while lately several commercial systems emerged (Riegl, 2017; Zoller+Fröhlich, 2017). A significant drawback of direct georeferencing is the extension of a scanner's error budget (Soudarissanane, 2016) due to the use of additional positioning and orientation sensors such as GNSS equipment or electronic compasses. With increasing scanning range, the impact on the relative rotation between two point clouds also increases, which is a result of the limited accuracy of compasses.

A vital prerequisite for deformation measurement is the repeated acquisition of a region of interest at different epochs. Consequently, the corresponding point clouds feature a large degree of overlap that can be used to transform both epochs into a common reference coordinate system. Apart from the aforementioned method of using radiometric features, geometric information can be used in the form of geometric primitives, such as planes (Gielsdorf et al., 2008; Previtali et al., 2014; technet, 2017). As a first step, planar segments have to be extracted from the original point clouds. Then, identical planes are computed instead of matching single points such as in radiometric approaches. By using planes instead of points, the precision of the resulting transformation parameters notably increases. However, the approach relies on the existence of planar areas within a region of interest and is hence mostly applied in urban environments. The most popular registration method uses redundantly captured regions of two point clouds and is called the iterative closest point algorithm (ICP) as proposed by Besl and McKay (1992). A substantial advantage of the last strategy over the aforementioned ones is the actual use of the information present in the point cloud. A drawback of these surface-based algorithms is their dependence to a sufficient pre-alignment of two datasets.

1.3 Motivation

A general aim in point cloud processing is to achieve a fully automatic workflow. Hence, the motivation arose to develop an automatic procedure for deformation monitoring that rules out potential biases provoked by different users. In addition, the desired solution should be versatile in terms of its input data. An algorithm that satisfies the latter aspect is the ICP algorithm that however relies on a satisfactory pre-alignment which leaves a methodical gap to the ideal of full automation. In order to achieve this prerequisite, several strategies appear to be suitable such as direct georeferencing, manual pre-alignment or pre-alignment algorithms. The desired solution should not rely on additional sensors or any form of user interaction. Hence, a two-tiered strategy was chosen that combines different algorithms for the pre-registration (Aiger et al., 2008) as well as the fine registration (Besl and McKay, 1992) of point clouds. It is well known that ICP-based approaches are sensitive to deformed regions, which yield erroneous registrations (Wujanz, 2012). Thus, it is of vital importance to reveal deformations within the captured datasets and to exclude these areas from the computation of transformation parameters. This aspect directly yields the following research questions:

- Is it possible to automatically reject deformed regions within unreferenced laser scans?
- Is the outcome biased by the geometric characteristics of the input as well as dependent on magnitude and distribution of deformations?

Kromer et al. (2017) approach the first research question by applying the Fischler and Bolles (1981) random sample consensus (RANSAC) in order to automatically exclude correspondences between two epochs which either arise as a consequence of deformation or false matches. Since Kromer et al. (2017) process data which have been captured by a permanent monitoring system at high temporal resolution, comparably low amounts of deformation can be expected between successive epochs. Thus, the risk of potential falsification of the outcome during the registration step by contaminated data should also be rather low, which otherwise may describe an insuperable hurdle for the parameter-dependent RANSAC. Since we assume to deal with deformation of larger magnitude and extent, an alternative solution is required that will be subject of Sect. 2.

This contribution demonstrates the automatic registration of multi-temporal point clouds as a vital prerequisite for the computation of geomorphometric measures. Therefore, a prototypical implementation of the iterative closest proximity algorithm (ICProx) that was integrated into the software Final Surface (GFaI 2017) was deployed. A brief description of the algorithm will be given in Sect. 2. Three regions of interest have been processed featuring different landforms such as ice glaciers, rock glaciers and a rockfall area. The study areas are described in detail in Sect. 3, while the computed results are the subject of Sect. 4. Section 5 summarises the findings presented in this article and gives an outlook on future endeavours.

2 Methods

In order to avoid falsification of the deformation measurement process and consequently the analysis of deformation, it is unavoidable to automatically identify deformed areas within point clouds and to exclude them from the computation of transformation parameters. For this task, the ICProx algorithm was deployed, which will be briefly introduced in the following. The ICProx is based on a spatial decomposition of the original datasets and identifies deformation and stable areas, respectively, by comparison of locally computed transformation parameters for individual segments (Wujanz et al., 2016b). Congruent respectively stable areas in terms of geometry are detected by a combinatorial approach termed the maximum subsample method (Neitzel, 2004, p. 109 ff.), which is, in terms of robustness against outliers/deformation, more reliable and hence superior to robust estimators or RANSAC. A detailed description can be found in Wujanz et al. (2016b). The algorithm consists of three essential phases:

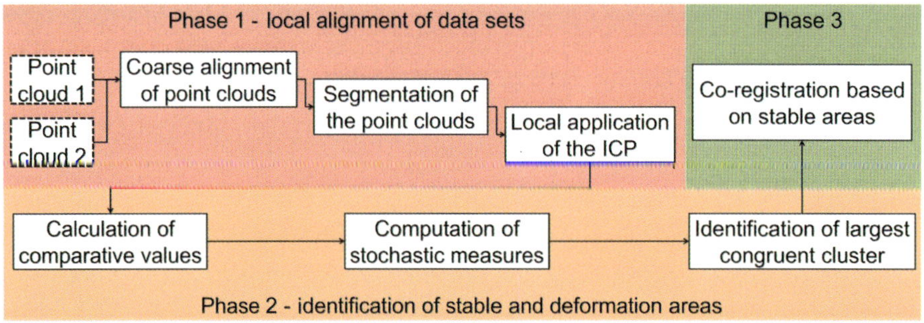

Figure 1. Workflow of the ICProx.

- phase 1: local alignment of datasets;
- phase 2: identification of both stable and deformation areas; and
- phase 3: computation of transformation parameters based on stable areas.

Figure 1 illustrates the workflow of the ICProx including the aforementioned phases and all related processing steps.

2.1 Local alignment of datasets

Phase 1 in turn contains three steps, while the first one automatically carries out the pre alignment of data. Therefore, the 4PCS algorithm as proposed by Aiger et al. (2008) is used in the first step. In the second step, the coarsely aligned data are decomposed in cubes of equal size that are also referred to as octree cells (Samet, 2006, p. 211 ff.). As already shown in Wujanz et al. (2016b), the choice of the octree cell size has notable effects on the outcome. In general, a smaller octree size leads to a higher degree of detecting stable regions, while the computational effort notably increases. Hence, setting the octree size is a balancing act between resolution and computation time. Subsequently, the ICP algorithm is locally applied within each octree cell, which represents the last step of phase 1. Therefore, so-called candidates are sampled in each cell of the reference coordinate system. For all candidates in the reference point cloud, the ICP will determine corresponding points in the second dataset. Thus, every cell receives an individual set of transformation parameters based on the established correspondences.

2.2 Identification of stable and deformation areas

The general idea in decomposing the original data into subsets is to locally increase or decrease the portion of deformation in order to identify geometric changes that occurred between scan acquisitions by means of suitable criteria in phase 2. Therefore, the individual sets of transformation parameters are applied to the corresponding centroid of a cell. Subsequently, distances between centroids before and after the application of the computed transformation parameters are compared. This strategy is in general comparable to the congruence analysis of geodetic networks by usage of combinatorial methods (Neitzel, 2004, p. 109 ff.). Congruent and thus stable regions share the following characteristic which helps to identify them: the distances between centroids before and after application of locally computed transformation parameters remain almost identical. Another unique feature of the algorithm is the automatic computation of uncertainty measures (ISO, 2008) that are calculated for the local registration in every octree cell instead of using fixed thresholds. This information is required in order to determine if an octree cell can be associated with a congruent set of cells or not. The general concept of this feature considers the fact that every scan of a stable object yields different point sampling (Wujanz et al., 2016b). In phase 3, the ICP is again applied to both epochs. This time, however, only stable regions, that were detected in phase 2, serve as input which is processed as a whole and not in segments. The result of phase 3 is a set of transformation parameters that forms the basis for the deformation estimation.

2.3 Some remarks on the spatial arrangement of deformation

Figure 2 depicts a test case on the left where stable areas are coloured in green, while deformation is highlighted in yellow. The aim of the ICProx algorithm is to identify these yellow areas and to exclude them from the final registration. The centre of the figure illustrates the decomposition of the input data. An individual set of transformation parameters is assigned to every cube by local application of the ICP algorithm. After congruence analysis, all segments are merged to one continuous dataset as depicted by green areas. Red regions have been classified as deformation and are thus excluded from the registration process.

From a geodetic perspective, the most desirable arrangement would be a more or less connected region that is subject to deformation which is surrounded by stable areas. In addition, the ratio between stable and deformed regions should be rather large in order to be able to compute transformation parameters with the highest possible redundancy. However, the

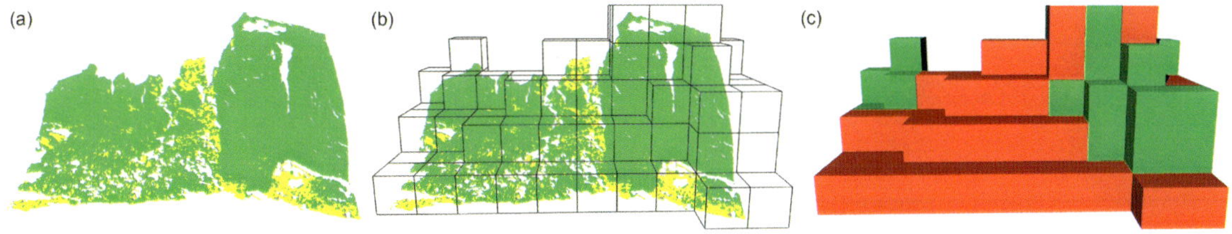

Figure 2. (**a**) Binary change map of a test case. (**b**) Octree structure that contains the scene. (**c**) Stable (green cubes) and unstable (red cubes) octree cells after congruence analysis.

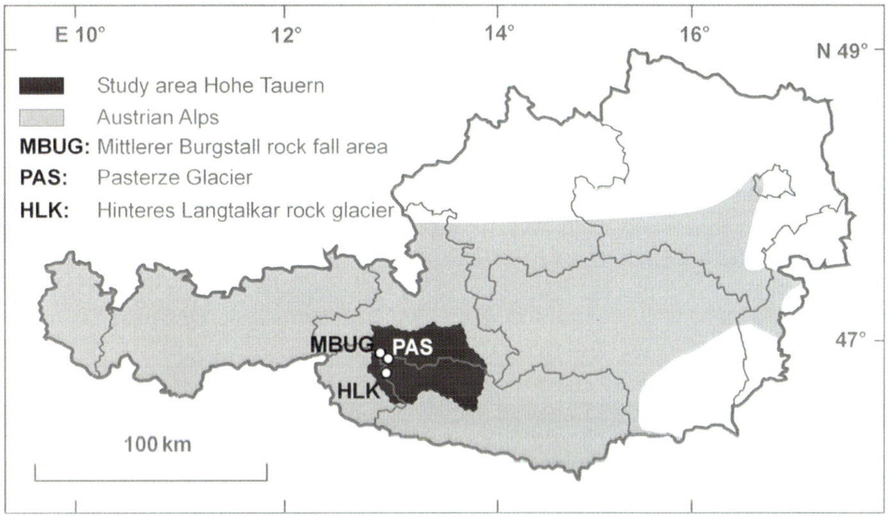

Figure 3. Study areas: the Pasterze Glacier, Hinteres Langtalkar rock glacier and Mittlerer Burgstall rockfall area within Austria.

distribution of stable areas, as well as their relative amount, is usually unpredictable in practice since every region of interest has its own individual characteristics. Countermeasures in order to receive favourable arrangements can be achieved by carrying out the surveys more frequently and/or to perform panorama scans which increases the chance of capturing additional stable areas. It is obvious that the result depicted on the right of Fig. 2 may yield imprecise transformation parameters due to the fact that a large amount of octree cells are subject to deformation, while the arrangement of stable cells is unfavourable.

3 Study areas

This study comprises three study areas in the Hohe Tauern area of the central Alps, Austria (see Fig. 3), monitoring three distinct geomorphic process groups:

- Hinteres Langtalkar: rock glacier movement representing permafrost dynamics;
- Pasterze Glacier: glacier retreat with subsequent proglacial processes; and
- Mittlerer Burgstall: rockfall activity.

All datasets were captured using a Riegl LMS Z620 laser scanner that applies a wavelength of 1550 nm and is able to measure distances of up to 2000 m; in applications in the study area, 1500 m were achieved. A summary of the characteristics of all epochs is given in Table 1.

3.1 The rock glacier Hinteres Langtalkar, Austria

The rock glacier Hinteres Langtalkar (46°59′ N, 12°47′ E) is located in the Schober Mountains at an elevation between 2455 and 2720 m a.s.l., and is ~900 m long and ~350 m wide. Using TLS, the rock glacier has been monitored since the year 2000 (Bauer et al., 2003). Extraordinary changes in surface elevation were detected around 2000 (Kaufmann and Ladstädter, 2002; Avian et al., 2005), leading to the assumption of acceleration most likely in 1994 (Avian et al., 2005). In order to geometrically quantify these surface elevation changes, a geodetic network was established in 1998 which has since been surveyed annually (Kienast and Kaufmann, 2004). The rock glacier itself shows distinct patterns of surface elevation changes and surface displacements. In addition, the rock glacier has reported to have a slower upper part and a rapidly moving lower part with maximum horizontal displacements of up to 2.5 m a^{-1} (Avian et al., 2009). The mean horizontal surface displacement rates vary

Table 1. Summary of all used measurements in every study area comprising date of acquisition, number of points (points per epoch) and ground sampling distance (GSD).

Year	Date of acquisition	No. of points (mio.)	Ground sampling distance (GSD) (cm)	Scanning increment (°)
Hinteres Langtalkar rock glacier				
2011	24 Aug 2011	1.6	6.6–41.8	0.040
2012	5 Sep 2012	7.8	4.8–33.1	0.029
Pasterze Glacier				
2010	13 Sep 2010	2.5	20–54	0.029
2012	10 Sep 2012	3.6	16–42	0.023
Burgstall rockfall area				
2014	8 Sep 2014	2.3	20–54	0.029
2015	12 Sep 2015	1.6	20–54	0.029

Figure 4. Scanner configuration at Hinteres Langtalkar rock glacier. The scanning position Hinteres Langtalkar (HLK) is indicated by a red ellipse with a white edge; the rock glacier outline is indicated by a white dashed line. White circles with green edges highlight reference points.

between 0.10 and 0.18 m year^{-1}, while maximum changes amount up to 3.6 m year^{-1} (Kaufmann and Ladstädter, 2009). Delaloye et al. (2008) estimate this rock glacier to currently be the fastest moving one in Europe, which underlines its importance for the scientific community.

Despite the multiple use of TLS for monitoring geomorphological processes (see Oppikofer et al., 2008), TLS was rarely used for analysis of periglacial effects such as rock glacier movement patterns until the end of the last decade. First annual TLS observations were undertaken by Avian et al. (2005, 2009) or Bodin et al. (2008). Data acquisition was carried out from one of two known scanning positions (Hinteres Langtalkar – HLK, 2454.709 m a.s.l., on the left in Fig. 4). The lowest and largest ground sampling distances (GSDs) for the 2011 campaign summed up to 6.6 cm and 41.8 cm, respectively, while the corresponding measures for the survey of 2013 are 4.8 and 33.1 cm; see Table 1. These GSD values were computed based on the closest and furthest distances between scanner and scene. In Fig. 4, all applied reference points are highlighted by white circles and the scanner position by a red ellipse with a white edge. These reference points were surveyed by geodetic means and are hence known in a superior coordinate system. It can notably be seen that the distribution is rather unfavourable from

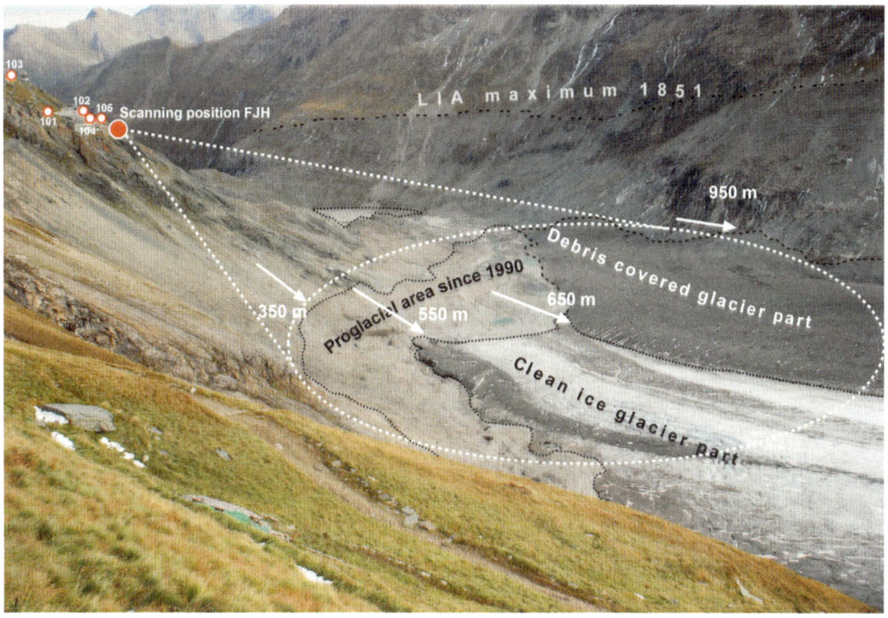

Figure 5. Scanning configuration at the Pasterze Glacier terminus area from the scanning position, Kaiser-Franz-Josefs-Höhe (FJH). Dots near the scanning position represent the location of the reference points. Numbers near the arrows indicate the distance from the scanning position.

a geodetic point of view as the active zone of the glacier, that is highlighted by a white dashed line, is not surrounded by reference points. This may lead to extrapolative and leverage effects, respectively. In order to provide trustable reference transformation parameters, a surface-based registration has been conducted after deformed regions were manually excluded. The scanning sector is marked by red dashed lines. The shown picture was taken in 2012.

3.2 The Pasterze Glacier, Austria

The Pasterze Glacier (47°04′ N, 12°44′ E) is located in the central part of the Austrian Alps. The geometric behaviour of this glacier has been of scientific interest since 1878 (Wakonigg and Lieb, 1996), which makes the Pasterze Glacier one of the longest-documented glaciers in the world. The glacier reached its latest maximum around 1852 in the Little Ice Age (Nicolussi and Patzelt, 2000) with a maximum area of 26.5 km^2. Since then, the glacier constantly lost substance leading to a recent area of approximately 17.3 km^2 in 2009 (Kaufmann et al., 2015). Accordingly, since the end of the Little Ice Age period, the Pasterze Glacier lost approximately 35 % of its area and about 60 % of its volume. The current retreat of the terminus area of the Pasterze Glacier has been observed using TLS since 2001, covering an area of approximately 0.9 km^2. On average, the annual elevation change within the debris-covered part sums up to approximately 3.7 m, while 6.3 m vertical loss was detected in the clean ice section between 2011 and 2012. A comprehensive view on the geomorphic consequences of the fast glacier retreat can be found in Avian et al. (2018). The scanning configuration at the Pasterze Glacier terminus area is depicted in Fig. 5, while the picture was taken in 2008.

The view at scanning position Kaiser-Franz-Josefs-Höhe (FJH; 2362.621 m a.s.l.) is excellent for geomorphic applications, whereas the availability of stable areas led to an unfavourable distribution of reference points that do not surround the ROI. As a consequence, the computed transformation parameters based on the reference points can only be seen as an approximate solution, as they can very likely cause extrapolative effects and hence have been refined by surface-based registration. The necessity of revealing deformed regions prior to performing co-registration has already been extensively stated, so the ICProx algorithm is deployed in the following to solve this task for a comparison among the datasets from 2010 and 2012. A summary of quality parameters of TLS measurements at the Pasterze Glacier terminus area is given in Avian et al. (2018).

3.3 The Mittlerer Burgstall rockfall area, Austria

The rockfall area Burgstall comprises the two former nunatak mountains (Kellerer-Pirklbauer et al., 2012) of Mittlerer and Hoher Burgstall which were encompassed by the Pasterze Glacier until the beginning of the 20th century. The Mittlerer Burgstall (47°06′ N, 12°42′ E; 2933 m a.s.l.) shows a distinct NW–SE trending ridge forming two rock faces which were exposed by glacier shrinkage in the last 100 years. In early summer 2007, a large rockfall at the top of the ridge, released approximately 430 000 m^3 (Viktor Kaufmann, Graz

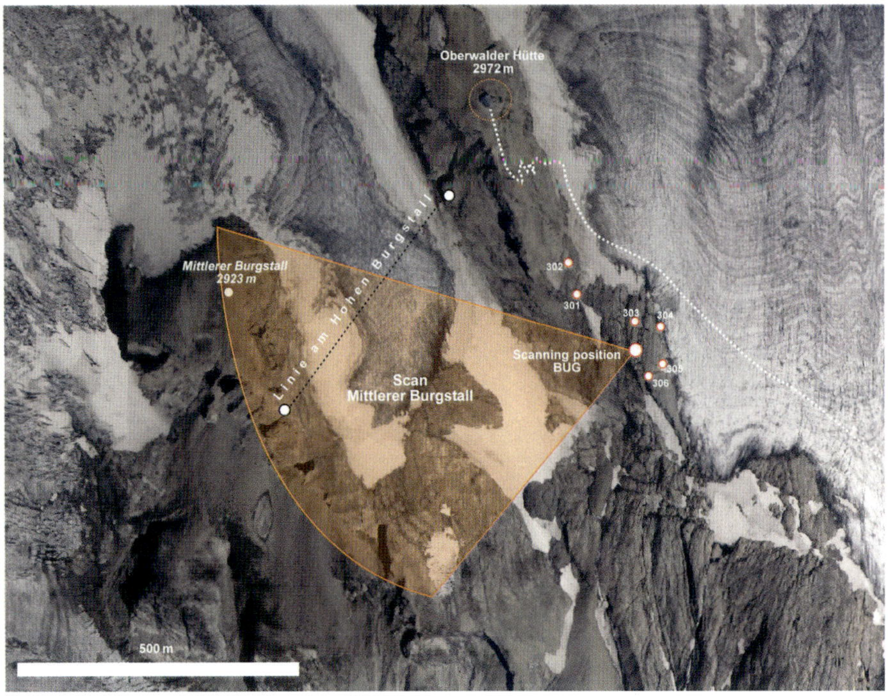

Figure 6. Configuration at scanning position Burgstall ("Scanning position BUG"). BUG is serving two objects, Mittlerer Burgstall and Hoher Burgstall rockfall area. Codes: 301–306: reference points; "Linie am Hohen Burgstall" cross section of traditional glacier surveying. Orthoimage from 1998, Hohe Tauern National Park©.

University of Technology, personal communication, 2013) from a detachment area of approximately 3100 m^2, transported debris to both sides of the ridge. Results from permafrost distribution modelling and ground temperature monitoring indicate (Kellerer-Pirklbauer et al., 2012) that the detachment area is located within or near the lower limit of permafrost. Consequently, in 2010, a geodetic network consisting of one main scanning position (BUG; 2794.958 m a.s.l.) and six reference points was established. From every reference point, each point can be viewed so that several scanning positions can be chosen to vary scanning geometry. The scanning configuration at scanning position Burgstall (BUG) can be seen in Fig. 6. Yet again, the sampling of reference points has to be rated as unfavourable. Hence, initial transformation parameters have been computed based on the reference points. Afterwards, deformed regions were manually excluded, while the remaining points were used to refine the relative alignment between point clouds based on the ICP.

4 Results and discussion

Every algorithm is sensitive to chosen settings as well as to the individual characteristics of the given data. In this section, the three aforementioned regions of interest have been analysed by using the ICProx algorithm. The input for the ICProx algorithm consisted of two unreferenced point clouds per ROI given in different local scanner coordinate systems. Pre-alignment of the epochs was carried out by the 4PCS algorithm, while subsequently the ICProx algorithm was deployed in order to reveal regions that were subject to deformation. The inherent challenges of the datasets presented in Sect. 3.1 and 3.2 can be associated with their sheer extent. A somewhat different degree of difficulty can be assigned to the ROI that was the subject of Sect. 3.3. This scene is demanding due to the scattered distribution of deformation and the comparably small magnitude of geometric changes as a consequence of processes after a huge rockfall.

Since the task of the ICProx algorithm is essentially a classification problem – to identify stable or deformed regions within a point cloud – the generated outcome can be numerically evaluated in the form of confusion matrices, while a visual inspection is carried out by so-called confusion maps. The reference data for the evaluation process are established by point clouds where deformed areas were manually removed, while the relative alignment of the remaining points was refined by surface-based registration. Based on these data, a point cloud was tinted in two colours representing the two classes of interest, namely deformation and stable region. The comparison between the reference and the results computed by the ICProx algorithm finally yield a confusion matrix and confusion maps, respectively, as depicted, for instance, on the left of Fig. 7. Therefore, the following colour scheme will be used and contains four elements, namely:

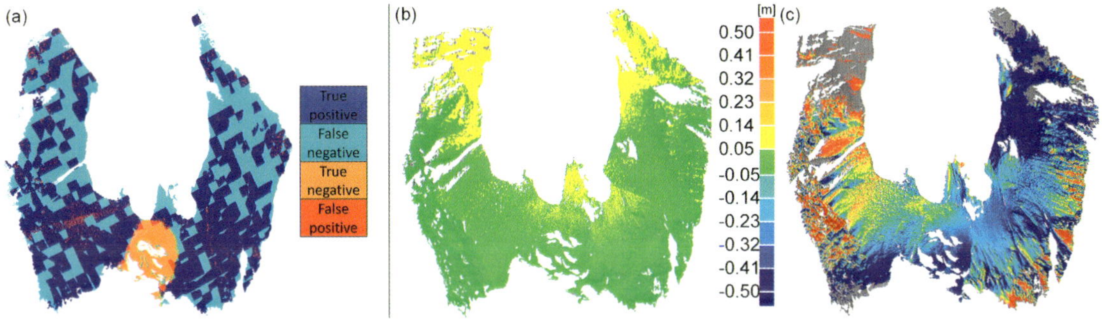

Figure 7. Colour-coded confusion map **(a)**, change map based on a comparison to the reference data outcome by usage of the ICProx **(b)** and the ICP **(c)** at Hinterer Langtalkar rock glacier.

- blue: area is stable, detected as stable (true positive);
- cyan: area is stable, detected as unstable (false negative);
- red: area is unstable, detected as stable (false positive); and
- orange: area is unstable, detected as unstable (true negative).

Note that a perfect result would only contain the first and last colours in the list. Yet, this aim is not achievable due to the applied octree structure.

4.1 Rock glacier Hinteres Langtalkar

After pre-alignment, the entire dataset was decomposed into octree cells of 30 m so that in total 661 cells emerged. In order to locally determine transformation parameters, candidates were sampled with 0.75 m on average. After congruence analysis, the largest subset contains 441 cells that were considered as being stable. Based on these areas, an ICP-based registration was carried out, leading to average residuals of 15.3 cm between the two point clouds. Afterwards, the computed results were compared to the reference, yielding a confusion matrix as well as a colour-coded confusion map which is depicted in Fig. 7. Figure 7a shows octree cells where orange represents the rock glacier and red cells are considered as being stable. This effect occurred predominately in the transition region between the glacier tongue and surrounding areas. The reason why these cells have been falsely assigned to the stable class can be found in the very low magnitude of deformation in the transition zone of the rock glacier which has a similar scale to the uncertainty measures (ISO, 2008). As a consequence, deformations of this size fall below the margin of detectability. The narrow red band on the left side of the scene was falsely assigned as being stable due to its limited spatial expansion.

The remaining parts of Fig. 7 show change maps based on comparisons between the reference dataset and their corresponding versions which were altered in position and orientation before application of the ICProx (Fig. 7b) and ICP (Fig. 7c). Hence, the motivation for conducting this analysis is to reveal differences between the two surface-based approaches against the reference data. A look at Fig. 7b shows that the majority of the scene complies within ±5 cm, while some parts feature deviations between 5 and 14 cm. These differences may be triggered by leverage effects due to an insufficient sampling of artificial targets for georeferencing. This suspicion substantiates if one has a look at Fig. 4b, where it can be seen that all targets do not surround the glacier tongue and are only located in the north westerly direction (above the glacier tongue in the figure). On average, the differences between the reference and the result computed based on the ICProx sum up to (i) 5.8 cm for points that are located above the reference surface and (ii) 4.5 cm for points below it, while the standard deviation among the two datasets measures 5.7 cm. Figure 7c again confirms that the ICP is not an appropriate algorithm for registration of point clouds that were subject to deformation. Again, the same statistical measures were computed as before where the mean deviation among the datasets sums up to 17.4 cm for points above the reference and 23.9 cm for points below the reference surface, respectively. The corresponding standard deviation of all residuals is 24.2 cm. It has to be mentioned that the listed statistical measures are somewhat whitewashed due to the fact that only deviations within the specified range between ±50 cm were considered but not the grey-coloured areas where the discrepancies exceed these limits.

The numerical assessment of the generated results is given in Table 2 and reveals that a comparably large part of stable areas was not correctly assigned, as indicated by the numbers in the cyan-coloured cell. The reason for this can be found in the weak geometric characteristics in the corresponding octree cells. The overall classification accuracy sums up to a moderate $60\,\% = 55\,\% + 5\,\%$ and can mostly be justified by the aforementioned geometric characteristics of the dataset. For the detection rate of stable areas, a value of $60\,\% = 55\,\%/(55\,\% + 37\,\%)$ was achieved. This result can be rated in two ways. On the one hand, it can be seen as a satisfactory result, while on the other hand, a substantial amount

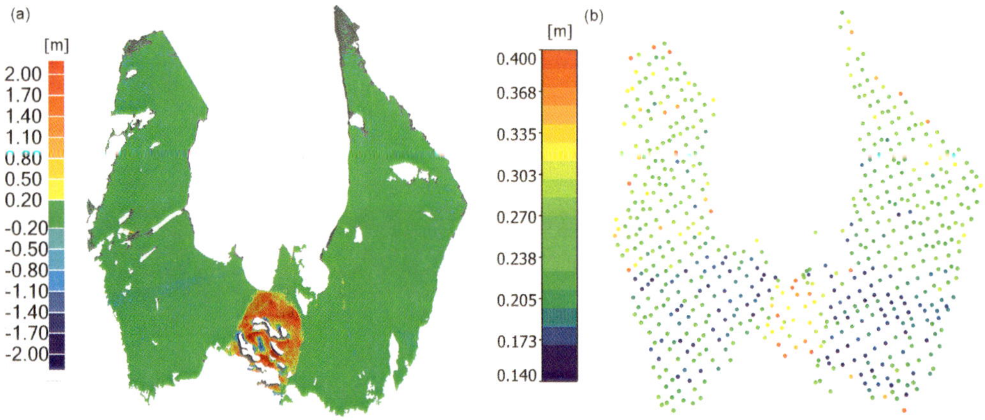

Figure 8. Change map of the Hinterer Langtalkar **(a)** and spatial uncertainties (m) for individual octree cells **(b)**.

Table 2. Confusion matrix for the classification of the rock glacier.

	Area is stable	Deformation
Classification: area is stable	55 %	3 %
Classification: deformation	37 %	5 %

of valid information was not used. The corresponding detection rate of deformation is 62.5 % = 5 %/(5 % + 3 %). After analysing the numbers in Table 2, it appears that the allegedly "simple" dataset with only 8 % of deformation has turned into a challenging task for the algorithm. However, a second look at Fig. 7 reveals that the stable areas surround the major area of deformation. Hence, a stable set of transformation parameters can be expected without extrapolative effects. The influence of 3 % of false positive cells can be expected to be little due to the fact that the magnitude of deformation is comparably low in these parts and due to the low ratio between true positive (55 %) and false positive (3 %) areas.

The change map based on the results of the ICProx algorithm is illustrated in Fig. 8a and describes the basis for deformation analysis by geologists. In Fig. 8b, the uncertainty values computed by the ICProx algorithm for individual octree cells (Wujanz et al., 2016b) can be seen forming the statistic basis to distinguish stable areas and deformation. Hence, small uncertainty values allow detecting deformations of small magnitude. It is obvious that the distribution of uncertainty values is quite heterogeneous, ranging from several centimetres to several decimetres. It can also be seen that the upper parts of the rock glacier have larger uncertainties. This can be explained by the fact that the local sampling rate is lower compared to the regions right below. The uncertainty values on the glacier tongue are also notably increased when compared to its surroundings. This is caused by the unsteady geometric characteristics of the glacier tongue in combination with the scanning configuration, as depicted in Fig. 4.

4.2 Pasterze Glacier terminus area

For this ROI, it can be expected that deformation occurred predominantly directly at the glacier's tongue, so that comparatively large octree cells were defined. Hence, the size of the cells was empirically set to 52.5 m, leading to 923 segments. On average, candidates were sampled every 6 m, which led to approximately 40 700 candidates. After application of the algorithm, a congruent set consisting of 526 octree cells emerged that was finally used to compute the transformation parameters for deformation monitoring. The residuals after registration based on stable regions were 28.8 cm on average. An explanation for this rather large quality measure can be found in the comparably low ground sampling distance in relation to the rugged terrain. Based on expert knowledge, a set of reference transformation parameters was computed featuring undoubtedly stable areas. This procedure is very time consuming, yet crucial in order to avoid falsification of the reference dataset by deformation.

After generating the aforementioned reference dataset, quality assessment was carried out. Figure 9a depicts the confusion map which was coloured according to the colour scheme proposed at the beginning of this section. In the centre of Fig. 9a, the glacier tongue is highlighted in orange, corresponding to large detected geometric changes. On the sides, stable areas can be found which have been predominantly detected by the algorithm. Cyan-coloured areas can be seen predominantly on the left of the tongue. These regions have not been detected as being stable due to the fact that their local geometric characteristics did not feature notable topographic changes. Consequently, the locally applied ICP converged into local minima, as the ICP requires a certain geometric contrast within the datasets. Red areas, however, have to be rated as being critical for the desired results. These regions can be found primarily in transition areas between glacier terminus and adjacent stable parts of the point cloud. The reason why these areas were falsely classified can be explained by their magnitude of deformation – they fall below

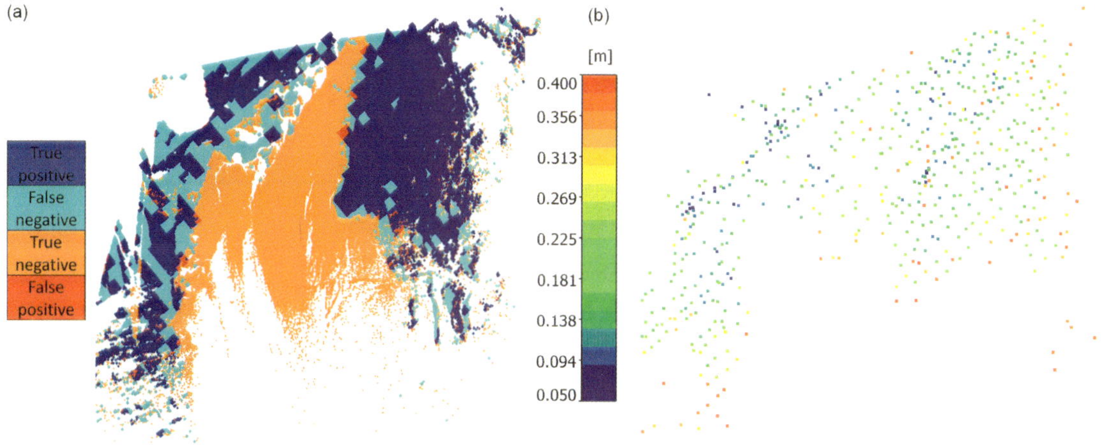

Figure 9. Confusion map **(a)** and spatial uncertainties for individual octree cells **(b)**.

Table 3. Confusion matrix for the classification of the ice glacier.

	Area is stable	Deformation
Classification: area is stable	43 %	2 %
Classification: deformation	19 %	36 %

the margin of detectability. Single red spots in the otherwise stable flanks left and right of the tongue were triggered by single-sliding or mass-wasting processes (Avian et al., 2018). These appearances also fall below the margin of detectability, yet not due to their magnitudes but their limited spatial extent. In Fig. 9b, spatial uncertainty measures for each octree cell are shown, which have been computed by the ICProx algorithm. A heterogeneous distribution is apparent, while an increasing characteristic can be seen towards the lower part of the image due to insufficient local sampling.

The numerical interpretation of the results can be found in Table 3. In total, 19 % of the dataset was falsely classified as deformation despite the fact that it actually remained geometrically stable. This effect does not provoke any substantial disadvantages, yet does simply mean that stable areas remain unused for the final registration. The portion of the critically false positive class, which is coloured in red in all tables and figures, sums up to 2 %. These regions undeniably have an influence on the registration, yet can be rated as uncritical due to their comparably low magnitudes and low fraction in relation to the stable areas.

All in all, the classification has to be rated quite positively with an overall accuracy of $79\,\% = 43\,\% + 36\,\%$. The detection rate for stable areas sums up to $69\,\% = 43\,\%/(43\,\% + 19\,\%)$; the one for deformation is even $95\,\% = 36\,\%/(36\,\% + 2\,\%)$. Consequently, the algorithm can be characterised as being "careful" – in case of doubt, areas are rather classified as being deformed.

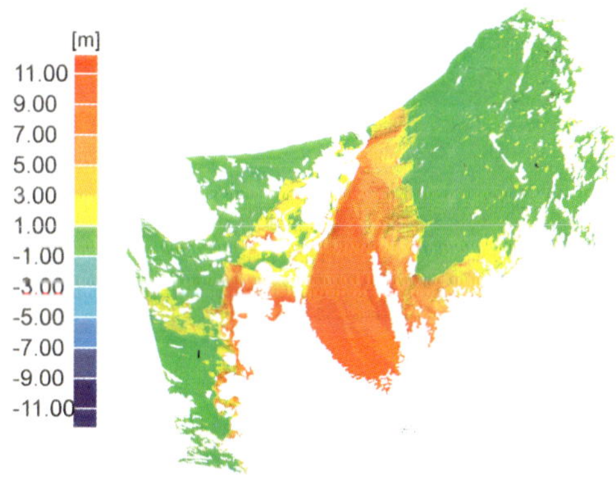

Figure 10. Resulting change map based on the detected stable regions.

Figure 10 features a point cloud that is coloured according to geometrical differences that occurred in between epochs. Large differences are notably visible at the glacier terminus area. Two regions need to be highlighted in particular. At first, the orange-to-red-coloured part has to be brought to attention with vertical surface elevation changes in the magnitude of -6.5 to $-10\,\mathrm{m\,a^{-1}}$. In this part of the glacier, the terminus area is only sparsely covered by debris. The second notable part is coloured in shades of yellow to green where the glacier tongue is covered by morainic debris. As a consequence, the ablation rates are significantly lower compared to the clean ice part with vertical surface elevation changes in the magnitude of -4 to $-6\,\mathrm{m\,a^{-1}}$ (Kellerer-Pirklbauer et al., 2008).

4.3 Mittlerer Burgstall rockfall area

Just as before, the input for the ICProx algorithm consisted of two unreferenced point clouds, given in different local scanner coordinate systems. In order to decompose the datasets, an octree size of 5 m in all cardinal directions was chosen

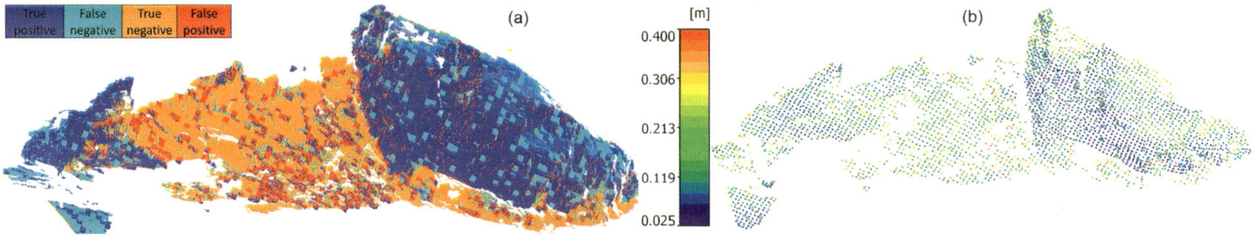

Figure 11. Confusion map of the Burgstall dataset based on the classification with the ICProx **(a)** and spatial uncertainties for individual octree cells **(b)**.

Table 4. Confusion matrix for the classification of the rockfall area.

	Area is stable	Deformation
Classification: area is stable	43 %	13 %
Classification: deformation	15 %	29 %

yielding 3494 cells. Candidates were sampled with an average spacing of 0.5 m so that about 120 000 correspondences had to be iteratively found by the ICP. After application of the ICProx algorithm, a congruent set of 2179 cells arose that was deployed to compute the desired transformation parameters for deformation monitoring. The residuals after registration based on stable regions were 17.2 cm on average. Again, the explanation for this rather large quality measure can be found in the comparably low ground sampling distance in relation to the rugged terrain. As before, expert knowledge was used to create a set of reference transformation parameters by manually selecting undoubtedly stable areas. Based on this information, a binary change map with the two classes, stable area and deformation, was created. These data served as a basis for the confusion map that is depicted in Fig. 11a. In contrast to the two previously shown examples, several red areas are notable, in which deformed regions were falsely classified as being stable. A close look at the red speckles which are located in the major rockfall area highlighted in orange shows that most of them contain some locally restricted blue regions. As a result, a mixture between stable and predominantly deformed areas occurs within the corresponding octree cells and hence lowers the algorithm's margin of detectability. The reverse case can be seen in Fig. 11b, which is mostly tinted in blue where small red spots are notable. Another reason for the comparably frequent occurrence of red regions is the low magnitude of deformations. The right part of the figure shows spatial uncertainties for individual octree cells given in metres. A heterogeneous distribution of uncertainty values is apparent, yet its variation appears to be weaker when compared to the other two examples. Note that the spatial uncertainties in large parts of the deformed region fall into the decimetre range.

The numerical assessment of the generated results is given in Table 4 and shows that a comparably large part of deformed areas was not correctly assigned, as indicated by the numbers in the red-coloured cell. The reason for this is twofold and can be found both in the ratio among the low magnitude of deformation and the corresponding local spatial uncertainties as well as the heterogeneous occurrence of geometric changes. The overall classification accuracy sums up to $72\,\% = 43\,\% + 29\,\%$. For the detection rate of stable regions, a value of $74\,\% = 43\,\%/(43\,\% + 15\,\%)$ was achieved that marks the best value in this contribution. However, the corresponding detection rate of deformation only sums up to $69\,\% = 29\,\%/(29\,\% + 13\,\%)$.

Based on the reference transformation parameters, a change map was generated that is depicted in Fig. 12. The majority of geometric changes occurred in the centre of the figure as well as on the bottom of the right part. Apart from larger connected regions, local nests of rockfall can be seen especially on the upper right part of the illustration.

A look at Table 4 shows that 13 % of all points were assigned to the false positive class. Hence, its influence onto the outcome is of interest in the following. Therefore, the computed set of transformation parameters based on the detected stable regions was applied to the second epoch and compared to the generated reference dataset. On average, the absolute deviation between the reference and computed result is 7.37 cm. The corresponding change map in Fig. 13 illustrates the differences in metres. In summary, the processed dataset proved to be challenging and revealed boundaries for the ICProx algorithm. Apart from magnitude and spatial distribution of deformation, the inherent relationship between detectability of geometric changes and local point density had a critical influence onto the outcome, which finally led to the comparably large amount of falsely classified stable regions. Thus, future surveys in this region should be carried out at higher spatial sampling rates.

5 Summary and outlook

In this contribution, an algorithm, termed the ICProx, was used to automatically identify and finally exclude deformation from point clouds captured at different points in time. If this diagnostic step is not performed, systematic falsification of the registration is inevitable, and consequently wrong conclusions are drawn during analysis of the deformation pat-

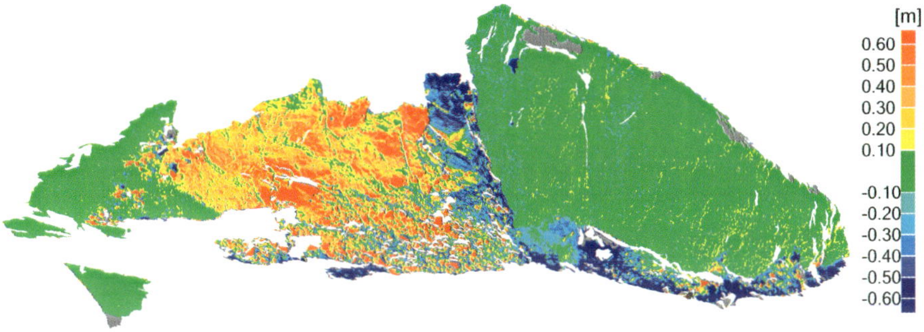

Figure 12. Change map based on the detected stable regions – comparison between the datasets from 2014 and 2015.

Figure 13. Comparison between the reference dataset and results generated based on the ICProx algorithm.

tern or computation of geomorphometric measures. Based on the detected stable regions, transformation parameters were computed as a vital prerequisite for deformation monitoring. The chosen study areas featured a glacier, a rock glacier, as well as a rockfall area. The datasets themselves proved to be demanding due to individual geometric characteristics, extent, relative amount and/or magnitude of deformations. In general, the algorithm performed well for the data presented in Sect. 3.1 and 3.2 where geometric changes occurred in a connected fashion. The most challenging task was represented by the data discussed in Sect. 3.3 due to the fact that deformation was scattered all over the datasets and the deformation rates were comparably low in comparison to the local uncertainty values. All results were numerically and visually evaluated against reference datasets. The mean classification accuracy for stable regions is 68 %, while the corresponding measure for deformation is 76 %. Thus, the results can be rated as quite promising. However, the importance of carefully acquiring data is particularly apparent in Sect. 4.3, as it clearly demonstrates the inherent link between local sampling density and potential detectability of deformation. It is important to point out that the effects provoked by the sampling uncertainty of TLS notably surpass their capabilities in terms of achievable 3-D precision.

Further research will focus on eliminating some existing disadvantages of the algorithm as well as the applied assessment procedure. One of them is the dependency of the outcome to the degree of decomposition of the point clouds by usage of an octree structure. Hence, an extension has to be implemented that is able to determine an optimal octree size under consideration of the local topography. This information could also be used to dismiss certain octree cells due to insufficient geometric properties that would otherwise occur in the algorithm's numerical and visual assessment. In addition, stable cells that border on deformed regions should be split and dismissed from the registration in order to avoid falsification in such transition zones as illustrated in Figs. 7 and 9. An alternative strategy could be to apply the ICProx algorithm in an iterative fashion where the octree size gradually decreases. With this, the boundaries of deformed regions would be detected at a higher resolution. Since geomorphometric monitoring is carried out frequently over long periods, strategies have to be developed in order to process time series that consist of several epochs.

Competing interests. The authors declare that they have no conflict of interest.

Special issue statement. This article is part of the special issue "4-D reconstruction of Earth surface processes: multi-temporal and multi-spatial high-resolution topography". It is not associated with a conference.

Acknowledgements. We warmly appreciate the support of several students of the Graz University of Technology and University of Graz within the scanning campaigns.

References

Adams, M., Gleirscher, E., and Gigele, T.: Automated Terrestrial Laser Scanner Measurements of Small-Scale Snow Avalanches, Proc. Intern. Snow Science Workshop Grenoble–Chamonix MontBlanc, 2013.

Aiger, D., Mitra, N. J., and Cohen-Or, D.: 4-points congruent sets for robust pairwise surface registration, ACM T. Graphic, 27, 85–95, https://doi.org/10.1145/1399504.1360684, 2008.

Al-Manasir, K. and Lichti, D.: Self–Calibration of a Leica HDS7000 Scanner, Proc., FIG Working Week, International Federation of Surveyors, Copenhagen, Denmark, 2015.

Anderson, J. and Bausch, C.: Climate Change and Natural Disasters: Scientific Evidence of a Possible Relation Between Recent Natural Disasters and Climate Change, Requested study on behalf of the European Parliament's Environment, Public Health and Food Safety Committee, 2006.

Avian, M., Lieb, G. K., and Kaufmann, V.: Recent and Holocene dynamics of a rock glacier system – the example of Langtalkar (Central Alps, Austria), Norsk Geogr. Tidsskr., 59, 1–8, 2005.

Avian, M., Kellerer-Pirklbauer, A., and Bauer, A.: LiDAR for monitoring mass movements in permafrost environments at the cirque Hinteres Langtal, Austria, between 2000 and 2008, Nat. Hazards Earth Syst. Sci., 9, 1087–1094, https://doi.org/10.5194/nhess-9-1087-2009, 2009.

Avian, M., Kellerer-Pirklbauer, A., and Lieb, G. K.: Geomorphic consequences of rapid deglaciation at Pasterze Glacier, Hohe Tauern Range, Austria, between 2010 and 2013 based on repeated terrestrial laser scanning data, Geomorphology, 310, 1–14, 2018.

Bauer, A., Paar, G., and Kaufmann, V.: Terrestrial laser scanning for rock glacier monitoring, 8th International Conference on Permafrost, Proceedings vol. 1, 55–60, 2003.

Bechtold, S. and Höfle, B.: HELIOS: A Multi-Purpose Lidar Simulation Framework For Research, Planning And Training Of Laser Scanning Operations With Airborne, Ground-Based Mobile And Stationary Platforms, ISPRS Ann. Photogramm. Remote Sens. Spatial Inf. Sci., III-3, 161–168, https://doi.org/10.5194/isprs-annals-III-3-161-2016, 2016.

Besl, P. J. and McKay, N. D.: Method for registration of 3-D shapes, Robotics-DL tentative, International Society for Optics and Photonics, 586–606, 1992.

Bodin, X. and Schoeneich, P.: High-resolution DEM extraction from terrestrial LIDAR topometry and surface kinematics of the creeping alpine permafrost: The Laurichard rock glacier case study (Southern French Alps), in: Ninth International Conference on Permafrost, edited by: Kane, D. L. and Hinkel, K. M., Institute of Northern Engineering, University of Alaska Fairbanks, 1, 137–142, 2008.

Böhler, W., Vicent, M. B., and Marbs, A.: Investigating laser scanner accuracy, ISPRS Arch. Photogramm. Remote Sens. Spatial Inf. Sci., 34, 696–701, 2003.

Böhm, J. and Becker, S.: Automatic marker-free registration of terrestrial laser scans using reflectance, Proceedings of 8th Conference on Optical 3-D Measurement Techniques, Zurich, Schweiz, 9–12 July 2007, 338–344, 2007.

Bradley, R. S., Vuille, M., Diaz, H. F., and Vergara, W.: Threats to water supplies in the tropical Andes, Science, 312, 1755–1756, 2006.

Cignoni, P., Rocchini, C., and Scopigno, R.: Metro: measuring error on simplified surfaces, Comput. Graph. Forum, 17, 167–174, 1998.

Cramer, M., Stallmann, D., and Haala, N.: Direct georeferencing using GPS/inertial exterior orientations for photogrammetric applications, Int. Arch. Photogramm. Remote Sens., 33, 98–205, 2000.

Delaloye, R., Perruchoud, E., Avian, M., Kaufmann, V., Bodin, X., Hausmann, H., Ikeda, A., Kääb, A., Kellerer-Pirklbauer, Krainer, K., Lambiel, C., Mihajlovic, D., Staub, B., Roer, I., and Thibert, E.: Recent interannual variations of rock glacier creep in the European Alps, in: Proceedings of the Ninth International Conference on Permafrost (NICOP), University of Alaska, United States of America, 343–348, 2008.

Fischler, M. A. and Bolles, R. C.: Random sample consensus: a paradigm for model fitting with applications to image analysis and automated cartography, Commun. ACM, 24, 381–395, https://doi.org/10.1145/358669.358692, 1981.

GFaI: Official website of the 3-D-Software Final Surface, available at: http://www.final-surface.com, last access: 17 May 2017.

Gielsdorf, F., Gründig, L., and Milev, I.: Deformation analysis with 3-D laser scanning, in: Proceedings of the 13th FIG Symposium, Lisbon, Portugal, 12–15 May, 2008.

Girardeau-Montaut, D., Roux, M., Marc, R., and Thibault, G.: Change detection on points cloud data acquired with a ground laser scanner, ISPRS Arch. Photogramm. Remote Sens. Spatial Inf. Sci., 36, 30–35, 2005.

Heunecke, O. and Welsch, W.: Terminology and classification of deformation models in engineering surveys, J. Geospat. Eng., 2, 35–44, 2000.

Höfle, B. and Pfeifer, N.: Correction of laser scanning intensity data: data and model-driven approaches, ISPRS J. Photogramm., 62, 415–433, 2007.

ISO: Evaluation of Measurement Data – Guide to the Expression of Uncertainty in Measurement, JCGM 100:2008(E), Joint Committee for Guides in Meterology (JCGM), 2008.

Jaboyedoff, M., Oppikofer, T., Abellán, A., Derron, M. H., Loye, A., Metzger, R., and Pedrazzini, A.: Use of LIDAR in landslide investigations: a review, Nat. Hazards, 61, 5–28, 2012.

Kaufmann, V. and Ladstädter, R.: Monitoring of active rock glaciers by means of digital photogrammetry, Proceedings of the ISPRS Commission III Symposium "Photogrammetric Computer Vision", 9–13 September, Graz, Austria, IAPRS, Vol. 34, Part 3B, 108–111, 2002.

Kaufmann, V. and Ladstädter, R.: Documentation and visualization of the morphodynamics of Hinteres Langtalkar rock glacier (Hohe Tauern Range, Austrian Alps) based on aerial photographs (1954–2006) and geodetic measurements (1999–2007), Proceedings of the 10th International Symposium on High Mountain Remote Sensing Cartography, Kathmandu, Nepal, 2009.

Kaufmann, V., Kellerer-Pirklbauer, A., Lieb, G. K., Slupetzky, H., and Avian, M.: Glaciological Studies at Pasterze Glacier (Austria) Based on Aerial Photographs, Monitoring and Modeling of Global Changes, Springer Netherlands, 173–198, 2015.

Kellerer-Pirklbauer, A., Lieb, G. K., Avian, M., and Gspurning, J.: The response of partially debris-covered valley glaciers to climate change: the example of the Pasterze Glacier (Austria) in the period 1964 to 2006, Geogr. Ann., 90, 269–285, 2008.

Kellerer-Pirklbauer, A., Lieb, G. K., Avian, M., and Carrivick, J.: Climate change and rock fall events in high mountain areas: numerous and extensive rock falls in 2007 at Mittlerer Burgstall, Central Austria, Geogr. Ann. A, 94, 59–78, 2012.

Kienast, G. and Kaufmann, V.: Geodetic measurements on glaciers and rock glaciers in the Hohe Tauern National park (Austria), Proceedings of the 4th ICA Mountain Cartography Workshop, Vall de Núria, Catalonia, Barcelona, Spain, 101–108, 2004.

Kromer, R. A., Abellán, A., Hutchinson, D. J., Lato, M., Chanut, M.-A., Dubois, L., and Jaboyedoff, M.: Automated terrestrial laser scanning with near-real-time change detection – monitoring of the Séchilienne landslide, Earth Surf. Dynam., 5, 293–310, https://doi.org/10.5194/esurf-5-293-2017, 2017.

Lague, D., Brodu, N., and Leroux, J.: Accurate 3-D comparison of complex topography with terrestrial laser scanner: application to the Rangitikei canyon (NZ), ISPRS J. Photogramm., 82, 10–26, 2013.

Lambrecht, A. and Kuhn, M.: Glacier changes in the Austrian Alps during the last three decades, derived from the new Austrian glacier inventory, Ann. Glaciol., 46, 177–184, 2007.

Lichti, D. D. and Licht, M. G.: Experiences with terrestrial laser scanner modelling and accuracy assessment, Int. Arch. Photogramm., 36, 155–160, 2006.

Lindenbergh, R. and Pietrzyk, P.: Change detection and deformation analysis using static and mobile laser scanning, Appl. Geomatics, 7, 65–74, 2015.

Lowe, D.: Object recognition from local scale-invariant features, The Proceedings of the seventh IEEE International Conference on Computer Vision, 2, 1150–1157, 1999.

Neitzel, F.: Identifizierung konsistenter Datengruppen am Beispiel der Kongruenzuntersuchung geodätischer Netze, German Geodetic Commission (DGK), Volume C, Nr. 565, Munich, Germany, 2004.

Neitzel, F.: Investigations of axes errors of terrestrial laser scanners, 5th International Symposium Turkish–German Joint Geodetic Days, Berlin, Germany, 2006.

Nicolussi, K. and Patzelt, G.: Discovery of early-Holocene wood and peat on the forefield of the Pasterze Glacier, Eastern Alps, Austria, Holocene, 10, 191–199, 2000.

Oppikofer, T., Jaboyedoff, M., and Keusen, H. R.: Collapse at the eastern Eiger flank in the Swiss Alps, Nat. Geosci., 1, 531–535, 2008.

Paffenholz, J. A., Alkhatib, H., and Kutterer, H.: Direct georeferencing of a static terrestrial laser scanner, J. Appl. Geodesy, 4, 115–126, 2010.

Previtali, M., Barazzetti, L., Brumana, R., and Scaioni, M.: Scan registration using planar features, Int. Arch. Photogramm., 40, 501, 2014.

Riegl: Riegl VZ-200i data sheet, available at: http://www.riegl.com/uploads/tx_pxpriegldownloads/RIEGL_VZ-2000i_Datasheet_2017-12-18_Preliminary.pdf (last access: 26 March 2018), 2017.

Salo, P., Jokinen, O., and Kukko, A.: On the calibration of the distance measuring component of a terrestrial laser scanner, in: Proc. in the XXIth ISPRS Congress, Silk Road for Information from Imagery, Vol. 37, p. B5, 2008.

Samet, H.: Foundations of Multidimensional and Metric Data Structures, Kaufmann, San Francisco, California, USA, 2006.

Schwalbe, E., Maas, H. G., Dietrich, R., and Ewert, H.: Glacier velocity determination from multi temporal terrestrial long range laser scanner point clouds, ISPRS Arch. Photogramm. Remote Sens. Spatial Inf. Sci., 37, 457–462, 2008.

Soudarissanane, S.: The geometry of terrestrial laser scanning; identification of errors, modeling and mitigation of scanning geometry, Doctoral dissertation, Delft University of Technology, 2016.

technet: Scantra – High-Precision Scan Registration without Targets, available at: http://technet-gmbh.de/index.php?id=1051&L=1 (last access: 17.05.2017), 2017.

Tirado, M. C., Clarke, R., Jaykus, L. A., McQuatters-Gollop, A., and Frank, J. M.: Climate change and food safety: a review, Food Res. Int., 43, 1745–1765, 2010.

van Veen, M., Hutchinson, D. J., Kromer, R., Lato, M., and Edwards, T: Effects of sampling interval on the frequency-magnitude relationship of rockfalls detected from terrestrial laser scanning using semi-automated methods, Landslides, 14, 1579–1592, 2017.

Wakonigg, H. and Lieb, G. K.: Die Pasterze und ihre Erforschung im Rahmen der Gletschermessungen, Wissenschaft im Nationalpark Hohe Tauern Kärnten – Kärntner 888 Nationalpark-Schriften, 8, 99–115, 1996.

Wujanz, D.: Towards Transparent Quality Measures in Surface Based Registration Processes: Effects of Deformation Onto Commercial and Scientific Implementations, Proceedings of the XXII Congress of the ISPRS, Melbourne, Australia, 2012.

Wujanz, D., Holst, C., Neitzel, F., Kuhlmann, H., Niemeier, W., and Schwieger, V.: Survey configuration for terrestrial laser scanning, Allgemeine Vermessungsnachrichten (AVN), 6, 158–169, 2016a.

Wujanz, D., Krueger, D., and Neitzel, F.: Identification of stable areas in unreferenced laser scans for deformation measurement, Photogramm. Rec., 31, 261–280, 2016b.

Wujanz, D., Burger, M., Mettenleiter, M., and Neitzel, F: An intensity-based stochastic model for terrestrial laser scanners, ISPRS J. Photogramm., 125, 146–155, 2017.

Zoller+Fröhlich: Data sheet, Z+F IMAGER 5010X, available at: http://www.zf-laser.com/fileadmin/editor/Datenblaetter/Z_F_IMAGER_5010X_System_Requirements_E_FINAL.pdf (last access: 21 May 2017), 2017.

4

Statistical modeling of the long-range-dependent structure of barrier island framework geology and surface geomorphology

Bradley A. Weymer[1,a], **Phillipe Wernette**[2], **Mark E. Everett**[1], **and Chris Houser**[3]

[1]Department of Geology and Geophysics, Texas A&M University, College Station, Texas 77843, USA
[2]Department of Geography, Texas A&M University, College Station, Texas 77843, USA
[3]Department of Earth and Environmental Sciences, University of Windsor, Windsor, Ontario N9B 3P4, Canada
[a]now at: GEOMAR – Helmholtz Center for Ocean Research Kiel, Wischhofstraße 1–3, 24148 Kiel, Germany

Correspondence: Bradley A. Weymer (brad.weymer@gmail.com)

Abstract. Shorelines exhibit long-range dependence (LRD) and have been shown in some environments to be described in the wave number domain by a power-law characteristic of scale independence. Recent evidence suggests that the geomorphology of barrier islands can, however, exhibit scale dependence as a result of systematic variations in the underlying framework geology. The LRD of framework geology, which influences island geomorphology and its response to storms and sea level rise, has not been previously examined. Electromagnetic induction (EMI) surveys conducted along Padre Island National Seashore (PAIS), Texas, United States, reveal that the EMI apparent conductivity (σ_a) signal and, by inference, the framework geology exhibits LRD at scales of up to 10^1 to 10^2 km. Our study demonstrates the utility of describing EMI σ_a and lidar spatial series by a fractional autoregressive integrated moving average (ARIMA) process that specifically models LRD. This method offers a robust and compact way of quantifying the geological variations along a barrier island shoreline using three statistical parameters (p, d, q). We discuss how ARIMA models that use a single parameter d provide a quantitative measure for determining free and forced barrier island evolutionary behavior across different scales. Statistical analyses at regional, intermediate, and local scales suggest that the geologic framework within an area of paleo-channels exhibits a first-order control on dune height. The exchange of sediment amongst nearshore, beach, and dune in areas outside this region are scale independent, implying that barrier islands like PAIS exhibit a combination of free and forced behaviors that affect the response of the island to sea level rise.

1 Introduction

Barrier island transgression in response to storms and sea level rise depends to varying degrees on preexisting geologic features. The traditional assumption of uniform sand at depth and alongshore cannot explain many observations. Models of barrier island evolution are required to ascertain the degree to which the island is either *free* (such as a large sand body) or *forced* (i.e., constrained) by the underlying geology. Despite growing evidence that the underlying geological structure, otherwise termed *framework geology*, of barrier islands influences nearshore, beach, and dune morphology (e.g., Belknap and Kraft, 1985; Houser, 2012; Lentz and Hapke, 2011; McNinch, 2004; Riggs et al., 1995), this variable remains largely absent from shoreline change models that treat the geology as being uniform alongshore (e.g., Dai et al., 2015; Plant and Stockdon, 2012; Wilson et al., 2015). Spatial variation in the height and position of the dune line impacts the overall transgression of the island with sea level rise (Sallenger, 2000). Transgression is accomplished largely through the transport and deposition of beach and dune sediments to the backbarrier as washover deposits during storms (Houser, 2012; Morton and Sallenger, 2003; Stone et al., 2004).

1.1 Framework geology controls on barrier island evolution

The dynamic geomorphology of a barrier island system is the result of a lengthy, complex, and ongoing history that is characterized by sea level changes and episodes of deposition and erosion (e.g., Anderson et al., 2015; Belknap and Kraft, 1985; Rodriguez et al., 2001). Previous studies demonstrate that the framework geology of barrier islands plays a considerable role in the evolution of these coastal landscapes (Belknap and Kraft, 1985; Evans et al., 1985; Kraft et al., 1982; Riggs et al., 1995). For example, antecedent structures such as paleo-channels, ravinement surfaces, offshore ridge and swale bathymetry, and relict transgressive features (e.g., overwash deposits) have been suggested to influence barrier island geomorphology over a wide range of spatial scales (Hapke et al., 2010, 2016; Houser, 2012; Lentz and Hapke, 2011; McNinch, 2004). In this study, the term framework geology is specifically defined as the topographic surface of incised valleys, paleo-channels, and/or the depth to ravinement surface beneath the modern beach.

As noted by Hapke et al. (2013), the framework geology at the regional scale (> 30 km) influences the geomorphology of an entire island. Of particular importance are the location and size of glacial, fluvial, tidal, and/or inlet paleo-valleys and paleo-channels (Belknap and Kraft, 1985; Colman et al., 1990; Demarest and Leatherman, 1985), and paleo-deltaic systems offshore or beneath the modern barrier system (Coleman and Gagliano, 1964; Frazier, 1967; Miselis et al., 2014; Otvos and Giardino, 2004; Twichell et al., 2013). At the regional scale, nonlinear hydrodynamic interactions between incident wave energy and nearshore ridge and swale bathymetric features can generate periodic alongshore variations in beach–dune morphology (e.g., Houser, 2012; McNinch, 2004) that are superimposed on larger-scale topographic variations as a result of transport gradients (Tebbens, et al., 2002). At the intermediate scale (10–30 km), feedbacks between geologic features and relict sediments of the former littoral system (e.g., Honeycutt and Krantz, 2003; Riggs et al., 1995; Rodriguez et al., 2001; Schwab et al., 2000) act as an important control on dune formation (Houser et al., 2008) and offshore bathymetric features (e.g., Browder and McNinch, 2006; Schwab et al., 2013). Framework geology at the local scale ($\leq$ 10 km) induces mesoscale ($\sim 10^1$–10^2 m) to microscale (< 1 m) sedimentological changes (e.g., Murray and Thieler, 2004; Schupp et al., 2006), variations in the thickness of shoreface sediments (Brown and Macon, 1977; Miselis and McNinch, 2006), and spatial variations in sediment transport across the island (Houser and Mathew, 2011; Houser, 2012; Lentz and Hapke, 2011).

To date, most of what is known regarding barrier island framework geology is based on studies performed at either intermediate or local scales (e.g., Hapke et al., 2010; Lentz and Hapke, 2011; McNinch, 2004), whereas few studies exist at the regional scale for United States coastlines (Hapke et al., 2013). The current study focuses on barrier islands in the United States and we do not consider work on barrier islands in other regions. Assessments of framework geology at regional and intermediate spatial scales for natural and anthropogenically modified barrier islands are essential for improved coastal management strategies and risk evaluation since these require a good understanding of the connections between subsurface geology and surface morphology. For example, studies by Lentz and Hapke (2011) and Lentz et al. (2013) at Fire Island, New York, suggest that the short-term effectiveness of engineered structures is likely influenced by the framework geology. Extending their work, Hapke et al. (2016) identified distinct patterns of shoreline change that represent different responses alongshore to oceanographic and geologic forcing. These authors applied empirical orthogonal function (EOF) analysis to a time series of shoreline positions to better understand the complex mul tiscale relationships between framework geology and contemporary morphodynamics. Gutierrez et al. (2015) used a Bayesian network to predict barrier island geomorphic characteristics and argue that statistical models are useful for refining predictions of locations where particular hazards may exist. These examples demonstrate the benefit of using statistical models as quantitative tools for interpreting coastal processes at multiple spatial and temporal scales (Hapke et al., 2016).

1.2 Statistical measures of coastline geomorphology

It has long been known that many aspects of landscapes exhibit similar statistical properties regardless of the length or timescale over which observations are sampled (Burrough, 1981). An often-cited example is the length L of a rugged coastline (Mandelbrot, 1967), which increases without bound as the length G of the ruler used to measure it decreases, in rough accord with the formula $L(G) \backsim G^{1-D}$, where $D \geq 1$ is termed the fractal dimension of the coastline. Andrle (1996), however, has identified limitations of the self-similar coastline concept, suggesting that a coastline may contain irregularities that are concentrated at certain characteristic length scales owing to local processes or structural controls. Recent evidence from South Padre Island, Texas (Houser and Mathew, 2011), Fire Island, New York (Hapke et al., 2010), and Santa Rosa Island, Florida (Houser et al., 2008), suggests that the geomorphology of barrier islands is affected to varying degrees by the underlying framework geology and that this geology varies, often with periodicities, over multiple length scales. The self-similarity of the framework geology and its impact on the geomorphology of these barrier islands was not examined explicitly.

Many lines of evidence suggest that geological formations in general are inherently rough (i.e., heterogeneous) and contain multiscale structure (Bailey and Smith, 2005; Everett and Weiss, 2002; Radliński et al., 1999; Schlager, 2004). Some of the underlying geological factors that lead to self-similar terrain variations are reviewed by Xu et al. (1993). In essence, competing and complex morphodynamic processes, influenced by the underlying geological structure, operate over different spatiotemporal scales, such that the actual terrain is the result of a complex superposition of the various effects of these processes (see Lazarus et al., 2011). Although no landscape is strictly self-similar on all scales, Xu et al. (1993) show that the fractal dimension, as a global morphometric measure, captures multiscale aspects of surface roughness that are not evident in conventional local morphometric measures such as slope gradient and profile curvature.

With respect to coastal landscapes, it has been suggested that barrier shorelines are scale independent, such that the wave number spectrum of shoreline variation can be approximated by a power law at alongshore scales from tens of meters to several kilometers (Lazarus et al., 2011; Tebbens et al., 2002). However, recent findings by Houser et al. (2015) suggest that the beach–dune morphology of barrier islands in Florida and Texas is scale dependent and that morphodynamic processes operating at swash (0–50 m) and surf-zone (< 1000 m) scales are different than the processes operating at larger scales. In this context, scale dependence implies that a certain number of different processes are simultaneously operative, each process acting at its own scale of influence, and it is the superposition of the effects of these multiple processes that shapes the overall behavior and shoreline morphology. This means that shorelines may have different patterns of irregularity alongshore with respect to barrier island geomorphology, which has important implications for analyzing long-term shoreline retreat and island transgression. Lazarus et al. (2011) point out that deviations from power-law scaling at larger spatial scales (tens of kilometers) emphasizes the need for more studies that investigate large-scale shoreline change. While coastal terrains might not satisfy the strict definition of self similarity, it is reasonable to expect them to exhibit long-range dependence (LRD). LRD pertains to signals in which the correlation among observations decays like a power law with separation, i.e., much slower than one would expect from independent observations or those that can be explained by a short-memory process, such as an autoregressive moving average (ARMA) with small (p, q) (Beran, 1994; Doukhan et al., 2003).

1.3 Research objectives

This study performed at Padre Island National Seashore (PAIS), Texas, United States, utilizes electromagnetic induction (EMI) apparent conductivity σ_a responses to provide insight into the relation between spatial variations in framework geology and surface morphology. Two alongshore EMI surveys at different spatial scales (100 and 10 km) were conducted to test the hypothesis that, like barrier island morphology, subsurface framework geology exhibits the LRD characteristic of scale independence. The σ_a responses, which are sensitive to parameters such as porosity and mineral content, are regarded herein as a rough proxy for subsurface framework geology (Weymer et al., 2015a). This assumes, of course, that alongshore variations in salinity and water saturation, and other factors that shape the σ_a response, can be neglected to first order. A corroborating 800 m ground-penetrating radar (GPR) survey, providing an important check on the variability observed within the EMI signal, confirms the location of a previously identified paleochannel (Fisk, 1959) at $\sim$5–10 m of depth. The overall geophysical survey design allows for a detailed evaluation of the long-range-dependent structure of the framework geology over a range of length scales spanning several orders of magnitude. We explore the applicability of autoregressive integrated moving-average (ARIMA) processes as models that describe the statistical connections between EMI and light detection and ranging (lidar) spatial data series. This paper utilizes a generalized fractional ARIMA $(0, d, 0)$ process (Hosking, 1981) that is specifically designed to model LRD for a given data series using a single differencing non-integer parameter d. The parameter d can be used in the present context to discriminate between forced, scale-dependent controls by the framework geology, i.e., stronger LRD ($d \to 0.5$), and free behavior that is scale independent, i.e., weaker LRD ($0 \leftarrow d$). In other words, it is the particular statistical characteristics of the framework geology LRD at PAIS that we are trying to ascertain from the EMI σ_a signal, with the suggestion that σ_a measurements can be used similarly at other sites to reveal the hidden LRD characteristics of the framework geology.

2 Background and regional setting

2.1 Utility of electromagnetic methods in coastal environments

Methods to ascertain the alongshore variability in framework geology, and to test long-range dependence, are difficult to implement and can be costly. Cores provide detailed point-wise geologic data; however, they do not provide laterally continuous subsurface information (Jol et al., 1996). Alternatively, geophysical techniques including seismic and GPR provide spatially continuous stratigraphic information (e.g., Buynevich et al., 2004; Neal, 2004; Nummedal and Swift, 1987; Tamura, 2012), but they are not ideally suited for LRD testing because the data combine depth and lateral information at a single acquisition point. Moreover, GPR signals attenuate rapidly in saltwater environments whereas seismic methods are labor-intensive and cumbersome. Conversely, terrain conductivity profiling is an easy-to-use alternative that has been used in coastal environments to investigate fundamental questions involving instrument performance characteristics (Delefortrie et al., 2014; Weymer et al., 2016), groundwater dynamics (Stewart, 1982; Fitterman and Stewart, 1986; Nobes, 1996; Swarzenski and Izbicki, 2009), and framework geology (Seijmonsbergen et al., 2004; Weymer et al., 2015a). Previous studies combining EMI with either GPR (Evans and Lizarralde, 2011) or coring (Seijmonsbergen et al., 2004) demonstrate the validity of EMI measurements as a means to quantify alongshore variations in the framework geology of coastlines.

In the alongshore direction, Seijmonsbergen et al. (2004) used a Geonics EM34™ terrain conductivity meter crossing a former outlet of the Rhine River, Netherlands, to evaluate alongshore variations in subsurface lithology. The survey was conducted in an area that was previously characterized by drilling and these data were used to calibrate the σ_a measurements. The results from the study suggest that coastal sediments can be classified according to σ_a signature and that high σ_a values occur in areas where the underlying conductive layer is thick and close to the surface. Although Seijmonsbergen et al. (2004) propose that EMI surveys are a rapid, inexpensive method to investigate subsurface lithology, they also acknowledge that variations in salinity as a result of changing hydrologic conditions, storm activity, and/or tidal influence confound the geological interpretation and should be investigated in further detail (see Weymer et al., 2016).

The challenge on many barrier islands and protected national seashores is obtaining permission for extracting drill cores to validate geophysical surveys. At PAIS, numerous areas along the island are protected nesting sites for the endangered Kemp's ridley sea turtle and migratory birds while other areas comprise historic archeological sites with restricted access. Thus, coring is not allowed and only noninvasive techniques, such as EMI–GPR, are permitted.

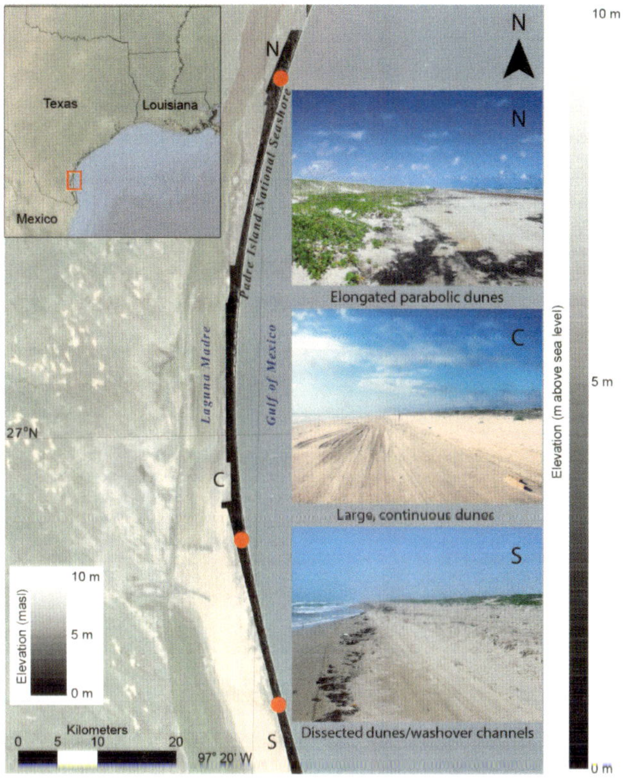

Figure 1. Location map and DEM of the study area at Padre Island National Seashore (PAIS), Texas, United States. Elevations for the DEM are reported as meters above sea level (m a.s.l.). Approximate locations of field images (red dots) from the northern (N), central (C), and southern (S) regions of the island showing alongshore differences in beach–dune morphology. Note that views are facing south for the central and southern locations, and the northern location view is to the north. Images taken in October 2014.

2.2 Regional setting

North Padre Island is part of a large arcuate barrier island system located along the Texas Gulf of Mexico coastline. The island is one of 10 national seashores in the United States and is protected and managed by the National Park Service, a bureau of the Department of the Interior. PAIS is 129 km in length, and is an ideal setting for performing EMI surveys because there is minimal cultural noise to interfere with the σ_a signal, which as stated earlier we regard as a proxy for alongshore variations in framework geology (Fig. 1). Additionally, there are high-resolution elevation data available from a 2009 aerial lidar survey. The island is not dissected by inlets or navigation channels (excluding Mansfield Channel separating the North and South Padre islands) or modified by engineered structures (e.g., groynes, jetties) that often interfere with natural morphodynamic processes (see Talley et al., 2003). The above characteristics make the study area an exceptional location for investigating the relationships between large-scale framework geology and surface morphology.

As described in Weymer et al. (2015a; Fig. 3), locations of several paleo-channels were established by Fisk (1959) based on 3000 cores and seismic surveys. More than 100 boreholes were drilled to the top of the late Pleistocene surface (tens of meters of depth) providing sedimentological data for interpreting the depth and extent of the various paleo-channels. These cores were extracted ~60 years ago, but the remnant Pleistocene and Holocene fluvial–deltaic features described in Fisk's study likely have not changed over decadal timescales.

Geologic interpretations based on the Fisk (1959) data suggest that the thickness of the modern beach sands is ~2–3 m, and they are underlain by Holocene shoreface sands and muds to a depth of ~10–15 m (Brown and Macon, 1977; Fisk, 1959). The Holocene deposits lie upon a Pleistocene ravinement surface of fluvial–deltaic sands and muds and relict transgressive features. A network of buried valleys and paleo-channels in the central segment of the island, as interpreted by Fisk (1959), exhibits a dendritic, tributary pattern. The depths of the buried valleys inferred from seismic surveys range from ~25 to 40 m (Brown and Macon, 1977). These channels have been suggested to have incised into the Pleistocene paleo-surface and became infilled with sands from relict Pleistocene dunes and fluvial sediments reworked by alongshore currents during the Holocene transgression (Weise and White, 1980). However, the location and cross-sectional area of each valley and paleo-channel alongshore is not well constrained. It is also possible that other channels exist other than those identified by Fisk (1959). As suggested in Weymer et al. (2015a), minima in the alongshore σ_a signal are spatially correlated with the locations of these previously identified geologic features. This observation provides an impetus for using EMI to map the known, and any previously unidentified, geologic features alongshore.

3 Methods

A combination of geophysical, geomorphological, and statistical methods are used in this study to quantify the relationships between framework geology and surface geomorphology at PAIS. A description of the EMI, GPR, geomorphometry and statistical techniques is provided in the following sections.

3.1 Field EMI and GPR surveys

Profiles of EMI σ_a responses are typically irregular and each datum represents a spatial averaging of the bulk subsurface electrical conductivity σ, which in turn is a function of a number of physical properties (e.g., porosity, lithology, water content, salinity). The "sensor footprint", or subsurface volume over which the spatial averaging is performed, is dependent on the separation between the transmitter–receiver (TX–RX) coils (1.21 m in this study) and the transmitter frequency. The horizontal extent, or radius, of the footprint can be more or less than the step size between subsequent measurements along the profile. The sensor footprint determines the volume of ground that contributes to σ_a at each acquisition point, and as will be discussed later, the radius of the footprint has important implications for analyzing LRD. The footprint radius depends on frequency and ground conductivity, but is likely to be of the same order as, but slightly larger than, the intercoil spacing. Two different station spacings were used to examine the correlation structure of σ_a as a function of spatial scale. An island-scale alongshore survey of ~ 100 km length was performed using a 10 m station spacing (station spacing ≫ footprint radius) such that each σ_a measurement was recorded over an independently sampled volume of ground. Additionally, a sequence of σ_a readings was collected at 1 m of spacing (station spacing < footprint radius) over a profile length of 10 km within the Fisk (1959) paleo-channel region of the island. This survey design allows for comparison of the long-range-dependent structure of the framework geology over several orders of magnitude (10^0–10^5 m).

The 100 km long alongshore EMI survey was performed during a series of three field campaigns, resulting in a total of 21 (each of length ~ 4.5 km) segments that were collected during 9–12 October 2014, 15–16 November 2014, and 28 March 2015. The EMI σ_a profiles were stitched together by importing GPS coordinates from each measurement into ArcGIS™ to create a single composite spatial data series. The positional accuracy recorded by a TDS Recon PDA equipped with a Holux™ WAAS GPS module was found to be accurate within ~ 1.5 m. To reduce the effect of instrument drift caused by temperature, battery, and other systematic variations through the acquisition interval, a drift correction was applied to each segment and the segments were then stitched together, following which a regional linear trend removal was applied to the composite dataset. An additional 10 km survey was performed along a segment of the same 100 km survey line in one day on 29 March 2015. This second composite data series consists of eight stitched segments.

The same multifrequency GSSI Profiler EMP-400™ instrument was used for each segment. All transects were located in the back-beach environment ~ 25 m inland from the mean tide level (MTL). This location was chosen to reduce the effect of changing groundwater conditions in response to nonlinear tidal forcing (see Weymer et al., 2016), which may be significant closer to the shoreline. As will be shown later, there is not a direct correlation between high tide and high σ_a values. Thus, we assume the tidal influence on the EMI signal can be neglected over the spatial scales of interest in the present study. Nevertheless, the duration and approximate tidal states of each survey were documented in order to compare with the EMI signal. Tidal data were accessed from NOAA's Tides and Currents database (NOAA, 2015b). Padre Island is microtidal and the mean tidal range within the study area is 0.38 m (NOAA, 2015a). A tidal signature in EMI signals may become more significant at other barrier islands with larger tidal ranges.

For all surveys, the EMI profiler was used in the same configuration and acquisition settings as described in Weymer et al. (2016). The transect locations were chosen to avoid the large topographic variations (see Santos et al., 2009) fronting the foredune ridge that can reduce the efficiency of data acquisition and influence the EMI signal. Measurements were made at a constant step size to simplify the data analysis; for example, ARIMA models require that data are taken at equal intervals (see Cimino et al., 1999). We choose herein to focus on data collected at 3 kHz, resulting in a depth of investigation (DOI) of ~ 3.5–6.4 m over the range of conductivities found within the study area (Weymer et al., 2016; Table 1). Because the depth of the modern beach sands is ~ 2–3 m or greater (see Brown and Macon, 1977; p. 56, Fig. 15), variations in the depth to shoreface sands and muds is assumed to be within the DOI of the profiler, which may not be captured at the higher frequencies also recorded by the sensor (i.e., 10 and 15 kHz) .

An 800 m GPR survey was performed on 12 August 2015 across one of the paleo-channels previously identified by Fisk (1959) located within the 10 km EMI survey for comparison with the σ_a measurements. We used a Sensors and Software pulseEKKO Pro™ system for this purpose. A survey-grade GPS with a positional accuracy of 10 cm was used to match the locations and measurements between the EMI–GPR surveys. Data were acquired in reflection mode at a nominal frequency of 100 MHz with a standard antenna separation of 1 m and a step size of 0.5 m. The instrument settings resulted in a DOI of up to 15 m. Minimal processing was applied to the data and includes a dewow filter and migration ($0.08\,\mathrm{m\,ns^{-1}}$), followed by automatic gain control (AGC) gain (see Neal, 2004). The theory and operational principles of GPR are discussed in many places (e.g., Everett, 2013; Jol, 2008) and will not be reviewed here.

3.2 Geomorphometry

Topographic information was extracted from aerial lidar data that were collected by the U.S. Army Corps of Engineers (U.S. ACE) in 2009 as part of the West Texas Aerial Survey project to assess post-hurricane conditions of the beaches and barrier islands along the Texas coastline. This dataset is the most recent publicly available lidar survey of PAIS and it provides essentially complete coverage of the island. With the exception of Hurricane Harvey, which made landfall near Rockport, Texas, as a category 4 storm in late August, 2017, Padre Island has not been impacted by a hurricane since July 2008, when Hurricane Dolly struck South Padre Island as a category 1 storm (NOAA, 2015a). The timing of the lidar and EMI surveys in this study precede the impacts of Hurricane Harvey, and it is assumed that the surface morphology across the island at the spatial scales of interest (i.e., 10^1–10^2 km) did not change appreciably between 2009 and 2015.

A 1 m resolution DEM was created from 2009 lidar point clouds available from NOAA's Digital Coast (NOAA, 2017). The raw point cloud tiles were merged to produce a combined point cloud of the island within the park boundaries of PAIS. The point clouds were processed into a continuous DEM using the ordinary kriging algorithm in SAGA GIS, which is a freely available open-source software (http://www.saga-gis.org, last access: 13 January 2018), and subsequent terrain analysis was conducted using an automated approach involving the relative relief (RR) metric (Wernette et al., 2016). Several morphometrics including beach width, dune height, and island width were extracted from the DEM by averaging the RR values across window sizes of 21 m × 21 m, 23 m × 23 m, and 25 m × 25 m. The choice of window size is based on tacit a priori knowledge and observations of the geomorphology in the study area. A detailed description of the procedure for extracting each metric is provided in Wernette et al. (2016).

Each DEM series is paired with the σ_a profile by matching the GPS coordinates (latitude and longitude) recorded in the field by the EMI sensor. Cross-sectional elevation profiles oriented perpendicular to the shoreline were analyzed every 10 m (y coordinate) along the EMI profile to match the same 10 m sampling interval of the σ_a measurements. The terrain variations along each cross-shore profile are summed to calculate beach and island volume based on the elevation thresholds mentioned above. Dune volume is calculated by summing the pixel elevations starting at the dune toe, traversing the dune crest, and ending at the dune heel. In total, six DEM morphometrics were extracted as spatial data series to be paired with the EMI data, each with an identical sample size ($n = 9694$), which is sufficiently large for statistical ARIMA modeling.

3.3 Statistical methods

Although the procedures for generating the EMI and lidar datasets used in this study are different, the intended goal is the same: to produce spatial data series that contain similar numbers of observations for comparative analysis using a combination of signal processing and statistical modeling techniques. The resulting signals comprising each data series represent the spatial averaging of a geophysical (EMI) or geomorphological elevation variable that contains information about the important processes that form relationships between subsurface geologic features and island geomorphology that can be teased out by means of comparative analysis (Weymer et al., 2015a). Because we are interested in evaluating these connections at both small and large spatial scales, our first approach is to determine the autocorrelation function and Hurst coefficient (self-similarity parameter) H and hence verify whether the data series are characterized by short- and/or long-range memory (Beran, 1992; Taqqu et al., 1995). LRD occurs when the autocorrelation within a series, at large lags, tends to zero like a power function, and so slowly that the sums diverge (Doukhan et al., 2003). LRD is often observed in natural time series and is closely related to self-similarity, which is a special type of LRD.

The degree of LRD is related to the scaling exponent H of a self-similar process, where increasing H in the range $0.5 < H \leq 1.0$ indicates an increasing tendency towards such an effect (Taqqu, 2003). Large correlations at small lags can easily be detected by models with short memory (e.g., ARMA, Markov processes) (Beran, 1994). Conversely, when correlations at large lags slowly tend to zero like a power function, the data contain long-memory effects and either fractional Gaussian noise (fGn) or ARIMA models may be suitable (Taqqu et al., 1995). The R/S statistic is the quotient of the range of values in a data series and the standard deviation (Beran, 1992, 1994; Hurst, 1951; Mandelbrot and Taqqu, 1979). When plotted on a log / log plot, the resulting slope of the best-fit line gives an estimate of H, which is useful as a diagnostic tool for estimating the degree of LRD (see Beran, 1994).

It has been suggested that R/S tends to give biased estimates of H, too low for $H > 0.72$ and too high for $H < 0.72$ (Bassingthwaigthe and Raymond, 1994), which was later confirmed by Malamud and Turcotte (1999). Empirical trend corrections to the estimates of H can be made by graphical interpolation, but are not applied here because of how the regression is performed. The R/S analysis in this study was performed using signal analysis software AutoSignal™ to identify whether a given signal is distinguishable from a random, white noise process and, if so, whether the given signal contains LRD. The H value is calculated by an inverse variance-weighted linear least-squares curve fit using the logarithms of the R/S and the number of observations, which provides greater accuracy than other programs that compute the Hurst coefficient.

Two of the simplest statistical time series models that can account for LRD are fGn and ARIMA. In the former case, fGn and its "parent" fractional Brownian motion (fBm) are used to evaluate stationary and nonstationary fractal signals, respectively (see Eke et al., 2000; Everett and Weiss, 2002). Both fGn and fBm are governed by two parameters: variance σ^2 and the scaling parameter, H (Eke et al., 2000). A more comprehensive class of time series models that has a similar capability to detect long-range structure is ARIMA. Because fGn and fBm models have only two parameters, it is not possible to model the short-range components. Additional parameters in ARIMA models are designed to handle the short-range component of the signal, as discussed by Taqqu et al. (1995) and others. Because the EMI data series presumably contain both short-range and long-range effects, we chose to use ARIMA as the analyzing technique.

ARIMA models are used across a wide range of disciplines in geoscience and have broad applicability for understanding the statistical structure of a given data series as it is related to some physical phenomenon (see Beran, 1992, 1994; Box and Jenkins, 1970; Cimino et al., 1999; Granger and Joyeux, 1980; Hosking, 1981; Taqqu et al., 1995). For example, Cimino et al. (1999) apply R/S analysis, ARIMA, and neural network analysis to different geological datasets including tree ring data, Sr isotope data of Phanerozoic sea water samples, and El Niño phenomena. The authors show that their statistical approach enables (1) recognition of qualitative changes within a given dataset, (2) evaluation of the scale (in)dependency of increments, (3) characterization of random processes that describe the evolution of the data, and (4) recognition of cycles embedded within the data series. In the soil sciences, Alemi et al. (1988) use ARIMA and Kriging to model the spatial variation in clay-cover thickness of a 78 km^2 area in northeastern Iran and demonstrate that ARIMA modeling can adequately describe the nature of the spatial variations. ARIMA models have also been used to model periodicity of major extinction events in the geologic past (Kitchell and Pena, 1984).

In all these studies, the statistical ARIMA model of a given data series is defined by three terms (p, d, q), where p and q indicate the order of the autoregressive (AR) and moving average (MA) components, respectively, and d represents a differencing or integration term (I) that is related to LRD. The AR element, p, represents the effects of adjacent observations and the MA element, q, represents the effects on the process of nearby random shocks (Cimino et al., 1999; De Jong and Penzer, 1998). However, in the present study our series are reversible spatial series that can be generated, and are identical, with either forward or backward acquisition, unlike a time series. Both p and q parameters are restricted to integer values (e.g., 0, 1, 2), whereas the integration parameter, d, represents potentially long-range structure in the data. The differencing term d is normally evaluated before p and q to identify whether the process is stationary (i.e., constant mean and σ^2). If the series is nonstationary, it is differenced to remove either linear ($d = 1$) or quadratic ($d = 2$) trends, thereby making the mean of the series stationary and invertible (Cimino et al., 1999), thus allowing determination of the ARMA p and q parameters.

Here, we adopt the definitions of an ARMA (p, q), and ARIMA (p, d, q) process following the work of Beran (1994). Let p and q be integers, where the corresponding polynomials are defined as

$$\phi(x) = 1 - \sum_{j=1}^{p} \phi_j x^j,$$

$$\psi(x) = 1 + \sum_{j=1}^{q} \psi_j x^j. \qquad (1)$$

It is important to note that all solutions of $\phi(x_0) = 0$ and $\psi(x) = 0$ are assumed to lie outside the unit circle. Additionally, let $\epsilon_t (t = 1, 2, \ldots)$ be independent, and identically distributed normal variables with zero variance σ_ϵ^2 such that an ARMA (p, q) process is defined by the stationary solution of

$$\phi(B) X_t = \psi(B) \epsilon_t, \qquad (2)$$

where B is the backward shift operator $BX_t = X_{t-1}, B^2 X_t = X_{t-1}, \ldots$ and, specifically, the differences can be expressed in terms of B as $X_t - X_{t-1} = (1 - B)X_t, (X_t - X_{t-1}) - (X_{t-1} - X_{t-2}) = (1 - B)^2 X_t \ldots$ Alternatively, an ARIMA (p, d, q) process X_t is formally defined as

$$\phi(B)(1 - B)^d X_t = \psi(B) \epsilon_t, \qquad (3)$$

where Eq. (3) holds for a dth difference $(1 - B)^d X_t$.

As mentioned previously, a more general form of ARIMA (p, d, q) is the fractional ARIMA process, or FARIMA, where the differencing term d is allowed to take on fractional values. If d is a non-integer value for some $-0.5 < d < 0.5$ and X_t is a stationary process as indicated by Eq.(3), then the model by definition is called a FARIMA process where d values in the range $0 < d < 0.5$ are of particular interest herein because geophysically relevant LRD occurs for $0 < d < 0.5$, whereas $d > 0.5$ means that the process is nonstationary but nonintegrable (Beran, 1994; Hosking, 1981).

A special case of a FARIMA process explored in the current study is ARIMA (0d0), also known as fractionally differenced white noise (Hosking, 1981), which is defined by Beran (1994) and others as

$$X_t = (1 - B)^{-d}\epsilon_t. \quad (4)$$

For $0<d<0.5$, the ARIMA (0d0) process is a stationary process with long-range structure and is useful for modeling LRD. As shown later, different values of the d parameter provide further insight into the type of causative physical processes that generate each data series. When $d<0.5$, the series X_t is stationary, which has an infinite MA representation that highlights long-range trends or cycles in the data. Conversely, when $d>-0.5$, the series X_t is invertible and has an infinite AR representation (see Hosking, 1981). When $-0.5<d<0$, the stationary, and invertible, ARIMA (0d0) process is dominated by short-range effects and is antipersistent. When $d=0$, the ARIMA (000) process is white noise, with zero correlations and a constant spectral density. Identification of an appropriate model is accomplished by finding small values of elements p, d, q (usually between 0 and 2) that accurately fit the most significant patterns in the data series. When a value of an element is 0, that element is not needed. For example, if $d=0$ the series does not contain a significant long-range component, whereas if $p=q=0$, the model does not exhibit significant short-range effects. If $p, d, q \neq 0$, the model contains a combination of both short- and long-memory effects.

4 Results

4.1 Spatial data series

4.1.1 EMI and GPR surveys

The unprocessed (raw) EMI σ_a responses show a high degree of variability along the island. High-amplitude responses within the EMI signal generally exhibit a higher degree of variability (multiplicative noise) compared to the low-amplitude responses. Higher σ_a readings correspond to a small sensor footprint and have enhanced sensitivity to small-scale near-surface heterogeneities (see Guillemoteau and Tronicke, 2015). Low σ_a readings suggest the sensor is probing greater depths and averaging over a larger footprint. In that case, the effect of fine-scale heterogeneities that contribute to signal variability is suppressed.

The 10 km alongshore survey is located within an inferred paleo-channel region (Fisk, 1959), providing some a priori geologic constraints for understanding the variability within the EMI signal (Fig. 2b). Here, the sample size is $n = 10\,176$, permitting a quantitative comparison with the 100 km long data series since they contain a similar number of observations. Unlike the 100 km survey, successive footprints of the sensor at each subsequent measurement point overlap along the 10 km survey. The overlap enables a fine-scale characterization of the underlying geological structure because the separation between the TX and RX coils (1.21 m), a good lower-bound approximation of the footprint, is greater than the step size (1 m).

The overall trend in σ_a for the 10 km survey is comparable to that of the 100 km survey, where regions characterized by high- and low-amplitude signals correspond to regions of high and low variability, respectively, implying that multiplicative noise persists independently of station spacing. The decrease in σ_a that persists between ~2.5 and 6 km along the profile (Fig. 2b) coincides in location with two paleo-channels, whereas a sharp reduction in σ_a is observed at ~8.2 km in close proximity to a smaller channel. Most of the known paleo-channels are located within the 10 km transect and likely contain resistive infill sands that should generate lower and relatively consistent σ_a readings (Weymer et al., 2015a). The low σ_a signal caused by the sand indirectly indicates valley incision since it is diagnostic of a thicker sand section, relatively unaffected by the underlying conductive layers. Thus, it is reasonable to assume that reduced variability in the signal is related to the framework geology within the paleo-channels, which we now compare with a GPR profile.

To corroborate the capability of the EMI data to respond to the variable subsurface geology, an 800 m GPR survey confirms the location of a previously identified paleo-channel (Fisk, 1959) at ~5–10 m depth (Fig. 3). A continuous undulating reflector from ~150 to 800 m along the profile is interpreted to be the surface mapped by Fisk (1959), who documented a paleo-channel at this location with a depth of ~8 m. Although the paleo-surface is within the detection limits of the GPR, it is likely that the DOI of the EMI data (~3–6 m) is not large enough to probe continuously along the contact between the more conductive ravinement surface and the more resistive infill sands. Along the transect at shallower depths highlighted by the red box in the lower radargram (Fig. 3), low EMI σ_a values correspond to fine stratifications in the GPR section, which is common for beach sands with little clay content that are not saline-saturated. The EMI highs between ~450 and 530 m coincide with parts of the GPR section that do not have the fine stratification and this may indicate the presence of clay or saline water. Here, the high conductivity zone for both the GPR and EMI is located within a recovering washover channel overlying the paleo-channel that is evident in the satellite imagery in the upper-left panel of Fig. 3. The overwash deposits consisting of a mix of sand and more finely grained backbarrier sediments likely mask the EMI sensors' ability to probe greater depths. Nonetheless, the high-conductivity zone represents a smaller ~100 m segment within the ~500 m wide paleo-channel, suggesting that variations in the EMI responses outside this zone are directly related to variations in the framework geology imaged by GPR.

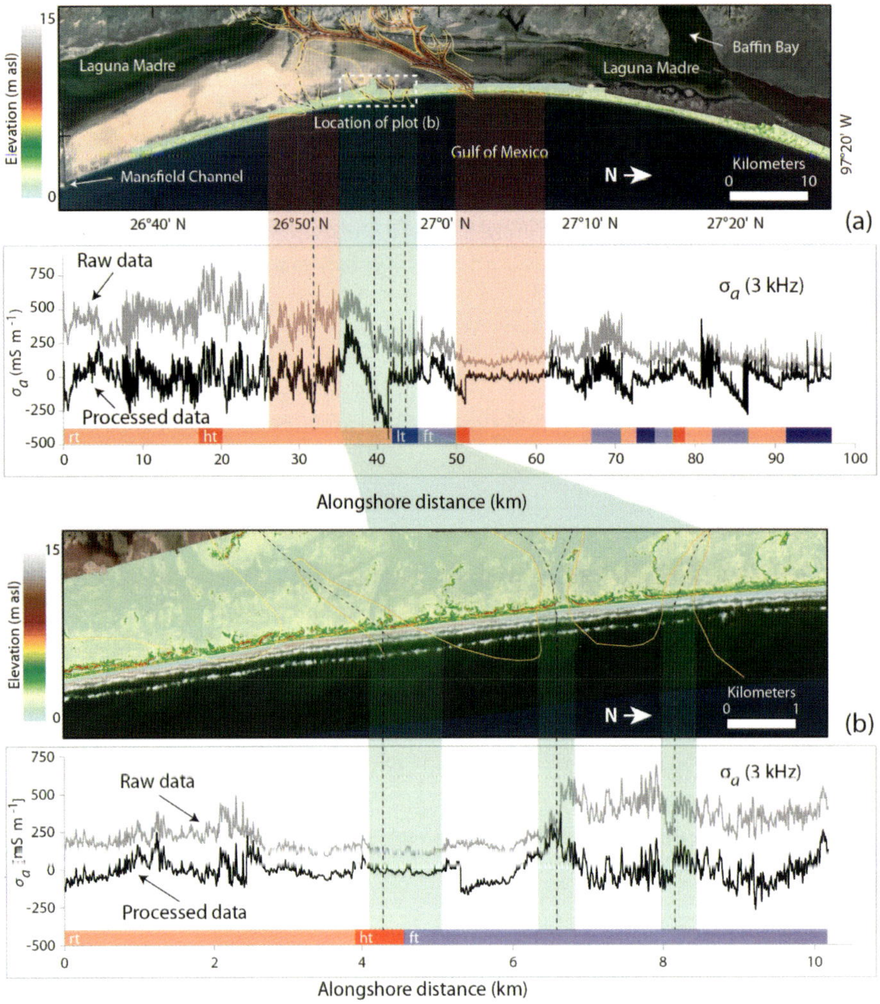

Figure 2. The 100 km **(a)** and 10 km **(b)** alongshore EMI surveys showing DEMs of the study area and previously identified paleo-channel region by Fisk (1959). Channels are highlighted in red and green, where the green region indicates the location of the 10 km survey. The 25 ft (7.6 m) contour intervals are highlighted with depths increasing from yellow to red and the center of the channels are represented by the black-dotted lines. For each survey, raw σ_a and zero-mean drift-corrected EMI responses are shown in gray and black, respectively. Tidal conditions during each EMI acquisition segment are shown below each panel. Low (lt) and falling tides (ft) are indicated by blue and light blue shades, respectively. High (ht) and rising tides (rt) are highlighted in red and light red, respectively.

4.1.2 Lidar-derived DEM morphometrics

The lidar-derived elevation data series along the 100 km transect are presented in Fig. 4. Each data series is shown with respect to the areal DEM of the study area where the approximate locations of each closely spaced paleo-channel are highlighted in gray. This visualization allows a qualitative analysis of the spatial relationships among paleo-channels, subsurface information encoded in the σ_a signal, and surface morphology over the entire length of the barrier island.

The morphology of the beach–dune system, as well as island width, changes substantially from north to south. In the paleo-channel region, beach width decreases in the central channel ($\sim$ 37–42 km) and is more variable outside this region. Beach width generally increases towards the northern section of the island. The volume of the beach tends to be lowest in the northern zone, varies considerably in the central part of the island, then stabilizes and gradually decreases towards the south. These zones correspond to the southern (0–30 km), central (30–60 km), and northern (60–100 km) sections of the island. Alongshore dune heights are generally greater in the south, become slightly more variable in the paleo-channel region, and decrease in the north except for the area adjacent to Baffin Bay. Dune volume is lowest in the northern section, intermittently increases in the central zone, and slightly decreases towards the south. The island is considerably narrower between Mansfield Channel and Baffin Bay (see Fig. 2a), increasing in width in the northern zone; island volume follows a similar trend. Overall, σ_a values are lower northward of the paleo-channel region compared to

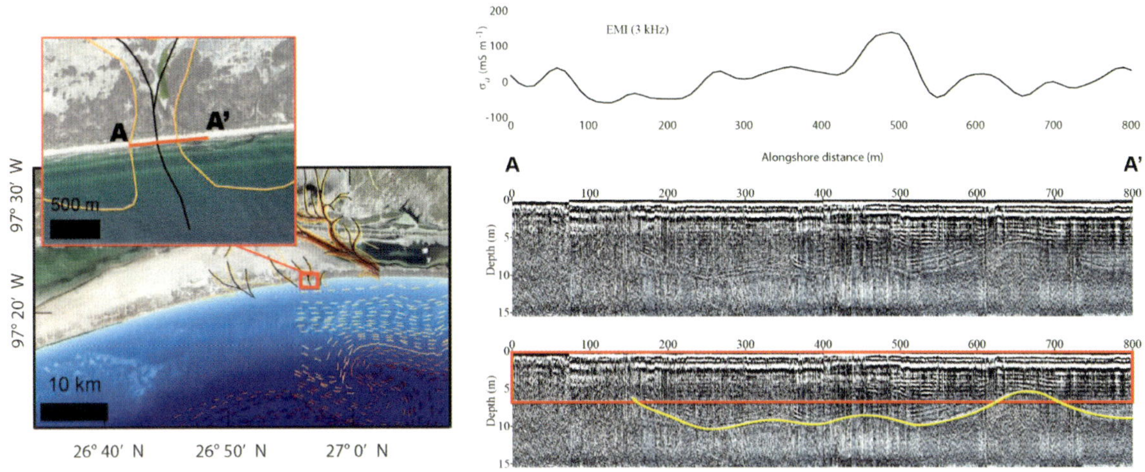

Figure 3. Comparison of EMI σ_a responses from the 100 km survey with 100 MHz GPR data within one of the Fisk (1959) paleo-channels. The 800 m segment (A–A′) crosses a smaller stream within the network of paleo-channels in the central zone of PAIS. The DOI of the 3 kHz EMI responses is outlined by the red box on the lower GPR radargram and the interpretation of the channel base (ravinement surface) is highlighted in yellow.

the southern zone where σ_a increases substantially. However, the lowest σ_a values are located within the region of paleo-channels inferred by Fisk (1959), supporting previous findings in the study area by Weymer et al. (2015a) and Wernette et al. (2018) that suggest a potential geologic control on alongshore geomorphic features.

Each spatial data series (Fig. 4a–g) represents a different superposition of effects caused by physical processes operating across a wide range of temporal and length scales (Weymer et al., 2015a). Short-range fluctuations represent small-scale heterogeneities, whereas long-range components capture variations in each metric at broader length scales. There is a high degree of variability within each signal that is directly related to the geological and geomorphological structure along the island. Within and outside the paleo-channel region, general associations between EMI σ_a responses and DEM metrics are visually subtle, motivating the statistics we now show with ARIMA modeling. To conduct the ARIMA analysis, we chose to divide the island into three zones based on the location of the known paleo-channels. As will be discussed later, the tripartite zonation allows for a quantitative analysis of LRD at three spatial scales (regional, intermediate, local) within and outside the area containing paleo-channels. It is important to note, however, that the framework geology is likely to exhibit LRD regardless of the length-scale over which it is observed.

4.2 Tests for LRD

4.2.1 Tests for LRD in EMI data series

Both EMI spatial data series appear to be nonstationary since the mean and variance of the data fluctuate along the profile. A closer visual inspection reveals, however, that cyclicity is present at nearly all spatial frequencies (Fig. 6), with the cycles superimposed in random sequence and added to a constant variance and mean (see Beran, 1994). This behavior is typical for stationary processes with LRD, and is often observed in various types of geophysical time series (Beran, 1992), for example records of Nile River stage minima (Hurst, 1951). A common first-order approach for determining whether a data series contains LRD is through inspection of the autocorrelation function, which we have computed in AutoSignal™ signal analysis software using a fast Fourier transform (FFT) algorithm (Fig. 5a, d). Both EMI signals exhibit large correlations at large lags (at kilometer and higher scales), suggesting the σ_a responses contain LRD, or long-memory effects in time series language. Results from a rescaled range R/S analysis (Fig. 5b, e) indeed show high H values of 0.85 ($r^2 = 0.98$) and 0.95 ($r^2 = 0.99$) for the 100 and 10 km surveys, indicating a strong presence of LRD at both regional and local spatial scales.

The manner in which different spatial frequency (i.e., wave number) components are superposed to constitute an observed EMI σ_a signal has been suggested to reveal information about the causative multiscale geologic structure (Everett and Weiss, 2002; Weymer et al., 2015a). For example, the lowest-wave-number contributions are associated with spatially coherent geologic features that span the longest length scales probed. The relative contributions of the various wave number components can be examined by plotting the σ_a signal power spectral density (PSD). A power law of the form $|\sigma_a(f)|^2 \sim f^\beta$ over several decades in spatial wave number is evident (Fig. 5c, f). The slope β of a power-law-shaped spectral density provides a quantitative measure of the LRD embedded in a data series and characterizes the heterogeneity, or roughness, of the signal. A value of $|\beta| > 1$ indicates a series

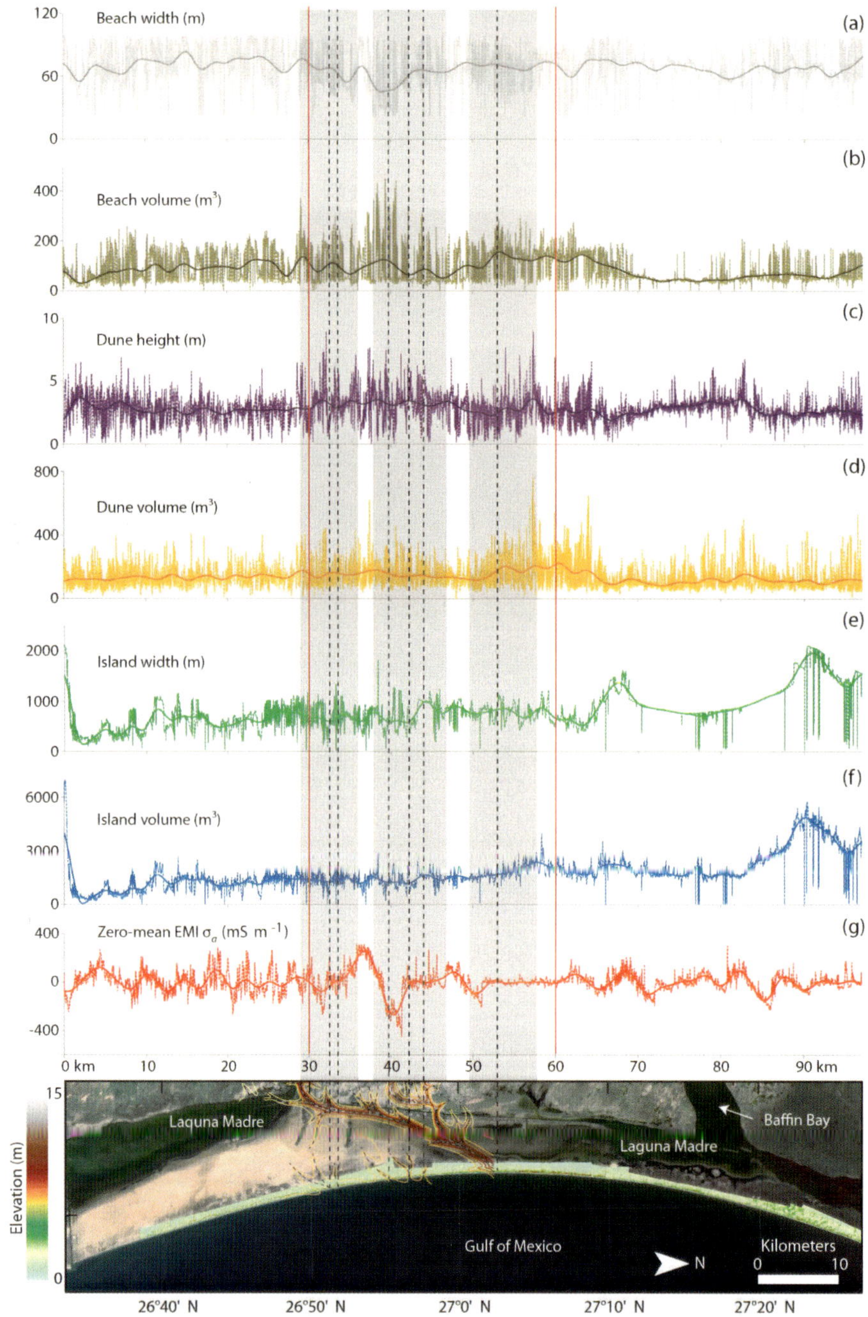

Figure 4. DEM metrics extracted from aerial lidar data. The sampling interval (step size) for each data series is 10 m and the coordinates are matched with each EMI acquisition point. Each panel corresponds to **(a)** beach width, **(b)** beach volume, **(c)** dune height, **(d)** dune volume, **(e)** island width, **(f)** island volume, and **(g)** EMI σ_a. The island is divided into three zones (red vertical lines) roughly indicating the locations within and outside the known paleo-channel region. A Savitzky–Golay smoothing filter was applied to all data series (lidar and EMI) using a moving window of $n = 250$ to highlight the large-scale patterns in each signal.

that is influenced more by long-range correlations and less by small-scale fluctuations (Everett and Weiss, 2002). For comparison, a pure white noise process would have a slope of exactly $\beta = 0$, whereas a slope of $\beta \sim 0.5$ indicates fractional Gaussian noise, i.e., a stationary signal with no significant long-range correlations (Everett and Weiss, 2002). The β values for the 100 and 10 km surveys are $\beta = -0.97$ and $\beta = -1.06$, respectively. These results suggest that both the 100 and 10 km EMI signals contain long-range correlations. However, there is a slightly stronger presence of LRD within the 10 km segment of the paleo-channel region compared to that within the segment that spans the entire length of the is-

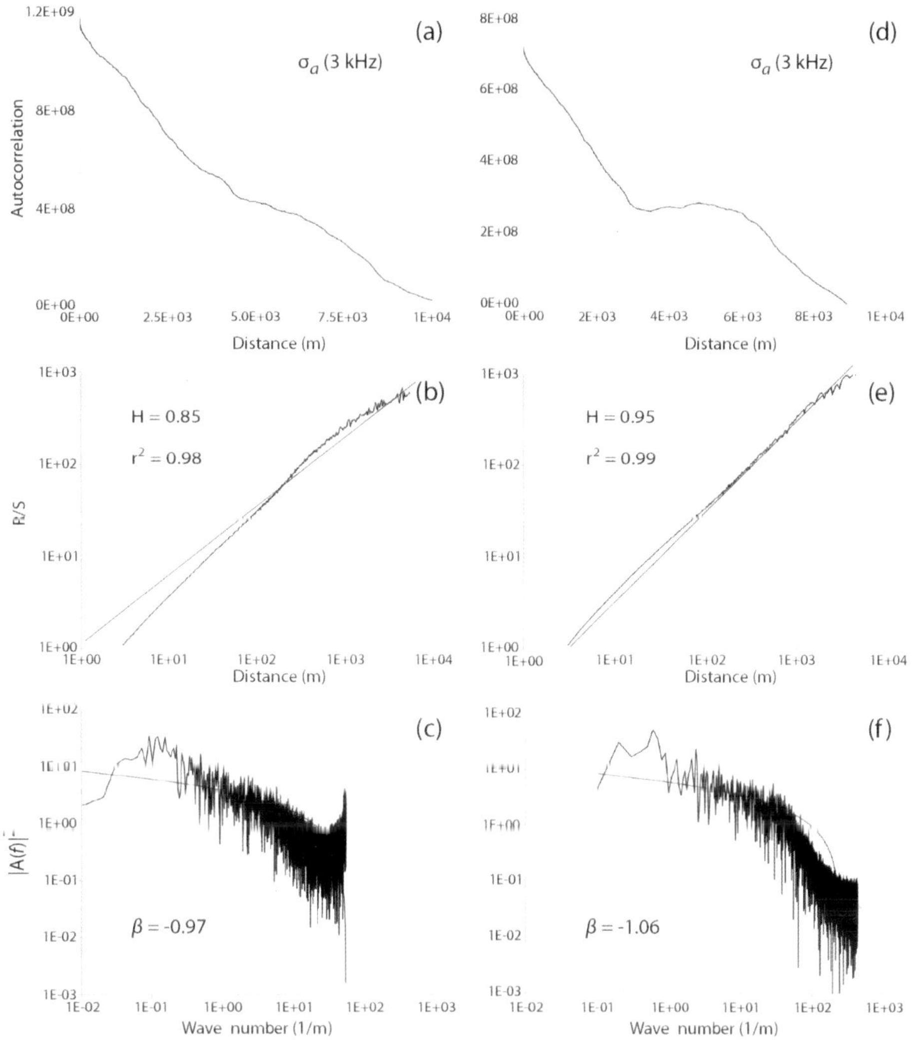

Figure 5. Autocorrelations of σ_a for the 100 km **(a)** and 10 km EMI surveys **(d)**. R/S analysis for the 100 km **(b)** and 10 km surveys **(e)**. PSD plots for the 100 km **(c)** and 10 km surveys **(f)**.

land. This indicates that long-range spatial variations in the framework geology are more important, albeit marginally so, at the 10 km scale than at the 100 km scale. It is possible that the variability within the signal and the degree of long-range correlation is also a function of the sensor footprint, relative to station spacing. This is critically examined in Sect. 4.3.

4.2.2 Tests for LRD in surface morphometrics

Following the same procedure as applied to the EMI data, we performed the R/S analysis for each beach, dune, and island metric. The calculated H values for the DEM morphometrics range between 0.80 and 0.95 with large values of $r^2 \sim 1$, indicating varying but relatively strong tendencies towards LRD. Beach width and beach volume data series have H values of 0.82 and 0.86, respectively. Dune height and dune volume H values are 0.83 and 0.80, whereas island width and island volume have higher H values of 0.95 and 0.92, respectively. Because each data series shows moderate to strong evidence of LRD, the relative contributions of short- and long-range structure contained within each signal can be further investigated by fitting ARIMA models to each dataset.

4.3 ARIMA statistical modeling of EMI

The results of the tests described in Sect. 4.2.1 for estimating the self-similarity parameter H and the slope of the PSD function suggest that both EMI data series, and by inference the underlying framework geology, exhibit LRD. The goal of our analysis using ARIMA is to estimate the p, d, and q terms representing the order, respectively, of AR, integrated (I), and MA contributions to the signal (Box and Jenkins, 1970) to quantify free vs. forced behavior along the island. For the analysis, the "arfima" and "forecast" statistical

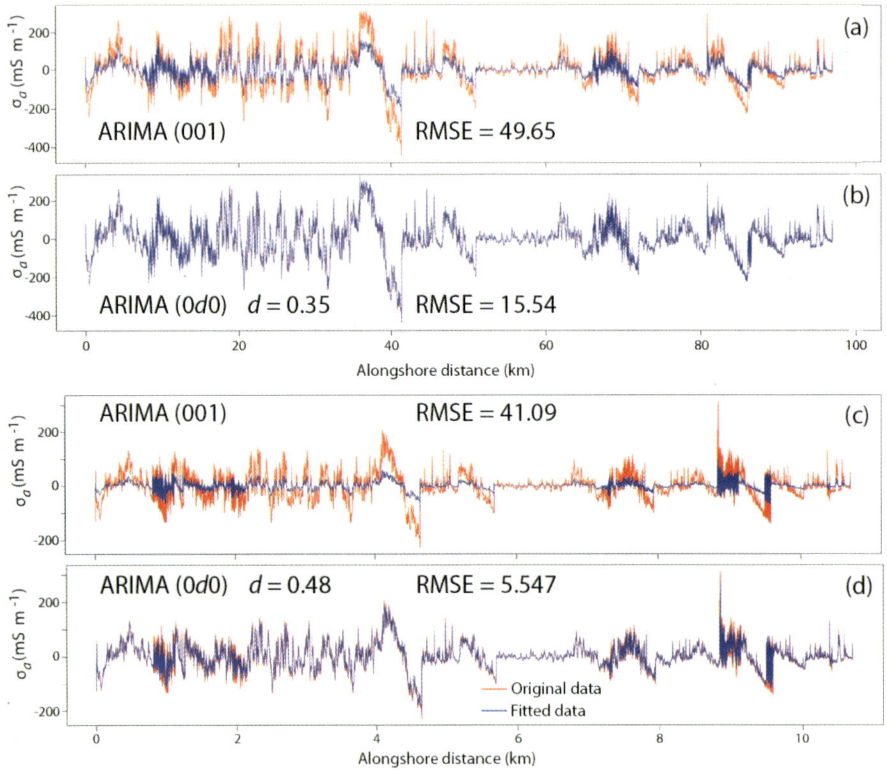

Figure 6. Examples of the worst-fit **(a, c)** and best-fit **(b, d)** ARIMA models for the 100 and 10 km EMI surveys. Model results are shown for the processed (drift-corrected) σ_a data. Residuals (RMSE) listed for each model give the standard deviation of the model prediction error. For each plot, original data are in red and fitted (model) data are in blue.

packages in R were used to fit a family of ARIMA (p, d, q) models to the EMI σ_a data and island morphometrics (Hyndman, 2015; Hyndman and Khandakar, 2008; Veenstra, 2013). Results of 10 realizations drawn from a family of ARIMA (p, d, q) models and their residuals (RMSE) are presented in Table 1. The worst fit (ARIMA 001) models are shown for the 100 and 10 km (Fig. 6a, c) surveys. The best fit (ARIMA 0d0) models for both the 100 and 10 km surveys are shown in Fig. 6b and d, respectively. For this analysis, the tests include different combinations of p, d, and q that model either short-range: ARIMA (100; 001; 101; 202; 303; 404; 505), long-range: ARIMA (010; 0d0), or composite short- and long-range processes: ARIMA (111). It is important to note that AR and MA are only appropriate for short-memory processes since they involve only near-neighbor values to explain the current value, whereas the integration (the I term in ARIMA) models long-memory effects because it involves distant values. Note that ARIMA was developed for one-way time series, in which the arrow of time advances in only one direction, but in the current study we are using it for spatial series that are reversible. Different realizations of each ARIMA (p, d, q) data series were evaluated, enabling physical interpretations of LRD at regional, intermediate, and local spatial scales. Determining the best-fitting model is achieved by comparing the residual score, or RMSE, of each predicted data series relative to the observed data series, where lower RMSE values indicate a better fit (Table 1).

Table 1. Comparison of residuals (RMSE) of each ARIMA model for the 100 and 10 km EMI surveys.

	EMI (100 km)	EMI (10 km)
ARIMA (100)	18.4	8.14
ARIMA (001)	49.7	41.1
ARIMA (101)	15.6	6.65
ARIMA (202)	40.6	7.31
ARIMA (303)	40.5	7.22
ARIMA (404)	40.3	7.22
ARIMA (505)	40.2	7.29
ARIMA (111)	15.8	5.72
ARIMA (010)	18.5	8.15
ARIMA (0d0)	15.5	5.55

Based on the residuals and visual inspection of each realization (Fig. 6), two observations are apparent: (1) both EMI data series are most accurately modeled by an ARIMA (0d0) process with non-integer d, and (2) the mismatch between the data and their model fit is considerably lower for the 10 km survey compared to the 100 km survey. The first observation

Table 2. Comparison of residuals (RMSE) of each ARIMA model for all spatial data series. Note that the residuals for each DEM metric correspond to the analysis performed at the regional scale (i.e., 100 km).

	ARIMA (100)	ARIMA (001)	ARIMA (101)	ARIMA (111)	ARIMA (010)	ARIMA (0*d*0)
Beach width	13.4	14.9	13.0	13.1	14.8	13.0
Beach volume	44.8	50.5	43.1	43.1	49.1	42.7
Dune height	0.7	0.8	0.7	0.7	0.8	0.7
Dune volume	60.6	63.9	59.7	59.2	69.03	58.9
Island width	138.4	253.2	121.3	121.1	140.8	120.9
Island volume	271.3	611.4	244.3	244.1	273.9	243.3

suggests that the data are most appropriately modeled by a FARIMA process, i.e., a fractional integration that is stationary ($0<d<0.5$) and has long-range dependence (see Hosking, 1981). This implies that spatial variations in framework geology at the broadest scales dominate the EMI signal and that small-scale fluctuations in σ_a caused, for example, by changing hydrological conditions over brief time intervals less than the overall data acquisition interval, or fine-scale lithological variations less than a few station spacings, are not as statistically significant. Regarding the second observation, the results suggest that a small station spacing (i.e., 1 m) is preferred to accurately model both short- and long-range contributions within the signal because large station spacings cannot capture short-range information. The model for the 10 km survey fits better because both p (AR) and q (MA) components increase with a smaller step size since successive volumes of sampled subsurface overlap. On the contrary, the sensor footprint is considerably smaller than the station spacing (10 m) for the 100 km survey. Each σ_a measurement in that case records an independent volume of ground, yet the dataset still exhibits LRD, albeit not to the same degree as in the 10 km survey.

4.4 ARIMA statistical modeling of island metrics compared with EMI

A sequence of ARIMA (p,d,q) models was also evaluated for the elevation morphometrics series to find best fits to the data. The analysis comprised a total of 36 model tests (Table 2). The RMSE values reveal that (1) all data series are best fit by an ARIMA (0*d*0) process with fractional d, i.e., a FARIMA process; (2) the ARIMA models, in general, more accurately fit the EMI data than the DEM morphometric data likely because the morphology is controlled by more than the framework geology alone; and (3) in all cases, the poorest fit to each series is the ARIMA (001) or MA process. This, in turn, means that the differencing parameter d is the most significant parameter amongst p, d, and q. It is important to note that different values of d were computed based on the best fit of each FARIMA model to the real data. A graphical representation of the FARIMA-modeled data series for each DEM metric is shown in Fig. 7, allowing a visual inspection of how well the models fit the observed data. Because each data series has its own characteristic amplitude and variability, it is not possible to compare RMSE among tests without normalization. The variance within each data series can differ by several orders of magnitude.

Instead of normalizing the data, a fundamentally different approach is to compare the EMI $\sigma_a d$ values with respect to each metric at regional, intermediate, and local scales (Table 3). Higher positive d values indicate a stronger tendency towards LRD. According to Hosking (1981), $\{x_t\}$ is called an ARIMA (0*d*0) process and is of particular interest in modeling LRD as d approaches 0.5 because in such cases the correlations and partial correlations of $\{x_t\}$ are all positive and decay slowly towards zero as the lag increases, while the spectral density of $\{x_t\}$ is concentrated at low frequencies. It is reasonable to assume that the degree of LRD may change over smaller intermediate and/or local scales, which implies a breakdown of self-similarity. For a self-similar signal, d is a global parameter that does not depend on which segment of the series is analyzed. In other words, the d values should be the same at all scales for a self-similar structure.

The results of the FARIMA analysis at the intermediate scale vary considerably within each zone of the barrier island (north, central, south) and for each spatial data series (Table 3). In the southern zone (0–30 km), EMI σ_a and beach volume have the strongest LRD ($d=0.44$), whereas the other metrics exhibit weak LRD (ranging from $d\sim 0$ to 0.2), which may be characterized approximately as a white noise process. Within the paleo-channel region (30–60 km), all of the island metrics show a moderate to strong tendency towards LRD ($0.3 \leq d \leq 4.2$); however, the EMI signal does not ($d=0.11$). In the northern zone (60–100 km) all data series contain moderate to strong LRD with the exception of beach and island width.

A FARIMA analysis was also conducted at the local scale by dividing the island into 10 km segments, starting at the southern zone (0–10 km) and ending at the northern zone of the island (90–100 km). A total of 70 FARIMA model realizations were evaluated and the resulting d values demonstrate that the EMI data segments show a stronger presence of LRD ($d>0.4$) within the paleo-channels (30–60 km) and

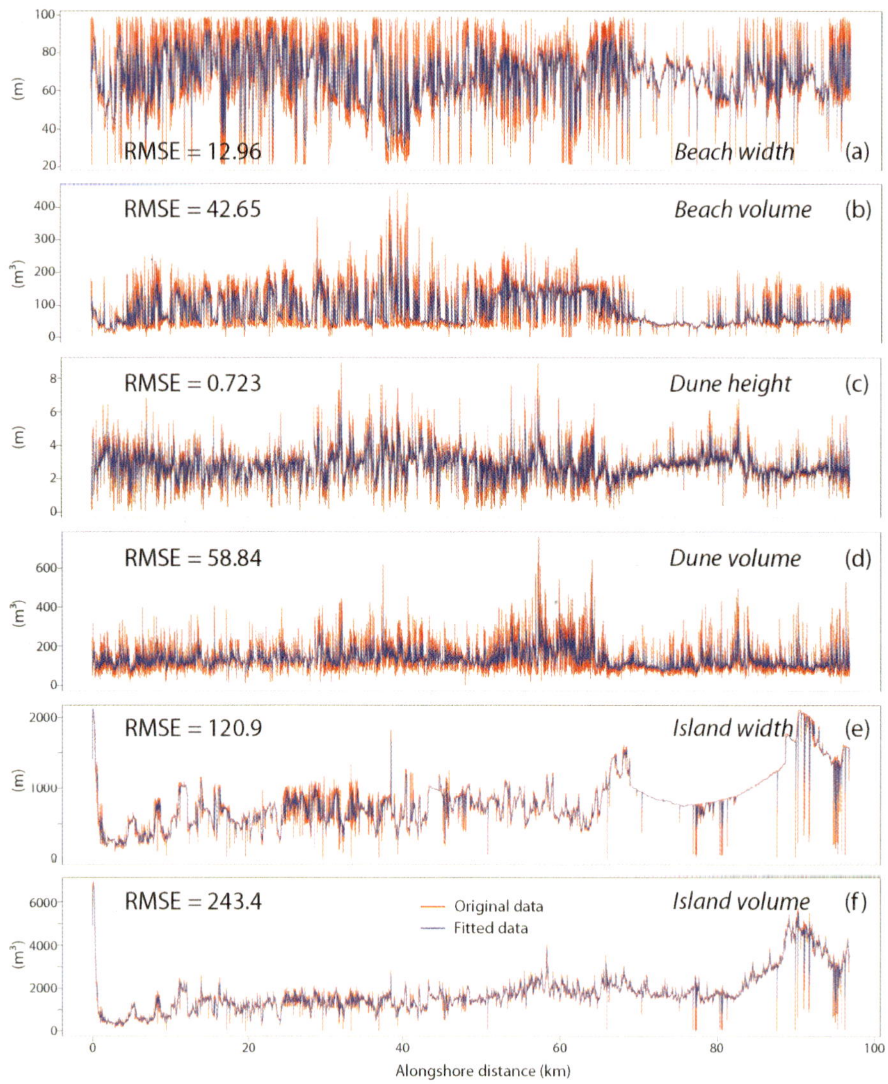

Figure 7. Example of the best fit ARIMA (0*d*0) models for each lidar-derived DEM metric: **(a)** beach width, **(b)** beach volume, **(c)** dune height, **(d)** dune volume, **(e)** island width, and **(f)** island volume.

further to the north (60–80 km) in close proximity to the ancestral outlet of Baffin Bay. These findings indicate that there may be local and/or intermediate geologic controls along different parts of the island, but that the framework geology dominates island metrics at the regional scale.

5 Discussion

Although it has long been known that processes acting across multiple temporal and length scales permit the shape of coastlines to be described by mathematical constructs such as power-law spectra and fractal dimension (Lazarus et al., 2011; Mandelbrot, 1967; Tebbens et al., 2002), analogous studies of the subsurface framework geology of a barrier island have not been carried out. This research supports previous studies demonstrating that near-surface EMI geophysical methods are useful for mapping barrier island framework geology and that FARIMA data series analysis is a compact statistical tool for illuminating the long and/or short-range spatial correlations between subsurface geology and geomorphology. The results of the FARIMA analysis and comparisons of the best-fitting d parameters show that beach and dune metrics closely match EMI σ_a responses regionally along the entire length of PAIS, suggesting that the long-range dependent structure of these data series is similar at large spatial scales. However, further evaluation of the d parameters over smaller data segments reveals that there are additional localized framework geology controls on island geomorphology that are not present at the regional scale.

At the intermediate scale, a low EMI d value ($d = 0.11$) suggests there is only a weak framework geologic control on barrier island morphometrics. A possible explanation is that

Table 3. Summary table showing the computed d parameters that most appropriately model each ARIMA ($0d0$) iteration (i.e., lowest RMSE).

Alongshore distance	Beach width	Beach volume	Dune height	Dune volume	Island width	Island volume	EMI σ_a
Regional							
0–100 km	0.38	0.42	0.34	0.32	0.13	$\sim$0.00	0.35
Intermediate							
0–30 km	$\sim$0.00	0.44	0.13	0.20	0.03	0.18	0.44
30–60 km	0.37	0.30	0.36	0.31	0.30	0.42	0.11
60–100 km	0.26	0.41	0.35	0.46	$\sim$0.00	0.50	0.49
Local							
0–10 km	0.41	0.39	0.20	0.21	0.09	0.18	0.36
10–20 km	0.30	0.42	0.20	0.26	0.37	$\sim$0.00	0.36
20–30 km	0.26	0.40	$\sim$0.00	$\sim$0.00	0.49	$\sim$0.00	$\sim$0.00
30–40 km	0.47	$\sim$0.00	0.41	0.25	0.29	0.28	$\sim$0.00
40–50 km	0.28	0.21	0.21	0.19	0.30	0.02	0.44
50–60 km	0.03	0.31	0.23	0.32	$\sim$0.00	0.33	0.48
60–70 km	0.16	0.37	0.29	0.34	$\sim$0.00	0.30	0.40
70–80 km	0.47	0.34	0.43	0.26	$\sim$0.00	0.42	0.49
80–90 km	0.27	0.19	0.42	0.39	0.01	0.02	$\sim$0.00
90–100 km	0.13	0.13	$\sim$0.00	0.06	0.44	0.47	0.41

the paleo channels, located within a ~30 km segment of the island, are not regularly spaced and on average are less than a few kilometers wide. This implies that the framework geology controls are localized (i.e., effective in shaping island geomorphology only at smaller spatial scales). At the local scale, relationships between the long-range dependence of EMI and each metric vary considerably, but there is a significant geologic control on dune height within the paleo-channel region ($d > 0.4$). It is hypothesized that the alongshore projection of the geometry of each channel is directly related to a corresponding variation in the EMI signal, such that large, gradual minima in σ_a are indicative of large, deep channel cross sections and small, abrupt minima in σ_a represent smaller, shallow channel cross sections. At shallower depths within the DOI probed by the EMI sensor, variability in the σ_a signal may correspond to changes in sediment characteristics as imaged by GPR (Fig. 3). Located beneath a washover channel, a zone of high-conductivity EMI σ_a responses between $\sim$450 and 530 m coincides with a segment of the GPR section where the signal is more attenuated and lacks the fine stratification that correlates much better with the lower σ_a zones. The contrasts in lithology between the overwash deposits and stratified infilled sands were detected by both EMI and GPR measurements.

It is argued herein that differences in the d parameter between EMI σ_a readings (our assumed proxy for framework geology) and lidar-derived surface morphometrics provide a new metric that is useful for quantifying the causative physical processes that govern island transgression across multiple spatial scales. All of the calculated d values in this study are derived from ARIMA ($0d0$) models that fit the observations, and lie within the range of $0 < d < 0.5$, suggesting that each data series is stationary but does contain long-range structure that represents randomly placed cyclicities in the data. For all models in our study, the d values range between $\sim$0 and 0.50, which enables a geomorphological interpretation of the degree of LRD and self-similarity at different spatial scales. In other words, the d parameter not only provides an indication of the scale dependencies within the data, but also offers a compact way of analyzing the statistical connections between forced (stronger $d \sim 0.5$) and free (weaker $d \sim 0$) behavior that may be more influenced by morphodynamic processes operating at smaller spatial scales.

Alongshore variations in beach width and dune height are not uniform at PAIS (Weymer et al., 2015b) and exhibit different spatial structure within and outside the paleo-channel region (Fig. 5). These dissimilarities may be forced by the framework geology within the central zone of the island but are influenced more by contemporary morphodynamic processes outside the paleo-channel region. This effect could be represented by higher-wave-number components embedded within the spatial data series. Beach and dune morphology in areas that are not controlled by framework geology (e.g., the northern and southern zones) exhibit more small-scale fluctuations representing a free system primarily controlled by contemporary morphodynamics (e.g., wave action, storm surge, wind).

Because variations in dune height exert an important control on storm impacts (Sallenger, 2000) and ultimately large-scale island transgression (Houser, 2012), it is argued here that the framework geology (or lack thereof) of PAIS acts as an important control on island response to storms and sea level rise. This study supports recent work by Wernette et al. (2018) suggesting that framework geology can influence barrier island geomorphology by creating alongshore variations in either oceanographic forcing and/or sediment supply and texture that controls smaller-scale processes responsible for beach–dune interaction at the local scale. The forced behavior within the paleo-channel region challenges shoreline change studies that consider only small-scale undulations in the dune line that are caused by natural randomness within the system. Rather, we propose that dune growth is forced by the framework geology, whose depth is related to the thickness of the modern shoreface sands beneath the beach. This depth is the primary quantity that is detected by the EMI sensor. With respect to shoreline change investigations, improving model performance requires further study of how the framework geology influences beach–dune morphology through variations in wave energy, texture, and sediment supply (e.g., Houser, 2012; McNinch, 2004; Schwab et al., 2013).

Our findings extend previous framework geology studies from Outer Banks, NC (e.g., Browder and McNinch, 2006; McNinch, 2004; Riggs et al., 1995; Schupp et al., 2006), Fire Island, NY (e.g., Hapke et al., 2010; Lentz and Hapke, 2011), and Pensacola, FL (e.g., Houser, 2012), where feedbacks between geologic features and relict sediments within the littoral system have been shown to act as an important control on dune growth and evolution. Nonetheless, most of these studies focus on offshore controls on shoreface and/or beach–dune dynamics at either local or intermediate scales because few islands worldwide exist that are as long and/or continuous as North Padre Island. To our knowledge, few framework geology studies have specifically used statistical testing to analyze correlations between subsurface geologic features and surface morphology. Two notable exceptions include Browder and McNinch (2006), and Schupp et al. (2006), both of which used chi-squared testing and cross-correlation analysis to quantify the spatial relationships among offshore bars, gravel beds, and/or paleo-channels at the Outer Banks, NC. Although these techniques are useful for determining spatial correlations among different datasets, they do not provide information about the scale (in)dependencies between the framework geology and surface geomorphology that FARIMA models are better designed to handle. The current study augments the existing literature in that (1) it outlines a quantitative method for determining free and forced evolution of barrier island geomorphology at multiple length scales, and (2) it demonstrates that there is a first-order control on dune height at the local scale within an area of known paleo-channels, suggesting that framework geology controls are localized within certain zones of PAIS.

Further study is required to determine how this combination of free and forced behavior resulting from the variable and localized framework geology affects island transgression. Methods of data analysis that would complement the techniques presented in this paper might include power spectral analysis, wavelet decomposition, and shoreline change analysis that implicitly includes variable framework geology. These approaches would provide important information regarding (1) coherence and phase relationships between subsurface structure and island geomorphology and (2) nonlinear interactions of coastal processes across large and small spatiotemporal scales. Quantifying and interpreting the significance of framework geology as a driver of barrier island formation and evolution and its interaction with contemporary morphodynamic processes is essential for designing and sustainably managing resilient coastal communities and habitats.

6 Conclusions

This study demonstrates the utility of EMI geophysical profiling as a new tool for mapping the length-scale dependence of barrier island framework geology and introduces the potential of FARIMA analysis to better understand the geologic controls on large-scale barrier island transgression. The EMI and morphometric data series exhibit LRD to varying degrees, and each can be accurately modeled using a non-integral parameter d. The value of this parameter diagnoses the spatial relationship between the framework geology and surface geomorphology. At the regional scale (~ 100 km), small differences in d between the EMI and morphometrics series suggest that the long-range-dependent structure of each data series with respect to EMI σ_a is statistically similar. At the intermediate scale (~ 30 km), there is a greater difference among the d values of the EMI and island metrics within the known paleo-channel region, suggesting a more localized geologic control with less contributions from broader-scale geological structures. At the local scale (10 km), there is a considerable degree of variability among the d values of the EMI and each metric. These results all point toward a forced barrier island evolutionary behavior within the paleo-channel region transitioning into a free, or scale-independent, behavior dominated by contemporary morphodynamics outside the paleo-channel region. In a free system, small-scale undulations in the dune line reinforce natural random processes that occur within the beach–dune system and are not influenced by the underlying geologic structure. In a forced system, the underlying geologic structure establishes boundary constraints that control how the island evolves over time. This means that barrier island geomorphology at PAIS is forced and scale dependent, unlike shorelines which have been shown at other barrier islands to be scale independent (Tebbens et al., 2002; Lazarus et al., 2011). The exchange of sediment amongst nearshore, beach, and dune in areas outside the paleo-channel region is scale independent, meaning that barrier islands like PAIS exhibit a combination of free and forced behaviors that will affect the response of the island to sea level rise and storms. We propose that our analysis is not limited to PAIS but can be applied to other barrier islands and potentially in different geomorphic environments, both coastal and inland.

Competing interests. The authors declare that they have no conflict of interest.

Acknowledgements. We are grateful to Patrick Barrineau, Andy Evans, Brianna Hammond Williams, Alex van Plantinga, and Michael Schwind for their assistance in the field. We thank two anonymous reviewers for their constructive comments during the open discussion. The field data presented in this paper were collected under the National Park Service research permit: no. PAIS-2013-SCI-0005. This research was funded in part by a Grants-in-Aid of Graduate Student Research Award by the Texas Sea Grant College Program to BW, and through a grant to CH from the Natural Science and Engineering Research Council of Canada (NSERC).

References

Alemi, M. H., Azari, A. S., and Nielsen, D. R.: Kriging and univariate modeling of a spatially correlated data, Soil Technol., 1, 133–147, 1988.

Anderson, J. B., Wallace, D. J., Simms, A. R., Rodriguez, A. B., Weight, R. W., and Taha, Z. P.: Recycling sediments between source and sink during a eustatic cycle: Systems of late Quaternary northwestern Gulf of Mexico Basin, Earth-Sci. Rev., 153, 111–138, 2015.

Andrle, R.: The west coast of Britain: Statistical self-similarity vs. characteristic scales in the landscape, Earth Surf. Proc. Land., 21, 955–962, 1996.

Bailey, R. J. and Smith, D. G.: Quantitative evidence for the fractal nature of the stratigraphie record: results and implications, P. Geologist. Assoc., 116, 129–138, 2005.

Bassingthwaighte, J. B. and Raymond, G. M.: Evaluating rescaled range analysis for time series, Ann. Biomed. Eng., 22, 432–444, 1994.

Belknap, D. F. and Kraft, J. C.: Influence of antecedent geology on stratigraphic preservation potential and evolution of Delaware's barrier systems, Mar. Geol., 63, 235–262, 1985.

Beran, J.: Statistical methods for data with long-range dependence, Stat. Sci., 7, 404–427, 1992.

Beran, J.: Statistics for long-memory processes, vol. 61, CRC Press, New York, USA, 1994.

Box, G. E. and Jenkins, G. M.: Time series analysis: forecasting and control, Holden-Day, San Francisco, CA, USA, 1970.

Browder, A. G. and McNinch, J. E.: Linking framework geology and nearshore morphology: correlation of paleo-channels with shore-oblique sandbars and gravel outcrops, Mar. Geol., 231, 141–162, 2006.

Brown, L. F. and Macon, J.: Environmental geologic atlas of the Texas coastal zone: Kingsville area, Bureau of Economic Geology, The University of Texas, Austin, USA, 1977.

Burrough, P.: Fractal dimensions of landscapes and other environmental data, Nature, 294, 240–242, 1981.

Buynevich, I. V., FitzGerald, D. M., and van Heteren, S., Sedimentary records of intense storms in Holocene barrier sequences, Maine, USA, Mar. Geol., 210, 135–148, 2004.

Cimino, G., Del Duce, G., Kadonaga, L., Rotundo, G., Sisani, A., Stabile, G., Tirozzi, B., and Whiticar, M.: Time series analysis of geological data, Chem. Geol., 161, 253–270, 1999.

Coleman, J. M. and Gagliano, S. M.: Cyclic sedimentation in the Mississippi River deltaic plain, Transactions of Gulf Coast Association of Geological Societies, 14, 67–80, 1964.

Colman, S. M., Halka, J. P., Hobbs, C., Mixon, R. B., and Foster, D. S.: Ancient channels of the Susquehanna River beneath Chesapeake Bay and the Delmarva Peninsula, Geol. Soc. Am. Bull., 102, 1268–1279, 1990.

Dai, H., Ye, M., and Niedoroda, A. W.: A Model for Simulating Barrier Island Geomorphologic Responses to Future Storm and Sea-Level Rise Impacts, J. Coastal Res., 31, 1091–1102, 2015.

De Jong, P. and Penzer, J.: Diagnosing shocks in time series, J. Am. Stat. Assoc., 93, 796–806, 1998.

Delefortrie, S., Saey, T., Van De Vijver, E., De Smedt, P., Missiaen, T., Demerre, I., and Van Meirvenne, M.: Frequency domain electromagnetic induction survey in the intertidal zone: Limitations of low-induction-number and depth of exploration, J. Appl. Geophys., 100, 14–22, 2014.

Demarest, J. M. and Leatherman, S. P.: Mainland influence on coastal transgression: Delmarva Peninsula, Mar. Geol., 63, 19–33, 1985.

Doukhan, P., Oppenheim, G., and Taqqu, M. S.. Theory and aplications of long-range dependence, Birkhauser, Boston, MA, USA, 2003.

Eke, A., Hermán, P., Bassingthwaighte, J., Raymond, G., Percival, D., Cannon, M., Balla, I., and Ikrényi, C.: Physiological time series: distinguishing fractal noises from motions, Pflüg. Arch., 439, 403–415, 2000.

Evans, M. W., Hine, A. C., Belknap, D. F., and Davis, R. A.: Bedrock controls on barrier island development: west-central Florida coast, Mar. Geol., 63, 263–283, 1985.

Evans, R. L. and Lizarralde, D.: The competing impacts of geology and groundwater on electrical resistivity around Wrightsville Beach, NC, Cont. Shelf Res., 31, 841–848, 2011.

Everett, M. E.: Near-surface applied geophysics, Cambridge University Press, New York, USA, 2013.

Everett, M. E. and Weiss, C. J.: Geological noise in near-surface electromagnetic induction data, Geophys. Res. Lett., 29, 10-11–10-14, 2002.

Fisk, H. N.: Padre Island and Laguna Madre Flats, coastal south Texas. Proceedings 2nd Coastal Geography Conference, Louisiana State University, Baton Rouge, LA, USA, 103–151, 1959.

Fitterman, D. V. and Stewart, M. T.: Transient electromagnetic sounding for groundwater, Geophysics, 51, 995–1005, 1986.

Frazier, D. E.: Recent deltaic deposits of the Mississippi river: their development and chronology, Transactions of Gulf Coast Association of Geological Societies, 17, 287–315, 1967.

Granger, C. W. and Joyeux, R.: An introduction to long-memory time series models and fractional differencing, J. Time Ser. Anal., 1, 15–29, 1980.

Guillemoteau, J. and Tronicke, J.: Non-standard electromagnetic induction sensor configurations: Evaluating sensitivities and applicability, J. Appl. Geophys., 118, 15–23, 2015.

Gutierrez, B. T., Plant, N. G., Thieler, E. R., and Turecek, A.: Using a Bayesian network to predict barrier island geomorphologic characteristics, J. Geophys. Res.-Earth, 120, 2452–2475, 2015.

Hapke, C. J., Lentz, E. E., Gayes, P. T., McCoy, C. A., Hehre, R., Schwab, W. C., and Williams, S. J.: A review of sediment budget imbalances along Fire Island, New York: can nearshore geologic framework and patterns of shoreline change explain the deficit?, J. Coastal Res., 510–522, 2010.

Hapke, C. J., Kratzmann, M. G., and Himmelstoss, E. A.. Geomorphic and human influence on large-scale coastal change, Geomorphology, 199, 160–170, 2013.

Hapke, C. J., Plant, N. G., Henderson, R. E., Schwab, W. C., and Nelson, T. R.: Decoupling processes and scales of shoreline morphodynamics, Mar. Geol., 381, 42–53, 2016.

Honeycutt, M. G. and Krantz, D. E.: Influence of the geologic framework on spatial variability in long-term shoreline change, Cape Henlopen to Rehoboth Beach, Delaware, J. Coastal Res., 38, 147–167, 2003.

Hosking, J. R.: Fractional differencing, Biometrika, 68, 165–176, 1981.

Houser, C.: Feedback between ridge and swale bathymetry and barrier island storm response and transgression, Geomorphology, 173, 1–16, 2012.

Houser, C. and Mathew, S.: Alongshore variation in foredune height in response to transport potential and sediment supply: South Padre Island, Texas, Geomorphology, 125, 62–72, 2011.

Houser, C., Hapke, C., and Hamilton, S.: Controls on coastal dune morphology, shoreline erosion and barrier island response to extreme storms, Geomorphology, 100, 223–240, 2008.

Houser, C., Wernette, P., Rentschlar, E., Jones, H., Hammond, B., and Trimble, S.: Post-storm beach and dune recovery: Implications for barrier island resilience, Geomorphology, 234, 54–63, 2015.

Hurst, H. E.: Long-term storage capacity of reservoirs, T. Am. Soc. Civ. Eng., 116, 770–808, 1951.

Hyndman, R. J.: Forecasting functions for time series and linear models. R package version 5.9., available at: http://github.com/robjhyndman/forecast (last access: 20 January 2018), 2015.

Hyndman, R. J. and Khandakar, Y.: Automatic time series forecasting: The forecast package for R, J. Stat. Softw., 27, 1–22, 2007.

Jol, H. M. (Ed.): Ground penetrating radar theory and applications, Elsevier, Amsterdam, the Netherlands, 2008.

Jol, H. M., Smith, D. G., and Meyers, R. A.: Digital ground penetrating radar (GPR): a new geophysical tool for coastal barrier research (Examples from the Atlantic, Gulf and Pacific coasts, USA), J. Coastal Res., 12, 960–968, 1996.

Kitchell, J. A. and Pena, D.: Periodicity of extinctions in the geologic past: deterministic versus stochastic explanations, Science, 226, 689–692, 1984.

Kraft, J., Belknap, D., McDonald, K., Maley, K., and Marx, P.: Models of a shoreface-nearshore marine transgression over estuarine and barrier systems and antecedent topography of the Atlantic coast, paper presented at the Geol. Soc. Am., 18 October 1982, New Orleans, LA, USA, Abstr. With Programs, 1982.

Lazarus, E., Ashton, A., Murray, A. B., Tebbens, S., and Burroughs, S.: Cumulative versus transient shoreline change: Dependencies on temporal and spatial scale, J. Geophys. Res.-Earth, 116, F02014, https://doi.org/10.1029/2010JF001835, 2011.

Lentz, E. E. and Hapke, C. J.: Geologic framework influences on the geomorphology of an anthropogenically modified barrier island: Assessment of dune/beach changes at Fire Island, New York, Geomorphology, 126, 82–96, 2011.

Lentz, E. E., Hapke, C. J., Stockdon, H. F., and Hehre, R. E.: Improving understanding of near-term barrier island evolution through multi-decadal assessment of morphologic change, Mar. Geol., 337, 125–139, 2013.

Malamud, B. D. and Turcotte, D. L.: Self-affine time series: I. Generation and analyses, Adv. Geophys., 40, 1–90, 1999.

Mandelbrot, B. B.: How long is the coast of Britain, Science, 156, 636–638, 1967.

Mandelbrot, B. B. and Taqqu, M. S.: Robust R/S analysis of long run serial correlation, Research report RC 7936, IBM, Yorktown Heights, NY, USA, 1979.

McNinch, J. E.: Geologic control in the nearshore: shore-oblique sandbars and shoreline erosional hotspots, Mid-Atlantic Bight, USA, Mar. Geol., 211, 121–141, 2004.

Miselis, J. L. and McNinch, J. E.: Calculating shoreline erosion potential using nearshore stratigraphy and sediment volume: Outer Banks, North Carolina, J. Geophys. Res.-Earth, 111, F02019, https://doi.org/10.1029/2005JF000389, 2006.

Miselis, J. L., Buster, N. A., and Kindinger, J. L.: Refining the link between the Holocene development of the Mississippi River Delta and the geologic evolution of Cat Island, MS: implications for delta-associated barrier islands, Mar. Geol., 355, 274–290, 2014.

Morton, R. A. and Sallenger Jr., A. H.: Morphological impacts of extreme storms on sandy beaches and barriers, J. Coastal Res., 19, 560–573, 2003.

Murray, A. B. and Thieler, E. R.: A new hypothesis and exploratory model for the formation of large-scale inner-shelf sediment sorting and "rippled scour depressions", Cont. Shelf Res., 24, 295–315, 2004.

Neal, A.: Ground-penetrating radar and its use in sedimentology: principles, problems and progress, Earth-Sci. Rev., 66, 261–330, 2004.

NOAA: National Hurricane Center, Data set, available at: http://www.nhc.noaa.gov/data/, last access: 29 April 2015a.

NOAA: Tides and Currents, available at: https://tidesandcurrents.noaa.gov, last access: 18 October 2015b.

NOAA: Digital Coast, available at: https://coast.noaa.gov/digitalcoast/, last access: 31 October 2017.

Nobes, D. C.: Troubled waters: Environmental applications of electrical and electromagnetic methods, Surv. Geophys., 17, 393–454, 1996.

Nummedal, D. and Swift, D. J.: Transgressive stratigraphy at sequence-bounding unconformities: some principles derived from Holocene and Cretaceous examples, in: Sea-level fluctuation and coastal evolution, Special Publication vol. 41, edited by: Nummedal, D., Pilkey, O. H., and Howard, J. D., Society of Economic Paleontologists and Mineralogists (SEPM), Tulsa, OK, USA, 241–260, 1987.

Otvos, E. G. and Giardino, M. J.: Interlinked barrier chain and delta lobe development, northern Gulf of Mexico, Sediment. Geol., 169, 47–73, 2004.

Plant, N. G. and Stockdon, H. F.: Probabilistic prediction of barrier-island response to hurricanes, J. Geophys. Res.-Earth, 117, F03015, https://doi.org/10.1029/2011JF002326, 2012.

Radliński, A., Radlińska, E., Agamalian, M., Wignall, G., Lindner, P., and Randl, O.: Fractal geometry of rocks, Phys. Rev. Lett., 82, 3078–3081, https://doi.org/10.1103/PhysRevLett.82.3078, 1999.

Riggs, S. R., Cleary, W. J., and Snyder, S. W.: Influence of inherited geologic framework on barrier shoreface morphology and dynamics, Mar. Geol., 126, 213–234, 1995.

Rodriguez, A. B., Fassell, M. L., and Anderson, J. B.: Variations in shoreface progradation and ravinement along the Texas coast, Gulf of Mexico, Sedimentology, 48, 837–853, 2001.

Sallenger Jr., A. H.: Storm impact scale for barrier islands, J. Coastal Res., 16, 890–895, 2000.

Santos, V. R., Porsani, J. L., Mendonça, C. A., Rodrigues, S. I., and DeBlasis, P. D.: Reduction of topography effect in inductive electromagnetic profiles: application on coastal sambaqui (shell mound) archaeological site in Santa Catarina state, Brazil, J. Archaeol. Sci., 36, 2089–2095, 2009.

Schlager, W.: Fractal nature of stratigraphic sequences, Geology, 32, 185–188, 2004.

Schupp, C. A., McNinch, J. E., and List, J. H.: Nearshore shore-oblique bars, gravel outcrops, and their correlation to shoreline change, Mar. Geol., 233, 63–79, 2006.

Schwab, W. C., Thieler, E. R., Allen, J. R., Foster, D. S., Swift, B. A., and Denny, J. F.: Influence of inner-continental shelf geologic framework on the evolution and behavior of the barrier-island system between Fire Island Inlet and Shinnecock Inlet, Long Island, New York, J. Coastal Res., 16, 408–422, 2000.

Schwab, W. C., Baldwin, W. E., Hapke, C. J., Lentz, E. E., Gayes, P. T., Denny, J. F., List, J. H., and Warner, J. C.: Geologic evidence for onshore sediment transport from the inner continental shelf: Fire Island, New York, J. Coastal Res., 29, 526–544, 2013.

Seijmonsbergen, A. C., Biewinga, D. T., and Pruissers, A. P.: A geophysical profile at the foot of the Dutch coastal dunes near the former outlet of the "Old Rhine", Neth. J. Geosci., 83, 287–291, 2004.

Stewart, M. T.: Evaluation of electromagnetic methods for rapid mapping of salt-water interfaces in coastal aquifers, Groundwater, 20, 538–545, 1982.

Stone, G. W., Liu, B., Pepper, D. A., and Wang, P.: The importance of extratropical and tropical cyclones on the short-term evolution of barrier islands along the northern Gulf of Mexico, USA, Mar. Geol., 210, 63–78, 2004.

Swarzenski, P. W. and Izbicki, J. A.: Coastal groundwater dynamics off Santa Barbara, California: Combining geochemical tracers, electromagnetic seepmeters, and electrical resistivity, Estuar. Coast. Shelf S., 83, 77–89, 2009.

Talley, D. M., North, E. W., Juhl, A. R., Timothy, D. A., Conde, D., deBrouwer, J. F. C., Brown, C. A., Campbell, L. M., Garstecki, T., Hall, C. J., Meysman, F. J. R., Nemerson, D. M., Souza, P. W., and Wood, R. J.: Research challenges at the land–sea interface, Estuar. Coast. Shelf S., 58, 699–702, 2003.

Tamura, T.: Beach ridges and prograded beach deposits as palaeoenvironment records, Earth-Sci. Rev., 114, 279–297, 2012.

Taqqu, M. S.: Fractional Brownian motion and long-range dependence, in: Long-range Dependence: Theory and Applications, edited by: Doukhan, P., Oppenheim, G., and Taqqu, M. S., Birkhauser, Boston, MA, USA, 2003.

Taqqu, M. S., Teverovsky, V., and Willinger, W.: Estimators for long-range dependence: an empirical study, Fractals, 3, 785–798, 1995.

Tebbens, S. F., Burroughs, S. M., and Nelson, E. E.: Wavelet analysis of shoreline change on the Outer Banks of North Carolina: An example of complexity in the marine sciences, P. Natl. Acad. Sci., 99, 2554–2560, 2002.

Twichell, D. C., Flocks, J. G., Pendleton, E. A., and Baldwin, W. E.: Geologic controls on regional and local erosion rates of three northern Gulf of Mexico barrier-island systems, J. Coastal Res., 63, 32–45, 2013.

Veenstra, J. Q.: Persistence and Anti-persistence: Theory and Software, PhD thesis, Electronic Thesis and Dissertation Repository, 1119, available at: https://ir.lib.uwo.ca/etd/1119 (last access: 20 January 2018), 2013.

Weise, B. R. and White, W. A.: Padre Island National Seashore: A guide to the geology, natural environments, and history of a Texas barrier island, vol. 17, Bureau of Economic Geology, The University of Texas, Austin, USA, 1980.

Wernette, P., Houser, C., and Bishop, M. P.: An automated approach for extracting Barrier Island morphology from digital elevation models, Geomorphology, 262, 1–7, 2016.

Wernette, P., Houser, C., Weymer, B. A., Everett, M. E., Bishop, M. P., and Reece, B.: Influence of a spatially complex framework geology on barrier island geomorphology, Mar. Geol., 398, 151–162, 2018.

Weymer, B. A., Everett, M. E., de Smet, T. S., and Houser, C.: Review of electromagnetic induction for mapping barrier island framework geology, Sediment. Geol., 321, 11–24, 2015a.

Weymer, B. A., Houser, C., and Giardino, J. R.: Poststorm Evolution of Beach-Dune Morphology: Padre Island National Seashore, Texas, J. Coastal Res., 31, 634–644, 2015b.

Weymer, B. A., Everett, M. E., Houser, C., Wernette, P., and Barrineau, P.: Differentiating tidal and groundwater dynamics from barrier island framework geology: Testing the utility of portable multi-frequency EMI profilers, Geophysics, 81, E347–E361, 2016.

Wilson, K. E., Adams, P. N., Hapke, C. J., Lentz, E. E., and Brenner, O.: Application of Bayesian Networks to hindcast barrier island morphodynamics, Coast. Eng., 102, 30–43, 2015.

Xu, T., Moore, I. D., and Gallant, J. C.: Fractals, fractal dimensions and landscapes – a review, Geomorphology, 8, 245–262, 1993.

5

Assessment of structural sediment connectivity within catchments: insights from graph theory

Étienne Cossart and Mathieu Fressard

Université de Lyon (Jean Moulin, Lyon 3), UMR 5600 CNRS – Environnement Ville Société, Lyon, 69007, France

Correspondence to: Étienne Cossart (etienne.cossart@univ-lyon3.fr)

Abstract. To describe the sedimentary signal delivered at catchment outlets, many authors now refer to the concept of connectivity. In this framework, the sedimentary signal is seen as an emergent organization of local links and interactions. The challenge is thus to open the black boxes that remain within a sediment cascade, which requires both accurate geomorphic investigations in the field (reconstruction of sequences of geomorphic evolution, description of sediment pathways) and the development of tools dedicated to sediment cascade modeling. More precisely, the development of tools devoted to the study of connectivity in geomorphology is still in progress, although graph theory offers promising perspectives (Heckmann and Schwanghart, 2013). In this paper, graph theory is applied to abstract the network structure of sediment cascades, keeping only the nodes (sediment sources, sediment stores, outlet) and links (linkage by a transportation agent), represented as vertices and edges. From the description of the assemblages of sedimentary flows, we provide three main indices to explore how small-scale processes may result in significant broad-scale geomorphic patterns. The main hypothesis guiding this work is that the network structure dictates how sediment inputs from various sources interact at tributary junctions and finally at the outlet of a cascading system. First, we use the flow index to assess the potential contribution of each node to the sediment delivery at the outlet. Second, we measure the influence of each node regarding how it is accessible from both sediment sources and the outlet (using the Shimbel index). Third, we propose a new connectivity index named "Network Structural Connectivity index" (NSC) revealing whether the potential contribution of a node is lower or higher than expected from its location within the network. These indices are first computed for a conceptual sediment cascade network and then applied to a catchment located in the southern French Alps. We demonstrate that this index may be used to simulate sediment transfer and help in identifying the hotspots of geomorphic change. In the present case, we try to predict how a sediment cascade may be impacted by an edge disruption or a reconnection.

1 Introduction

The concept of connectivity now provides an overarching framework in geosciences to explore better how catchments function. Connectivity was first defined in ecology to assess the spatial coherence of a system of landscape elements, a coherence that is necessary to maintain or restore ecological functions (Bennett, 2004). Following these pioneering contributions, connectivity has been increasingly used by hydrologists to model hydrological connection patterns (Delahaye et al., 2001; Douvinet et al., 2008). For instance, Ali and Roy (2009) stated that hydrological connectivity can be quantified as a function of the available water volume (calculated from a hydrological balance) and the rate of transfer. More recently, connectivity has appeared as a fruitful conceptual framework in geomorphology (Brierley et al., 2006; Wainwright et al., 2011; Fryirs, 2013; Hoffmann, 2015). It helps in studying the spatiotemporal unsteadiness of sediment transport within catchments and why sediment cascades can be considered a "jerky conveyor belt" (Ferguson,

1981). Unsteadiness patterns in sediment transfers are indeed a major field of research for geomorphologists, and refer to the "spatial and temporal paradox" described by McGuiness et al. (1971): in a catchment, sediment delivery from sources on hillslopes is not correlated with sediment delivery at the outlet. Consequently, sediment cascades are not necessarily efficient in transferring sediments, highlighting a "sediment delivery problem" (Walling, 1983). Finally, geomorphic signals, especially sediment delivery, cannot be interpreted easily (e.g., in terms of climate change, anthropogenic influences) and may rather reveal a "sedimentological anarchy" (Walker, 1990; Bravard, 1998; Schumm, 2005). At the catchment scale, geomorphic processes may be alternately coupled to create a sediment impulse or may act to create a blockage.

Recently, many authors have sought a complex-systems approach to conceptualize the continuum of sediment transfer: how processes, at local scales, may be combined to understand the functioning of the whole sediment cascade (Fryirs et al., 2007; Borselli et al., 2008; Fryirs, 2013; Bracken et al., 2015). Such a multiscale framework was conceptualized by Heckmann and Schwanghart (2013), who clearly distinguished the coupling of processes and connectivity. On the one hand, geomorphic coupling is "the linkage of distinct landforms or landscape units by sediment transport" (Harvey, 2001), it refers to "elementary interactions at the relatively small scale" (Faulkner, 2008). On the other hand, "the degree of coupling, the combined effect of lateral (hillslope to channel) and longitudinal (from one river reach to another) linkages between system components, is termed (sediment) connectivity" (Heckmann and Schwanghart, 2013). Shifting from the local to the catchment scale remains the main issue in explaining how erosion and sediment transfer at small spatial scale interact and result in broad scale geomorphic patterns and processes (Bracken et al., 2015). Addressing this issue requires the development of numerical methods to acquire an exhaustive inventory of all the local linkages within the sediment cascades, to assess their properties, and then to predict the result of their combination at catchment scale. One promising field of research has been opened up by the application of graph theory, which offers mathematical tools to analyze statistically the assemblages of all the components of a sediment cascade. It is a spatially explicit analysis since nodes and edges are spatial objects and since distance plays a role in the modeling (Heckmann and Schwanghart, 2013; Heckmann et al., 2015). This mathematical approach is based on a graph representation of the hydrological network that has been used for decades to describe watercourses shape and organization (Strahler, 1957; Chorley and Kennedy, 1971; Shreve, 1974). This methodological framework focuses particularly on structural connectivity, i.e., the influence of the spatial patterns formed by the linkages on sediment delivery. One main objective is to provide a quantitative index that would help in comparing the skeleton of the sediment cascades in both space and time. It could also be used to estimate the contribution of a given part of the catchment to provide sediment at the outlet and to predict where local erosion should be monitored (Cavalli et al., 2013).

In this paper, we propose a new connectivity index named "Network Structural Connectivity index" (NSC). Following a brief review of the literature on connectivity indices, we explore the main mathematical tools provided by graph theory to measure structural sediment connectivity. We particularly aim to complement the indices already published in order to develop a framework in which geomorphic simulations of changes can be carried out. The proposed connectivity index (NSC) is first described through a conceptual sediment cascade network (fictive and simple). Then, it is applied to the Celse-Nière study area (French Alps) to explore the spatial patterns of (dis)connectivity within the catchment based on a scenario analysis. Finally, we discuss the main applications and interpretations of the proposed connectivity index.

2 State of the art

By stating that catchments are inefficient at supplying sediment to the outlet, Walling (1983) pointed out a problem that arises from (dis)connectivity. He showed that catchments (especially larger ones) tend to be characterized by a low sediment delivery ratio (SDR) defined as

$$\mathrm{SDR} = \frac{V_\mathrm{o}}{V_\mathrm{h}}, \tag{1}$$

where V_o is the volume of sediment delivered at the outlet of the catchment and V_h is the volume of sediment eroded from hillslopes. The SDR is a synthetic index that provides a proxy to assess the connectivity of a catchment and allows comparisons in both space and time. Recently, it has been demonstrated that SDR (and connectivity) decreases with increasing landscape morphological complexity (Baartman et al., 2013). One main criticism regarding this index is that catchments remain a black box: no attention is paid to the geomorphic linkages present at the local scale, nor to the feedbacks among geomorphic processes (Gumiere et al., 2011; Viel, 2012; Fryirs, 2013). In this respect, the SDR has been interpreted as a simple "performance" factor to relate erosion measured at the plot scale to sediment yields observed at the larger scale. Its usefulness has been critically discussed (Hoffmann, 2015).

To open such black boxes, the concept of connectivity was subdivided into two distinct parts (With et al., 1997; Tischendorf and Fahrig, 2000; Turnbull et al., 2008). On the one hand, structural connectivity refers to spatial patterns in the landscape, such as the spatial distribution of landscape units, which influence sediment transfer patterns and sediment paths. On the other hand, functional connectivity focuses on how geomorphic processes may activate or block sediment transfer along the spatial links within the sediment cascade (Kimberly et al., 1997; With and King, 1997;

Belisle, 2005; Uezu et al., 2005). The latter is also often called process-based sediment connectivity and was documented in depth in a recent review (Bracken et al., 2015). Here, we focus on structural connectivity, whose quantification is required to explore and understand the responses of geomorphic systems (Wainwright et al., 2011).

2.1 Assessment of structural sediment connectivity

Borselli et al. (2008) and Cavalli et al. (2013) developed a connectivity index (Eq. 2) that refers to structural connectivity. It estimates the connectivity at one location within the catchment as the ratio between an upslope (Eq. 3) and a downslope component (Eq. 4):

$$\mathrm{IC} = \log_{10}\left(\frac{D_{\mathrm{up}}}{D_{\mathrm{dn}}}\right), \quad (2)$$

$$D_{\mathrm{up}} = \overline{W}\,\overline{S}\sqrt{A}, \quad (3)$$

$$D_{\mathrm{dn}} = \sum_i \frac{d_i}{W_i S_i}, \quad (4)$$

where W is the average weighting factor of the upslope contributing area, S is the average slope gradient of the upslope contributing area (m m^{-1}), A is the upslope contributing area (m^2), d_i is the length of the flow path along the ith cell according to the steepest downslope direction (m), and W_i and S_i are the weighting factor and the slope gradient of the ith cell, respectively. In Borselli et al. (2008), the weighting factor W corresponded to the C factor in the USLE-RUSLE models (Wischmeier and Smith, 1978; Renard et al., 1997), which represents friction hindering sediment transfer. More recently, it has been demonstrated that topographic surface roughness can provide a good estimation of the weighting factor (Cavalli et al., 2013; Baartman et al., 2013): a high roughness value is seen as impeding sediment transfer. This index opens up a fruitful field of research to assess structural connectivity. First, it opens the black boxes within a catchment: the IC can be calculated for each cell of the catchment, highlighting those cells that may efficiently route the sediment flux at the outlet. Second, this index takes into account all the links between a cell and all other components of the catchment, in agreement with the definition of connectivity. Third, the index can be mapped enabling comparisons between various locations (a specific tool has been developed in ArcGIS), and the calculation of maps of connectivity evolution over time (see, for example, case study applications in, among others, Foerster at al., 2014 and López-Vicente et al., 2016). Nevertheless, this index remains empirical and comparisons between catchments should be made carefully. More specifically, while such a method helps in identifying specific configurations that are highly susceptible to sediment delivery, it remains difficult to provide scenarios of change (e.g., the consequences of land use change, retention basin creation or removal).

2.2 Graph theory applications to structural connectivity

Another promising field of research is the application of graph theory, which provides a robust mathematical framework for describing networks such as sediment cascades (Heckmann and Schwanghart, 2013; Heckmann et al., 2015; Cossart, 2016). Graph theory is applied to model a network structure as nodes (representing sediment sources, sediment stores and the outlet) connected by edges (representing linkages by a geomorphological process).

Two nodes i and j are joined or adjacent if there is an edge from i to j. A directed graph with n nodes can be represented by an $n \times n$ adjacency matrix $\mathbf{A}$ and constructed as follows: if there is an edge from node i to node j, then we enter 1 in row i, column j of the matrix $\mathbf{A}$.

Both spatial and topological configurations of the network and the fluxes associated with the respective edges are responsible for the sediment delivery at the outlet (Heckmann and Schwanghart, 2013; Heckmann et al., 2015). The goal is thus to obtain a pattern that can be described by algebraic tools (typology of linkages, identification of local sinks, etc.) to show the overall structure of the sedimentary cascade. Graph theory enables an objective description of the connections of sedimentary flows, and thus an estimation of the potential influence of the network on the amount of sediment load. Indices provided by graph theory have been developed to describe the properties of single landscape units (nodes), sediment pathways (edges) and sediment cascades (edge sequences = paths). The nodes can be characterized by the number and type of links that may provide or transport sediments. In graph theory, the so-called "source nodes" define nodes characterized by the absence of input links, sinks are characterized by no output link, and other nodes correspond to connectors whose importance is revealed by their degree (number of input and/or output links). It is important to note that the meaning of "source" differs from graph theory (node with no upstream contributor: starting points of the network) to geomorphology (node that can potentially supply sediments, which can be any node of the graph). The links may be characterized by the geomorphic process that carries sediments. Regarding the edge sequences, their main characteristic is whether or not they contribute to the sediment delivery at the outlet: do they correspond to an independent subcascade or not?

Another application is to carry out "flow analyses": in a directed graph (such as sediment cascades), each edge has a capacity and each edge is assigned a flow. A flow must satisfy the restriction that the amount of flow into a node equals the amount of flow out of it, unless it is a source, which has only an outgoing flow, or a sink, which has only an incoming flow. At each node, the sediments are supplied from the upstream flow and the potential local sediment delivery. The latter can be considered homogeneous on the whole catchment, or a potential sediment delivery map can be integrated to the simulation. This simulation is based on an assumption of flow conservation, and is a complementary approach to the assessment of sediment connectivity. In the case of sources with no incoming links, they can be assigned a default common value. The "network effect" (Pumain and Saint-Julien, 2010; Czuba and Foufoula-Georgiou, 2015) describes how the network structure may affect the "potential for creation, persistence, or dispersion of sediment waves" (Gran and Czuba, 2017) and finally the total delivery at the outlet (Czuba and Foufoula-Georgiou, 2015; Cossart, 2016).

Nevertheless, these indices do not predict the role of sediment connectivity on sediment delivery, especially in the case of scenarios of changes. New methodological procedures from graph theory need to be developed, so we propose here a brief description of the state of the art of previous studies in which the influence of a spatial network structure on material or immaterial fluxes has been thoroughly explored using graph theory. Although such studies have focused on geography (Cole and King, 1968; Gleyze, 2008), social networks (Freeman, 1979) or, more recently, ecology (Ludwig et al., 2002; Belisle, 2005), they have developed metrics and discussed concepts particularly relevant to geomorphology: the relationship between connectivity and the total amount of fluxes passing through the system, and the identification of local hotspots where any change may have an impact on the whole system (Marra et al., 2014; Czuba and Foufoula-Georgiou, 2014, 2015; Masselink et al., 2017). In such studies, one key requirement is to provide a hierarchy of the influence of nodes within the network. Nodes characterized by high connectivity have a considerable influence within a network as they control the fluxes passing between many other nodes. Such high-connectivity nodes are also those where a disruption would lead to more serious damage of the network's functioning (Haggett and Chorley, 1969; Newman, 2010) and may be considered hotspots. They lie on the largest number of possible paths within the network. Many indices provided by graph theory have been applied to undirected graphs (for a review, see Rodrigue, 2017). Such indices are the result of a long procedure of research held by geographers to formalize spatial networks and spatial interactions (Pumain and Saint-Julien, 2010). From their experience we can avoid difficulties and understand how structural connectivity can be measured in directed graphs such as sediment cascades.

The betweenness centrality index (B) measures the extent to which a node i lies on paths between other nodes (Eq. 5):

$$B_i = \frac{\sum n_{ijk}}{n_{jk}}, \tag{5}$$

where n_{ijk} is the number of paths that exist from a node j to a node k and that pass through i, and n_{jk} is the total number of paths within the network, from j to k. Such simulation provides a good evaluation of the potential volume that may pass through the nodes and is helpful in interpreting the real fluxes observed in each node of the network. One main criticism is that this index enhances the role of nodes close to the center of gravity of the network and is not especially efficient in ranking the influence of eccentric nodes (Rodrigue, 2017). However, such eccentric nodes are close to the sources of the network, so that they should be ranked in relation to their importance in sediment transfer. Furthermore, spatial patterns are taken into account in a simplistic way: the distance (and the friction effect associated to the distance to hinder fluxes) is not considered. We note that distance can be calculated in different ways: it can represent the length of the path, the time necessary to travel along the path, the cost necessary to travel, etc. Consequently, various properties of the edges can be used to approximate the velocity of sediment transfer (Czuba and Foufoula-Georgiu, 2015).

The Shimbel index (Shi) takes into account the distance between nodes and considers whether the location of the node increases or decreases the total length of all possible paths within the network (Eq. 6) (Newman, 2010; Rodrigue, 2017). For one node i, it corresponds to the sum of the length of all the shortest paths connecting all other nodes j in the graph (d_{ij}). To facilitate comparisons in both space and time, this index should be normalized, i.e., divided by the sum of the length of all the paths in the network, from j to k (d_{jk}):

$$\mathrm{Shi}_i = \frac{\sum d_{ij}}{\sum d_{jk}}. \tag{6}$$

If the Shimbel index is high, then the node contributes to creating long paths within the network (and thus attenuates the compactness of the network). If the Shimbel index is low, then the node maximizes the compactness of the network. This index is much more efficient at ranking the influence of eccentric nodes on the network and can be enriched by considering various types of distance (geodesic, time, etc.). It is considered a very good proxy of accessibility and is thus sometimes called an "accessibility index". Nevertheless, we should point out that the lower the Shimbel index, the higher the accessibility (and thus the connectivity) of the nodes: while counterintuitive, this feature is now well accepted within the scientific community (Rodrigue, 2017).

Both indices enable an in-depth description of the skeleton of a network and highlight the potential impacts of the network structure on flux patterns. They can thus provide conceptual and mathematical frameworks to explore the structure of sediment cascades. Nevertheless, they cannot be applied directly to measure sediment connectivity as sediment cascades are directed graphs and thus more complicated in terms of mathematical conceptualization.

3 Methodology: the Network Structural Connectivity index (NSC)

From both recent developments on sediment connectivity metrics and past studies on the applications of graph theory to spatial networks, we now develop a methodological framework focusing on the analysis of sediment transfer within a

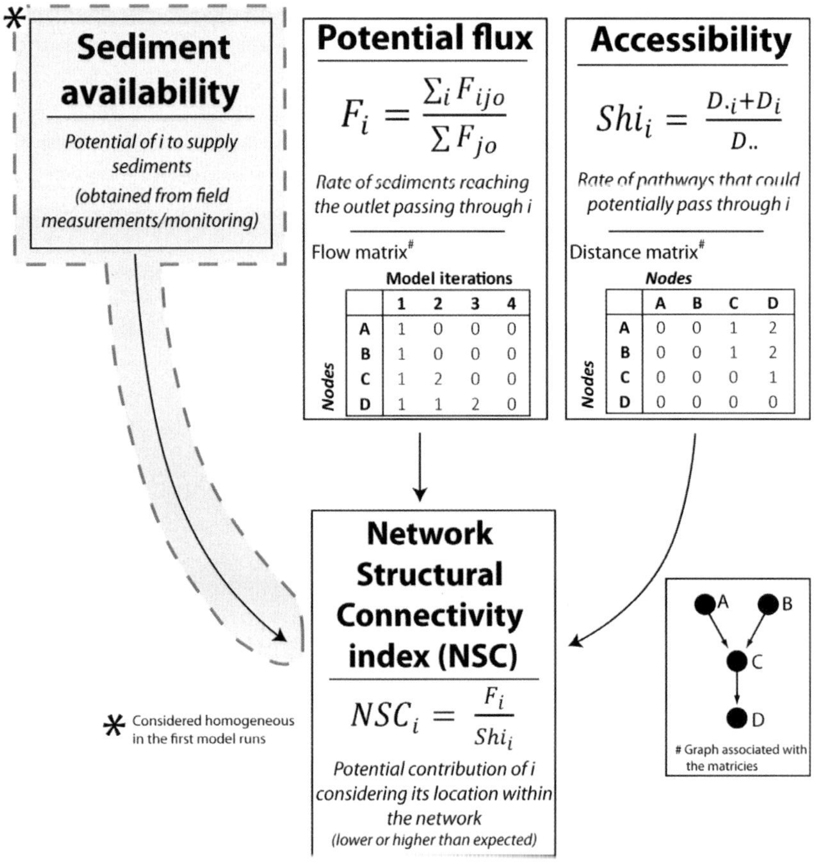

Figure 1. Overview of the main parameters used to build the network structural connectivity index (NSC) (equation details can be found in Sect. 3).

network. At the catchment scale, graph theory helps in identifying specific assemblages and, in particular, the subcascades that are not connected to each other. It corresponds to connected components, i.e., a subgraph in which nodes are all connected to each other (Newman, 2010). Their identification may reveal the spatial fragmentation of the sediment cascade and thus highlight the extent of the active contributing area in terms of sediment delivery at the outlet.

At a more local scale, to provide indices that can measure sediment connectivity within the sediment cascade, we have developed the NSC, which is based on (1) potential sediment fluxes and (2) accessibility (Fig. 1). The objective is to assess spatially the respective influence of each node on sediment connectivity inside the catchment area.

3.1 Potential flows in directed graphs (F)

As in undirected graphs, the first issue is to quantify the "network effect" (according to Pumain and Saint-Julien, 2010) to highlight how the spatial structure of paths influences the amount of sediment transferred to the outlet. In sediment cascades, only the paths that come from a node j to the outlet o have to be considered, so that, for each node i, we have to count the number of paths from j to o that include i (F_{ijo}). This measurement is divided by the total number of paths that come from all nodes j to o (F_{jo}) to reveal the proportion of the total number of paths that lies on i (Eq. 7):

$$F_i = \frac{\sum_i F_{ijo}}{\sum F_{jo}}. \tag{7}$$

F_{ijo} and F_{jo} can be calculated by reconstructing the sediment pathways throughout the cascade. Under the hypothesis of "all things being equal", a virtual volume of sediments (1 unit) is set on each node so that a spatially uniform sediment input is considered. As suggested by Gran and Czuba (2017) such a theoretical simplification allows the simulation of sediment wave generation and movement and exhibits the specific influence of the spatial structure of the network on the sediment wave pattern. In the case of converging flows, the progressive increase in the fluxes is exhibited. In the case of diverging flows, the flow can be subdivided between edges (equally or according to a weighting factor, e.g., slope gradient) so that the complex pattern of sediment wave generation and movement can also be described. In terms of mathematical procedure, the evacuation of sediment can be simulated by a multiplication of the adjacency matrix by a matrix representing the sediment variability ($\mathbf{S}_n$) (Eq. 8). This is a one-column matrix, where each row represents a node of the cascade. In the initial conditions (first column, $\mathbf{S}_0$), a value of 1 is considered for each row to represent the virtual volume of sediments (one unit per node) at the beginning of the transfer. Each multiplication corresponds to an iteration, in which each sediment unit is transferred along one edge, according to the links described by the adjacency matrix (Eq. 8 and Fig. 1). The result provides a matrix $\mathbf{S}_n$, highlighting where the sediments are after one single iteration:

$$\mathbf{S}_n = \mathbf{S}_{n-1} \times A. \tag{8}$$

The operation is repeated until all the virtual sediments are evacuated, and the results can be represented within a synthetic matrix ($\mathbf{S}.$), concatenating $\mathbf{S}_0$, $\mathbf{S}_1$,... , $\mathbf{S}_n$ matrices, obtained during the calculation (Fig. 1).

3.2 Accessibility from sources to sinks (Shi)

Within a sediment cascade, the influence of geomorphic units (sources, stores, and sinks) on sediment delivery can be assessed by considering their location inside the cascades. A node whose centrality is high (i.e., characterized by a low summation of all distances between the other nodes and itself) has potentially greater influence on the overall sediment cascade. A first objective is to hierarchize the influence of nodes that correspond to confluences. In other words, if a strategic confluence is disconnected from the outlet (i.e., if an edge connecting the strategic node is disrupted) the spatial pattern of the transport capacity will be affected. Not only would the total amount of sediment fluxes change, but the "mutual interferences" that occur at such geomorphic hotspots would also be modified (Benda et al., 2004a, b). A second objective is to compare the potential influence of the network sources: the smaller the distance to the outlet, the greater the influence is.

Characterizing the nodes by their location within the network refers to the concept of accessibility and is thus very similar to the calculation of the Shimbel index in the case of undirected graphs. In directed graphs, the calculation of the accessibility (Shi) of each node i can be made from a distance matrix $\mathbf{D}$ (Eq. 9 and Fig. 1):

$$\overline{\mathrm{Shi}_i} = \frac{\mathbf{D}_{.i} + \mathbf{D}_{i.}}{\mathbf{D}_{..}}, \tag{9}$$

where $\mathbf{D}_{.i}$ is the total of the distances between i and the nodes (sources and stores) that feed i, $\mathbf{D}_{i.}$ is the distance between i and the nodes located downstream, and $\mathbf{D}_{..}$ is the total of the distances of all the paths within the network.

3.3 Combination of indices: NSC

The indices F and Shi provide a quantitative and complementary description of the cascade skeleton (e.g., both river network and hillslope processes): the first reveals the potential proportion of flux discharge passing through each node (e.g., water, sediments, and other types of constituents); the second measures the ratio of the length of paths that lie on i (from all points within the catchment and to the outlet) and the total length of path within the network. Classically, the potential flux (e.g., sediment) discharge at a node increases in relation to the number of paths that come from sources (i.e., the active contributing area is larger; cf. the upslope component of Borselli's IC). Nevertheless, due to the geometry of paths, confluences and, more generally, the network structure, there can be some interference in terms of sediment movement through the system: the sediment discharge at a node can be higher or lower than expected from the single number of paths supplying sediments. To estimate this possible under- or overrepresentation of potential sediment volume at each node, a ratio between F and Shi can be calculated (Eq. 10 and Fig. 1):

$$\mathrm{NSC}_i = \frac{F_i}{\mathrm{Shi}_i}. \tag{10}$$

The expected results can be considered a normalization of the potential sediment fluxes F_i.

3.4 Implementation

From a geomorphological map, a graph can be digitized in GIS software (QGIS), which consists of depicting a regular network of nodes. Each node can be characterized by the geomorphic unit to which it belongs, and the linkages between the nodes can be digitized from geomorphological expertise. The links correspond to directed fluxes, driven by gravity, and we only consider converging flow. To simplify the network structure, each node cannot have two output links. The "Network" QGIS tools can be used to generate the adjacency matrix (as an edge list matrix) and exported to R software. In the latter, the matrix can be converted into an origin-to-destination matrix, and the distance matrix is automatically created (for simplicity, the distance between adjacent pairs of geomorphic units is unity and not the Euclidean distance) using the "igraph" package (Csardi and Nepusz, 2006). All calculations on matrices were conducted in R, and the results were exported to QGIS to be mapped.

4 Results and interpretation

4.1 Implementation on a virtual catchment

To address the role of network structure in sediment fluxes, a conceptual sediment cascade network is first used as an example. The indices are calculated on a small virtual catchment of seven nodes and six links (Fig. 2a), from the adjacency matrix and the distance matrix (Tables 1 and 2). The first model run is parameterized in a simplistic way so that

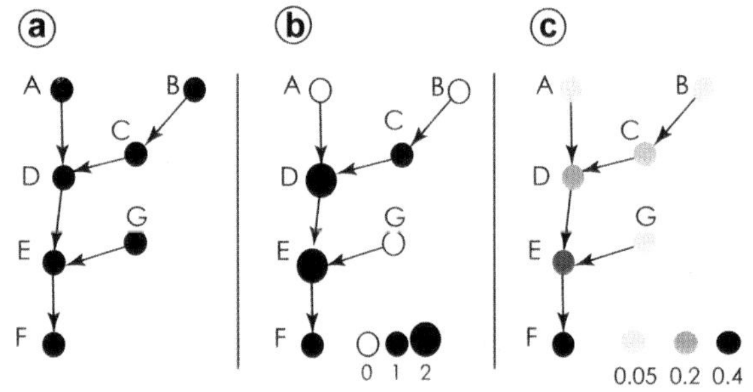

Figure 2. The virtual sediment cascade. **(a)** The structure of the cascade, represented by a graph; all nodes are supposed equals in terms of sediment availability (spatially uniform sediments availability; virtual units = 1). **(b)** The potential flow of sediments after one iteration during the simulation (see also Table 3, column "1"); **(c)** map potential flux index (F_i) values (see also Table 3, column "F_i").

Table 1. Adjacency matrix of the virtual sediment cascade.

		Destination node						
		A	B	C	D	E	F	G
Origin node	A	1	0	0	1	0	0	0
	B	0	1	1	0	0	0	0
	C	0	0	1	1	0	0	0
	D	0	0	0	1	1	0	0
	E	0	0	0	0	1	1	0
	F	0	0	0	0	0	1	0
	G	0	0	0	0	1	0	1

all sources are assumed to be of equal importance, using spatially uniform sediment inputs (volume availability equals 1 at each node at time 0) and a topological distance (each edge corresponds to a distance of 1 unit).

4.1.1 Potential fluxes (F)

First, a map of the potential fluxes (F_i) within the sediment cascade can be drawn (Fig. 2b and c, Table 3). Such a result can be useful for defining a local monitoring strategy (e.g., survey locations) for sediment transfer and then interpolating local measurements at the catchment scale. Moreover, the F_i index may provide a hierarchy between the nodes by assessing the increase in sediment upstream and downstream of the node. For instance, in our virtual study case, the amount of sediment classically increases downstream, as there is no interruption of the cascade and we only consider converging flow. Nevertheless, the main increase occurs at node D, pointing out that this node is a junction that receives input from various tributary systems. D may correspond to a zone of sediment persistence within the catchment, so that any disruption of this node (blockage due to sediment surplus that leads to aggradation, anthropogenic action, etc.) would significantly modify the development of the sediment pulse.

Table 2. Distance matrix (origin-to-destination) of the virtual sediment cascade. On the right, the rows detail the calculation of the accessibility index.

		Destination node								
		A	B	C	D	E	F	G	$\mathbf{D}_{i}$	Shi$_i$
Origin node	A	0	0	0	1	2	3	0	6	0.21
	B	0	0	1	2	3	4	0	10	0.35
	C	0	0	0	1	2	3	0	6	0.24
	D	0	0	0	0	1	2	0	3	0.24
	E	0	0	0	0	0	1	0	1	0.35
	F	0	0	0	0	0	0	0	0	0.52
	G	0	0	0	0	1	2	0	3	0.10
	$\mathbf{D}_{\cdot i}$	0	0	1	4	9	15	0	29	

Table 3. Analysis of the potential sediment flow within the sediment cascade. The first rows correspond to the iterations simulating the evacuation of sediments. On the right, the rows detail the calculation of the flow index.

		Model iteration							
		0	1	2	3	4	5	F_{ijo}	F_i
Node	A	1	0	0	0	0	0	1	0.05
	B	1	0	0	0	0	0	1	0.05
	C	1	1	0	0	0	0	2	0.09
	D	1	2	1	0	0	0	4	0.18
	E	1	2	2	1	0	0	6	0.27
	F	1	1	2	2	1	0	7	0.32
	G	1	0	0	0	0	0	1	0.05
							Total	22	

However, one main criticism is that the F_i index pays little attention to the sediment sources (here A, B and G) (Fig. 2c). As the latter correspond to the initiation zones of the sediment cascade, the events that happen there may influence long pathways to the outlet. As mentioned for the betweenness index (B), it is necessary to discriminate better the potential influence of sources and stores located next to sources.

4.1.2 Accessibility (Shi)

The accessibility (Shi) map can then be computed based on the distance matrix (Fig. 3 and Table 2). Here, G is characterized by better accessibility, greater than both A and B: indeed the short distance between G and the outlet suggests it would have a higher influence on sediment delivery. From a geomorphic point of view, it reveals two distinct consequences without causal relation: first, the time of sediment transfer is short; second, the possibility of intermediate storage is low. Figure 3 thus illustrates such a hierarchy of the influence of sediment sources on sediment delivery at the outlet. In terms of management, it shows the sources that can be activated to

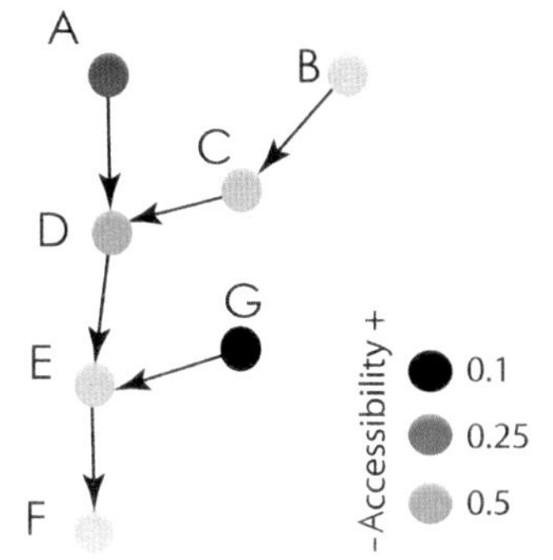

Figure 3. Assessment of the accessibility index (Shi) within the virtual sediment cascade.

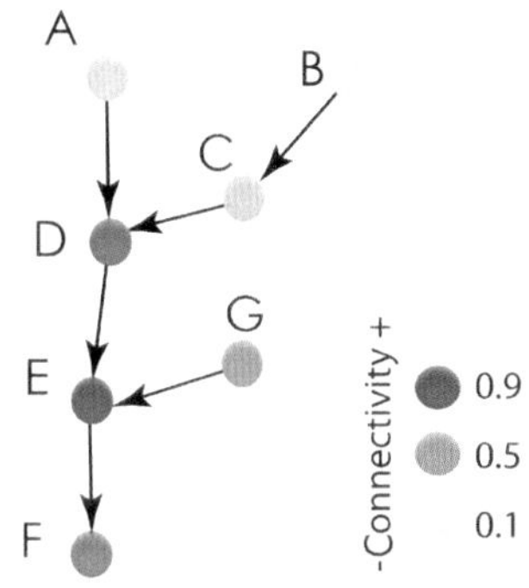

Figure 4. Assessment of the connectivity within the virtual sediment cascade using the NSC index.

cope with sediment exhaustion at the outlet or, conversely, sources where protection strategies should be applied in the case of sediment overflow. Because of this issue, this index is not a good proxy of connectivity as it underestimates the role of the outlet and all nodes close to the outlet, and does not pay attention to the confluences between various pathways inside the sediment cascade. At the catchment scale, the roles of D and E are not shown although they are important connectors between pathways developed from sources A and B. The indices of both nodes have to be carefully compared to note that D is closer to different sources and the outlet than E.

4.1.3 NSC

The NSC provides synthetic metrics, suggesting that E and D are the most important geomorphic hotspots (Fig. 4), with very similar values (0.8 and 0.71, respectively). E and D are at confluences and thus lie on various sediment paths from distinct sources. Their potential influence on the whole sediment cascade is high, so that any disruption of these nodes would considerably alter the elementary interactions between many nodes and sediment paths. Consequently, D and E may significantly modify the ability of the cascade to provide sediments and should be further studied in depth to document the functional connectivity or to assess erosional rates (local monitoring, field observations). The outlet F has quite a high index (0.66) but lower than E and D. This value highlights that the connectivity of a node is not proportional to the total volume of sediment that passes through it. In the case of F, the NSC is partly influenced by the high potential sediment volume that passes through it, but we point out that any disruption at this node would be ambiguous. For example, it would interrupt the sediment delivery but the organization of the three tributary subcascades from sources A, B and G would not be modified. Furthermore, the patterns of mutual interferences at the confluences (E and D) would also remain unmodified, so that sediment pulse generation and movement would not be modified. Consequently, the spatial structure of the cascade network would be roughly unchanged under the hypothesis of F disruption. Regarding the sources (A, B and D), a hierarchy is evidenced: the source G has a greater influence on the sedimentary signal at the outlet due to its proximity (NSC = 0.53), which is greater than A and B (NSC equals 0.27 and 0.12, respectively).

4.1.4 Sensitivity of NSC to sediment availability

To investigate how sensitive the indices are to parameterization, we modify the initial conditions of our virtual sediment cascade. Regarding sediment availability, we consider G exhausted (volume equals 0) and B overflowing (volume equals 2). All other nodes remain unchanged. Regarding the distance between E and F, it is now twice the initial value (D_{EF} equals 2) (Fig. 5a and Table 4).

To interpret the result we first differentiate nodes whose connectivity increases and nodes whose connectivity decreases. As expected, the potential flow F_i is mainly modified at B, whose influence increases (F_B shifts from 0.05 to 0.07), and at G, whose influence becomes null (Fig. 5b and Table 4). At other nodes, an increase is observed at D (F_D shifts from 0.18 to 0.21), while F_E and F_F remain roughly unchanged (shifting from 0.80 to 0.85 and 0.66 to 0.56, respectively). While node D was already strategic in the first simulation, the increase in sediment availability at B reinforces its influence on the whole sediment cascade. Downstream, the potential flow at E and F is not reinforced by the amount of sediment delivered at B because of the exhaustion of G.

Considering the accessibility (Fig. 5c), the higher eccentricity of F has an impact on $\mathbf{A}_F$ but, more generally, alters the accessibility of all nodes. The accessibility coefficient decreases significantly at B: the subcascade organized from B is the longest and all the sediment paths that may exist along this subcascade are impacted by the greater distance between E and F. As a consequence, the outlet is here significantly less accessible from the source B than from the sources A and G (the latter remaining the closer). It can be noticed that the accessibility of D is not impacted by the higher eccentricity of F. $\mathbf{A}_D$ remains roughly stable, and even suggests a slight improvement in accessibility. All nodes that are characterized

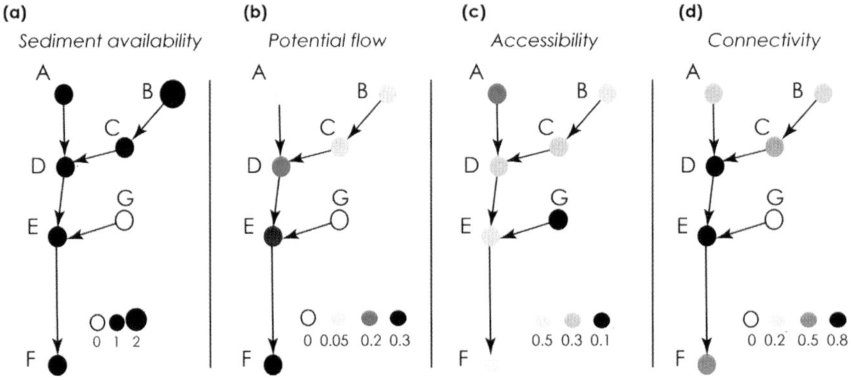

Figure 5. Flow, accessibility and connectivity indices following a modified parameterization. Note how the connectivity of node D is reinforced while the connectivity of F is reduced.

Table 4. Analysis of the potential flow and calculation of connectivity following a new parameterization. The rows indicate the patterns of sediment evacuation at each iteration of the simulation. Source B provides twice as many sediments and the distance between E and F is twice that during the initial conditions.

		Model iteration									
		0	1	2	3	4	5	F_{ijo}	F_i	Shi$_i$	NSC$_i$
Node	A	1	0	0	0	0	0	1	0.03	0.18	0.20
	B	2	0	0	0	0	0	2	0.07	0.35	0.20
	C	1	2	0	0	0	0	3	0.10	0.28	0.38
	D	1	2	3	0	0	0	6	0.21	0.25	0.83
	E	1	2	2	3	0	0	8	0.28	0.33	0.85
	F	1	1	2	2	3	0	9	0.31	0.55	0.56
	G	0	0	0	0	0	0	0	0.00	0.08	0.00
							Total	29			

by great centrality, and that are closer to both the sources and the outlet, are not affected by an increasing eccentricity at the margins of the cascade (if the distance from sources, or to the outlet, increases).

Finally, regarding the connectivity index (Fig. 5d), the new parameterization modifies the hierarchy of nodes. First, the influence of the confluence nodes increases and, concomitantly, the influence of F decreases. As F is characterized by eccentricity, its influence on the overall system decreases. A disruption would modify the sediment cascade structure less (interplay at tributary junctions, organization of tributary subcascades) than in the previous model. D here appears as a node of high connectivity as it is close to two main sources and to the outlet. The node E is also of prime importance in terms of connectivity, but its NSC value is lower than expected from its strategic location. In fact, it is connected to an exhausted source (G). Looking at the sources, a hierarchy is clearly observed: the influence of B increases due to its main contribution to the sediment flow, while the influence of G becomes null as it is exhausted.

The NSC thus reveals the degree of coupling to both the sources and the outlet for each unit of a sediment cascade. More precisely, it reflects the structural connectivity as it enhances the role of spatial patterns (distance, confluences, etc.) of the network. To make the interpretation easier, a high-connectivity node may modify the overall spatial network structure in the case of disruption (e.g., modifications of flux interplay at junctions, creation of independent sediment subcascades unconnected to the outlet). Moreover, the parameterization could be gradually enriched with geomorphic expertise to pay more attention to sediment availability or to the ability of geomorphic processes to transfer sediment along the paths (i.e., the edges).

4.2 Application to a real sediment cascade

The NSC is now applied to a real sediment cascade, whose functioning has already been conceptualized and quantified (Cossart and Fort, 2008; Cossart, 2016).

4.2.1 The Celse-Nière catchment

The Celse-Nière catchment is located in the French Southern Alps, on the eastern flank of the Massif des Écrins (Fig. 6). The current glaciation of this catchment is limited (about 6 km^2), in spite of high altitudes (summit approaching 4000 m at Ailefroide), because this area is partly sheltered from oceanic influences due to its eastward location. For instance, precipitation is about 995 mm yr^{-1} at Pelvoux (1280 m), which is significantly lower than on the western flank of the Massif des Écrins (1195 mm yr^{-1} at Valjouffrey, 1160 m). The equilibrium line altitude ranges from 3000 to 3200 m (Cossart, 2004), which is 200 m higher than in the western part of the massif.

We focus here on the headwater (about 10 km^2, from 2500 to 3850 m a.s.l.), which is still occupied by glaciers. Some small tributaries converge into the upper part of the catchment (Ailefroide, Coup de Sabre) as their mouth is blocked by the Sélé valley glacier tongue. Such small catchments are

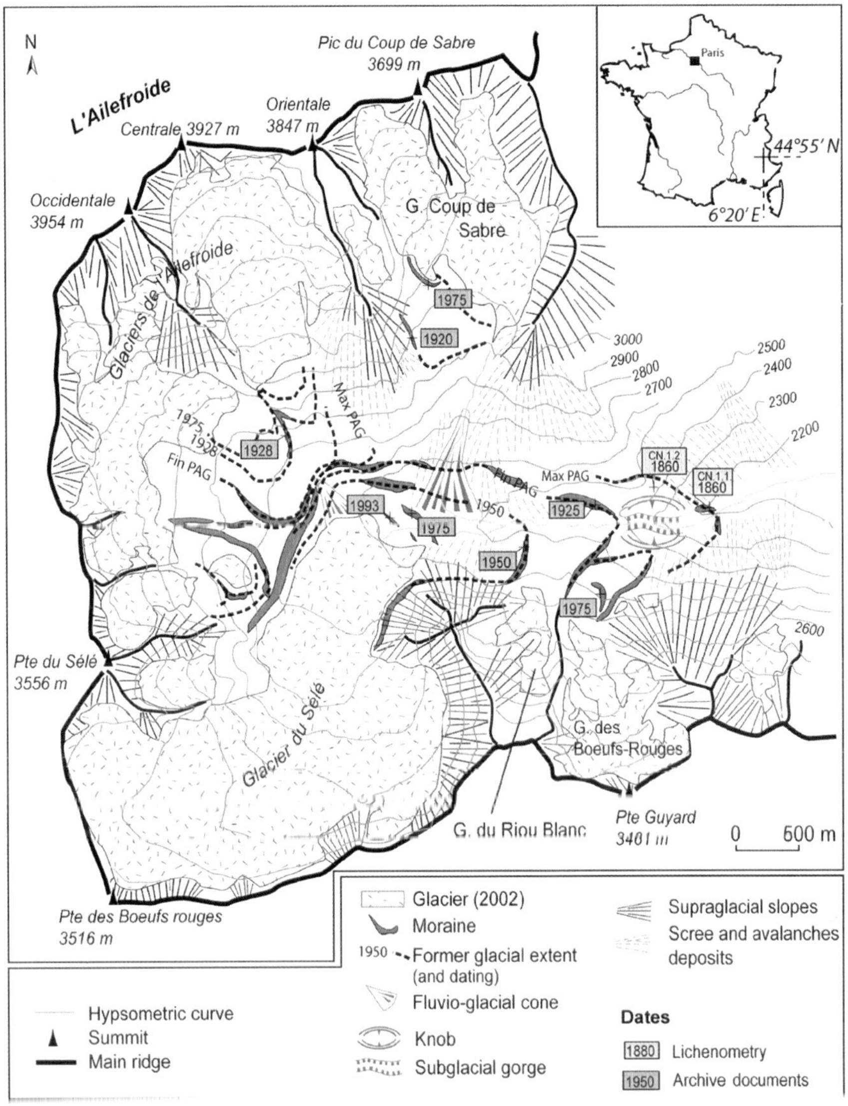

Figure 6. Geomorphological map of the study area: Celse-Nière catchment and sediment cascade.

occupied by cirque glaciers, which lie just below granitic and gneissic free faces and provide both sediments and meltwater to the valley floor. Special attention has already been given to the linkages between the glacial margins and the glacio-fluvial systems (Cossart, 2004, 2016). The presence of morainic ridges still interrupts the sedimentary cascade system, thus forcing local aggradation and change in the glacio-fluvial pattern (Fig. 6). Such a complex assemblage makes this area particularly suitable for assessing connectivity and simulating the impacts of new blockages or, conversely, some reconnections.

4.2.2 The structure of the network

As in the virtual case study, a reduced-complexity network model is considered to address the role of the network's spatial structure (Fig. 7). The initial conditions of run 1 of the model are as follows: (1) sediment input (volume availability equals 1 at each node at time 0) is spatially uniform and (2) topological distance is considered (each edge corresponds to a distance of 1 unit).

First, it can be noticed that only 56 % of all the paths are connected to the outlet, while the others are connected to permanent sinks. Twenty-five connected components are identified, highlighting a high fragmentation of the system: only 6 of these encompass more than 10 nodes (including the main connected component connected to the outlet) and 11 encompass fewer than 5 nodes. By applying the typology established by Fryirs et al. (2007), disconnections are due to barriers, buffers and blankets (Fig. 7a). Barriers are here mostly due to moraines: lateral moraines may affect the longitudinal coupling of processes, especially the coupling between glacio-fluvial streams from the cirque glaciers and

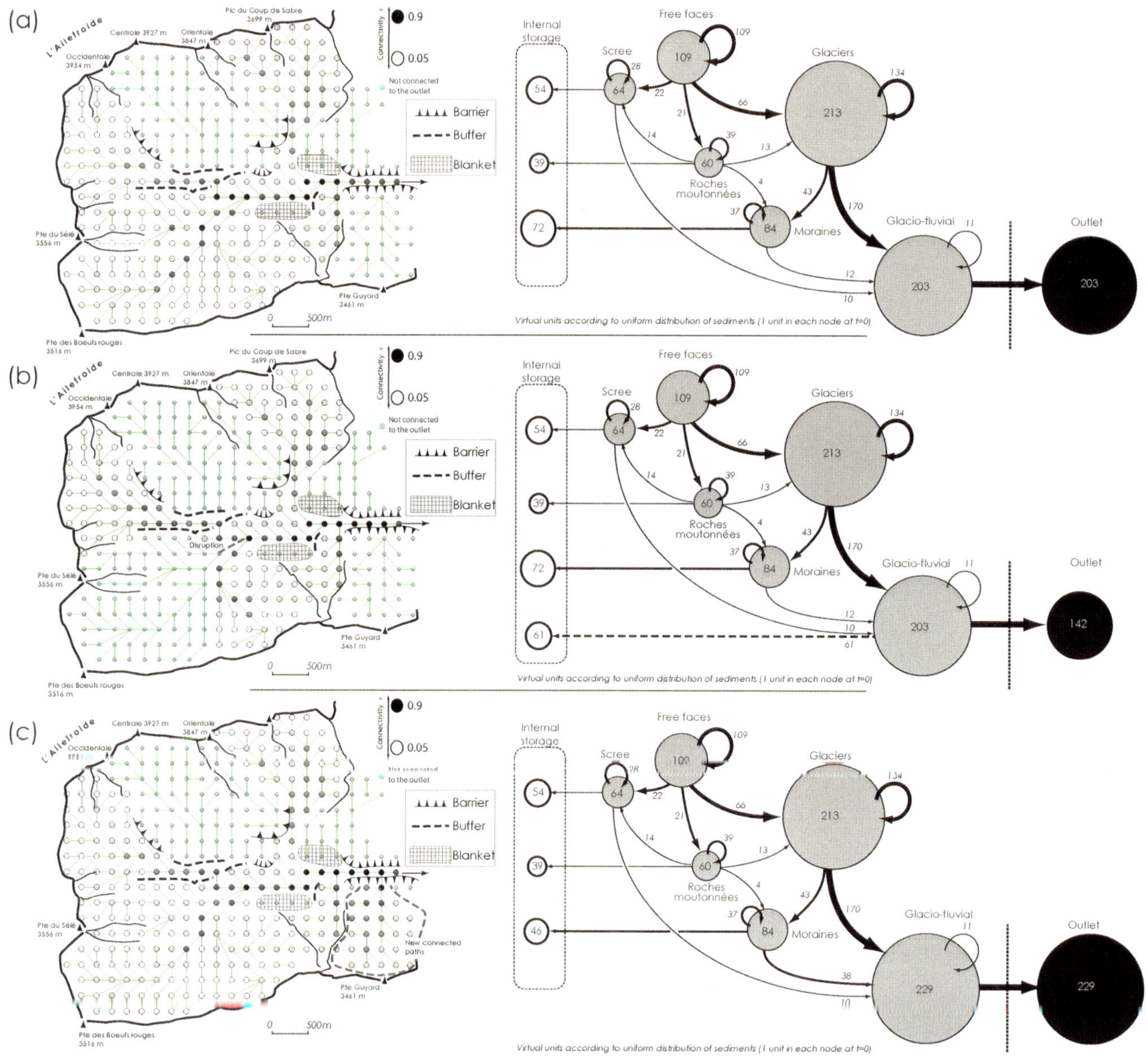

Figure 7. Assessment of connectivity. **(a)** Current structure of the cascade. **(b)** Connectivity map after the simulation of a disruption at Sélé toe. **(c)** Connectivity map after the simulation of a reconnection at Guyard outlet.

the mainstem. In that case, the node that corresponds to the upper part of the moraine is a sink, the node corresponding to the downslope part of the moraine is a source, but there is no link between these both nodes.

Buffers correspond not only to moraines, where they decouple scree deposits from the mainstem (scree deposition is forced by the morainic ridge), but also roches moutonnées and glacio-fluvial terraces. Blankets correspond to scree deposits made of large grain-size boulders that cannot be removed by fluvial processes. Second, the NSC highlights the influence of the trunk valley located between the margin of the Glacier-du-Sélé and the confluence with the Coup-de-Sabre proglacial river, where high-connectivity nodes are observed (Fig. 6a). This means that potential sediment fluxes are higher than expected from the active contributing areas upstream, so that the network may develop here zones of potential sediment persistence within the catchment. Furthermore, such nodes correspond to tributary junctions between the main subcascades of the system. Here the mutual interference may have an influence on the generation of sediment pulse, and thus on its evolution (transfer to the outlet), as well as on the tributary subcascades located upstream. Sediment persistence, an aggradation pattern in these zones, may generate a retrogressive aggradation that may modify the functioning of subcascades (e.g., decrease in sediment transfer along subcascades in the case of impoundment). For these reasons, such nodes can be considered hotspots of geomorphic change, which can propagate a perturbation along the whole cascade due to a geomorphic change and thus modify the overall functioning of the system. A significant input of sediments (due, for instance, to a hydro-meteorological event) in these areas would increase the sediment delivery at the outlet and may also alter the ability of tributaries to deliver sediments. The NSC also exhibits a hierarchy between the sources. As they are closer to the outlet, all the sources located in the Coup-de-Sabre subcatchment have a greater influence on sediment delivery than the sources located in the Ailefroide, Sélé or Bœufs-Rouges areas.

Thus, the map of the NSC helps to conceptualize the continuum of sediment transfer and to predict the downstream transfer and delivery of sediment fluxes measured at one point (not necessarily at the outlet). Node connectivity may need to be examined to establish sampling strategies for small-scale measurements of erosion in the field. Furthermore, this first examination highlights that the potential impacts of external drivers (anthropogenic impact, hydro-meteorological event and, more generally, climate change) are space-dependent: the impacts may be greater and efficiently propagated if they affect high-connectivity areas.

4.2.3 What if...?

The connectivity hierarchy between nodes can be interpreted as the potential influence of the node on sediment delivery and the overall functioning of the cascade. The NSC and, more generally, tools provided by graph theory enable the simulation of scenarios to predict which events would have more impact on the cascade. We applied two other model scenarios in R to simulate the consequences of a local disconnection and a local reconnection. Run 2 corresponds to an algorithm that tests which edge removal would have the greatest impact on sediment connectivity (i.e., the largest decrease in connectivity at the node characterized by the highest connectivity) (Fig. 7b). This simulation could reflect the possible impact of an anthropogenic feature (e.g., a dam) or a hillslope process (e.g., a dam created by a landslide mass or a debris flow). Run 3 (Fig. 7c) corresponds to an algorithm that tests which creation of a new edge would lead to the greatest improvement in connectivity (i.e., where can we create the highest increase in the NSC value at a node?). This simulation could reflect the disruption of a barrier, the removal of a blanket or the overwhelming of a buffer, for instance following a high-magnitude geomorphic event.

Run 2 highlights that the greatest impact would occur if the edge located at the toe of the Glacier du Sélé was disrupted. It would lead to the disconnection from the outlet of the main subcascade (fed by Sélé sediment sources) so that only 40 % of the nodes would remain connected to the outlet (compared to 56 % in the initial stage). Nevertheless, the modification of the whole system is not just a question of sediment delivery at the outlet. First, we point out that it may

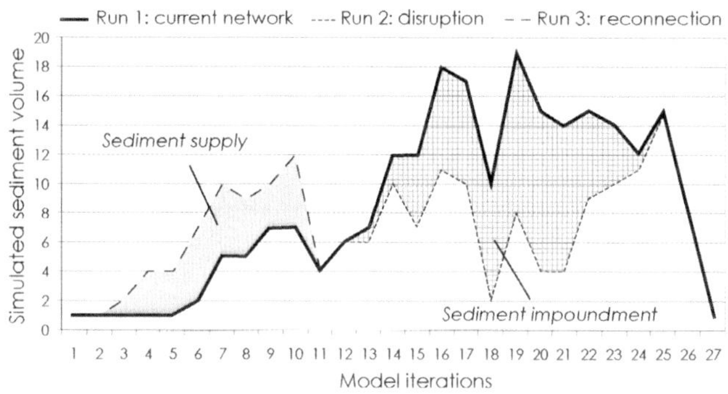

Figure 8. Sedimentographs associated with runs 1, 2 and 3 on Celse-Nière catchment. Run 1 shows the current effect of the spatial network structure on sediment waves, run 2 the potential impact of an edge disruption in a hotspot zone, and run 3 the potential impact of a reconnection. Runs 1, 2 and 3 correspond to the graphs in Fig. 7a, b and c, respectively.

provoke an attenuation of the peak of the sediment wave if we compare the simulated sedimentographs between runs 1 and 2 (Fig. 8). Second, the disruption creates a subdivision of the sediment cascade connected to the outlet into two main connected components of quite similar size. The connected component that becomes disconnected from the outlet after run 1 (Bœufs-Rouges area) encompasses 72 nodes, and the other (still connected to the outlet) includes 142 nodes. Any other edge removal would lead to a new connected component of a smaller size (< 72 nodes). The disruption would be more significant than that of an edge located at the confluence with the Coup-de-Sabre proglacial river. In this latter case, many nodes would be disconnected from the outlet, but the three subcascades of Ailefroide, Sélé and Bœufs-Rouges would be less impacted and would still be self-organized within a large connected component. As a result, the structure of the sediment cascade would be less fragmented. Finally, the NSC shows where a local disruption may split two connected components of quite similar size, which may have a strong impact in terms of geomorphic functioning. As many geomorphic processes are scale-dependent, some variables influencing sediment transfer that are interdependent at one scale may well be independent at another (Schumm, 2005). A corollary is that the split of a cascade into two connected components (whose sizes are approximately half that of the initial cascade) may modify the scale at which geomorphic processes or controls act and markedly change the functioning of the system.

For run 3, a new edge is added to improve the overall sediment connectivity (Fig. 6c). In this case, a link between the Guyard subcatchment and the trunk valley would create the highest NSC value at the confluence. Such an increase is due to the large number of nodes (64 %) that would become connected to the outlet. Furthermore, these nodes (especially the sources) are relatively close to the outlet, so that the sediment wave exhibits an increase at the beginning of the pulsation

(Fig. 8). A reconnection of the subcascade in the Ailefroide area would have a lesser impact because of its eccentricity. It can be noticed that the reconnection of the Guyard subcascade would decrease the influence of the Coup-de-Sabre subcascade on the overall network: in this scenario of reconnection, all the sources of this area are affected by a decrease in NSC. According to this new cascade structure, the hierarchy of sources would be modified: the sources of the Guyard area would have a greater influence than the Coup-de-Sabre sources, which would have a greater influence than the Ailefroide, Sélé and Bœufs-Rouges sources.

The NSC provides an exploration of the cascade structure and may explain to what extent a small-scale modification (disruption of a node, creation of a linkage) may result in significant changes in broad-scale geomorphic patterns and processes. In our examples, it can also predict the potential impacts on the sediment wave pattern (Fig. 8). More generally, the NSC enables comparisons between different states of connectivity within the same catchment.

5 Discussion

Several papers have already addressed the assessment of connectivity (Borselli et al., 2008; Cavalli et al., 2013; Czuba and Foufoula-Georgiu, 2015; Gran and Czuba, 2017; Hoffmann, 2015), so we focus here on the potential improvements offered by graph theory. Following Heckmann and Schwanghart (2013), a graph provides both a geometric and an algebraic description of the sediment cascade network. By identifying the connected components within the system, the examination of the geometry of the graph first helps to assess the SDR at the catchment scale and to identify the recurrent types of process coupling. In detail, it aims to determine the interaction of processes that are (in)efficient within the cascade at supplying sediments to the outlet. From the case studies developed here, we also highlight how an algebraic framework may help in exploring how the cascading system functions. Reduced-complexity network models associated with simple hypotheses (spatially uniform sediment inputs, simple assessment of the distance between nodes based on a topological distance) highlight the zones of greater sediment persistence, where sediment may accumulate in the network. Consequently, the role of the network's spatial structure in the generation of a sediment wave can be simulated and the associated sedimentograph can help in deciphering what is really due to external boundary conditions in sediment delivery variability at the outlet (Fryirs, 2017). Another result is the rank assessment of the geomorphic importance of the nodes, considering their location within the catchment. Following previous authors (Cavalli et al., 2013; Czuba and Foufoula-Georgiu, 2015; Gran and Czuba, 2017), the NSC identifies some specific zones that correspond to hotspots of geomorphic change due to their strategic location closer to the sources and the outlet. One main improvement in the graph theory framework is that geomorphic expertise can be integrated within algebraic formalization. First, sediment availability can be seen as $\mathbf{S}_0$ matrix. Second, the influence of barriers, buffers and blankets on network structure (connections and disconnections patterns) can be taken into account within the adjacency matrix. Furthermore, graph theory helps in developing simulation scenarios, while the associated algorithms show the properties of such geomorphic hotspots. Hotspots not only influence the total amount of sediment that is delivered at the outlet but also reveal where disconnections or reconnections may have a strong impact on the organization of the sediment cascade. A disconnection occurring at a hotspot may split the cascade into connected components of quite similar size, modifying the sedimentograph pattern, and may break the organization of path sequences.

Such a focus on the spatial dependence of geomorphic processes should be complemented, and the graph can be gradually enriched to take into account more complexity. A geomorphic hierarchy of nodes (in terms of sediment supply) can be parameterized: for instance, if a storage landform overflows or, conversely, is exhausted. The matrix representing the sediment sources can then be adjusted. Second, distance is an important parameter that can modify the results of Ai, and thus NSC. Distance involves friction, which hampers the sediment transfer: the greater the distance, the higher the potential friction opposing sediment delivery (i.e., the transfer time increases). Many other kinds of distance can be taken into account, such as the Euclidian distance, but in geomorphology other types of distance may be more relevant. For example, a distance expressed as a time, to reveal the "virtual velocity" of sediment transfer from one unit to another, can be particularly appropriate, although difficult to assess. A cost distance can also be relevant. By revealing how hampered (or efficient) the sediment transfer is along the edge, a Manning coefficient, or more generally a roughness index (Broselli et al., 2008; Cavalli et al., 2013; Baartman et al., 2013), can be a good proxy of the friction that hampers the sediment transfer. Such parameters can be calculated from high-resolution digital elevation models, combined with the edge characteristics through GIS procedures, and finally integrated within the matrices necessary for the calculations. In this way, graph theory provides a methodological framework that can be extensively enriched by various parameters to reveal the transport capacity, and consequently the potential interplay between network geometry and spatial patterns of transport. One perspective of research is to complement this approach with a more dynamic modeling of the sediment cascade network structure. It is well known that the assemblages of links and nodes may evolve in accordance with various external forces (e.g., climate, human practices, tectonics). For instance, agent-based models can be suitable to predict the possible evolution of the structure by considering negative and/or positive feedbacks along the edges (Reulier et al., 2016).

6 Conclusion

This paper seeks to develop an original methodology dedicated to the study of sedimentary cascades under the hypothesis that the influence of connectors and paths on sediment delivery is space-dependent. The methods rely on graph theory to assess structural connectivity: the sediment cascade is described as a network and consequently as a graph. Inspired by indices developed in other disciplines (transportation science, sociology, ecology), a potential flow and an accessibility of geomorphic units (i.e., accessibility to sediment sources and to the outlet) can be measured throughout the sediment cascade. Both indices are combined to estimate a connectivity index, which reveals how influential a node is within a sediment cascade. Specific applications were implemented in GIS software (QGIS) as well as in software dedicated to data analysis and matrix calculations (R).

The application on a simple virtual catchment and then on a real catchment shows how the network spatial structure may lead (or not) to sediment mobilization and exportation, from the upper slopes to the outlet of watersheds. The behavior of sediment cascades appears space-dependent: the geometry of paths and the location of nodes have a direct influence on structural connectivity and thus on the ability of the sediment cascade to deliver sediments. As a consequence, some hotspots of geomorphic change can be identified within the catchment. The impact of an external force on the sediment cascade depends on the location of its action: the higher the connectivity of the node, the higher the impact on the cascade. Moreover, some simulations can be conducted to predict how local perturbations may have an impact on the overall cascade.

This issue (characterizing sediment connectivity at the catchment scale) is one of the main challenges in geomorphology and may help in understanding how a sediment wave develops and moves downstream. In detail, it may be possible to decipher better whether sediment pulses reveal the spatial network structure or external boundary conditions (e.g., climate change, anthropogenic pressure, tectonics). Such results are of importance for management issues, for instance in the discussion of the location of a retention basin for sediments in a context of severe erosion, or where sediment continuity should be restored to cope with sediment exhaustion in some rivers.

Competing interests. The authors declare that they have no conflict of interest.

Acknowledgements. The authors acknowledge with thanks the anonymous reviewers for their fruitful comments that significantly improved the quality of the manuscript.

References

Ali, G. V. and Roy, A. G.: Revisiting hydrologic sampling strategies for an accurate assessment of hydrologic connectivity in humid temperate systems, Geography Compass, 3, 350–374, 2009.

Baartman, J. E. M., Masselink, R., Keesstra, S. D., and Temme, A. J. A. M.: Linking landscape morphological complexity and sediment connectivity, Earth Surf. Proc. Land., 38, 1457–1471, 2013.

Belisle, M.: Measuring landscape connectivity: the challenge of behavioural landscape ecology, Ecology, 86, 1988–1995, 2005.

Benda, L. E. E., Andras, K., Miller, D., and Bigelow, P.: Confluence effects in rivers: interactions of basin scale, network geometry, and disturbance regimes, Water Resour. Res., 40, 1–15, 2004a.

Benda, L. E. E., Poff, N. L., Miller, D., Dunne, T., Reeves, G., Pess, G., and Pollock, M.: The network dynamics hypothesis: how channel networks structure riverine habitats, BioScience, 54, 413–427, 2004b.

Bennett, G.: Integrating biodiversity conservation and sustainable use. Lessons learned from ecological networks, World Conservation Union (IUCN), Gland, Switzerland, 2004.

Borselli, L., Cassi, P., and Torri, D.: Prolegomena to sediment and flow connectivity in the landscape: a GIS and field numerical assessment, Catena, 75, 268–277, 2008.

Bracken, L. J., Turnbull, J., Wainwright, J., and Bogaart, P.: Sediment connectivity: a framework for understanding sediment transfer at multiple scales, Earth Surf. Proc. Land., 40, 177–188, 2015.

Bravard, J. P.: Le temps et l'espace dans les systèmes fluviaux, deux dimensions spécifiques de l'approche géomorphologique, Ann. Géogr., 107, 3–15, 1998.

Brierley, G., Fryirs, K., and Jain, V.: Landscape connectivity: the geographic basis of geomorphic applications, Area, 38, 165–174, 2006.

Cavalli, M., Trevisani, S., Comiti, F., and Marchi, L.: Geomorphometric assessment of spatial sediment connectivity in small Alpine catchments, Geomorphology, 188, 31–41, 2013.

Chorley, R. J. and Kennedy, B. A.: Physical Geography: A Systems Approach, Prentice-Hall International, London, UK, 370 pp., 1971.

Cole, J. P. and King, C. A. M.: Quantitative Geography: Techniques and theories in geography, Wiley, New York, USA, 471 pp., 1968.

Cossart, É.: La recrudescence de l'activité torrentielle dans un bassin versant en cours de déglaciation au cours du XXe siècle, Géomorphologie, 3, 225–240, 2004.

Cossart, É.: L'(in)efficacité géomorphologique des cascades sédimentaires en question: les apports d'une analyse réseau, Cybergeo: European Journal of Geography [Online], document 778, doi:10.4000/cybergeo.27625, 2016.

Cossart, É. and Fort, M.: Sediment release and storage in early deglaciated areas: Towards an application of the exhaustion model from the case of Massif des Écrins French Alps since the Little Ice Age, Norsk Geogr. Tidsskr., 62, 115–131, 2008.

Csardi, G. and Nepusz, T.: The igraph software package for complex network research, InterJournal. Complex Systems, 1695, 1–9, 2006.

Czuba, J. A. and Foufoula-Georgiou, E.: A network-based framework for identifying potential synchronizations and amplifications of sediment delivery in river basins, Water Resour. Res., 50, 3826–3851, 2014.

Czuba, J. A. and Foufoula-Georgiou, E.: Dynamic connectivity in a fluvial network for identifying hotspots of geomorphic change, Water Resour. Res., 51, 1401–1421, 2015.

Delahaye, D., Guermond, Y., and Langlois, P.: Spatial interaction in the run-off process, Cybergeo: European Journal of Geography [Online], document 213, doi:10.4000/cybergeo.3795, 2001.

Douvinet, J., Delahaye, D., and Langlois, P.: Modélisation de la dynamique potentielle d'un bassin versant et mesure de son efficacité structurelle, Cybergeo: European Journal of Geography [Online], document 412, doi:10.4000/cybergeo.16103, 2008.

Faulkner, H.: Connectivity as a crucial determinant of badland geomorphology and evolution, Geomorphology, 100, 91–103, 2008.

Ferguson, R. I.: Channel forms and channel changes, in: British Rivers, edited by: Lewin, J., Allen and Unwin, London, UK, 90–125, 1981.

Foerster, S., Wilczok, C., Brosinsky, A., and Segl, K.: Assessment of sediment connectivity from vegetation cover and topography using remotely sensed data in a dryland catchment in the Spanish Pyrenees, J. Soils Sediments, 14, 1982–2000, 2014.

Freeman, L. C.: Centrality in social networks conceptual clarification, Soc. Networks, 1, 215–239, 1979.

Fryirs, K.: DisConnectivity in catchment sediment cascades: a fresh look at the sediment delivery problem, Earth Surf. Proc. Land., 38, 30–46, 2013.

Fryirs, K., Brierley, G. J., Preston, N. J., and Kasai, M.: The disconnectivity of catchment-scale sediment cascades, Catena, 70, 49–67, 2007.

Fryirs, K. A., River sensitivity: A lost foundation concept in fluvial geomorphology, Earth Surf. Proc. Land., 42, 55–70, 2017.

Gleyze, J. F.: Effets spatiaux et effets réseau dans l'évaluation d'indicateurs sur les nœuds d'un réseau d'infrastructure, Cybergeo: European Journal of Geography [Online], document 370, doi:10.4000/cybergeo.5532, 2008.

Gran, K. B. and Czuba, J. A.: Sediment pulse evolution and the role of network structure, Geomorphology, 277, 17–30, 2017.

Gumiere, S. J., Le Bissonnais, Y., Raclot, D., and Cheviron, B.: Vegetated filter effects on sedimentological connectivity of agricultural catchments in erosion modelling: a review, Earth Surf. Proc. Land., 36, 3–19, 2011.

Haggett, P. and Chorley, R. J.: Network analysis in geography, edited by: Arnold, E., Science, London, UK, 348 pp., 1969.

Harvey, A. M.: Coupling between hillslopes and channels in upland fluvial systems: implications for landscape sensitivity, illustrated from the Howgill Fells, northwest England, Catena, 42, 225–250, 2001.

Heckmann, T. and Schwanghart, W.: Geomorphic coupling and sediment connectivity in an alpine catchment – exploring sediment cascades using graph theory, Geomorphology, 182, 89–103, 2013.

Heckmann, T., Schwanghart, W., and Phillips, J. D.: Graph theory: Recent developments of its application in geomorphology, Geomorphology, 243, 130–146, 2015.

Hoffmann, T.: Sediment residence time and connectivity in non-equilibrium and transient geomorphic systems, Earth Sci. Rev., 150, 609–627, 2015.

Kimberly, A. W., Gardner, R. H., and Turner, M. G.: Landscape connectivity and population distributions in heterogeneous environments, Oikos, 78, 151–169, 1997.

López-Vicente, M., Nadal-Romero, E., and Cammeraat, E. L.: Hydrological connectivity does change over 70 years of abandonment and afforestation in the Spanish pyrenees, Land Degrad. Dev., doi:10.1002/ldr.2531, 2016.

Ludwig, J. A., Eager, R. W., Bastin, G. N., Chewings, V. H., and Liedloff, A. C.: A leakiness index for assessing landscape function using remote sensing, Landscape Ecol., 17, 157–171, 2002.

Marra, W. A., Kleinhans, M. G., and Addink, E. A.: Network concepts to describe channel importance and change in multichannel systems: test results for the Jamuna River, Bangladesh, Earth Surf. Proc. Land., 39, 766–778, 2014.

Masselink, R. J., Heckmann, T., Temme, A. J., Anders, N. S., Gooren, H., and Keesstra, S. D.: A network theory approach for a better understanding of overland flow connectivity, Hydrol. Process., 31, 207–220, 2017.

McGuinness, J. L., Harrold, L. L., and Edwards, W. M.: Relation of rainfall energy streamflow to sediment yield from small and large watersheds, J. Soil Water Conserv., 26, 233–235, 1971.

Newman, M. E. J.: Networks: An Introduction, Oxford University Press, Oxford, UK, 2010.

Pumain, D. and Saint-Julien, T.: Analyse spatiale: les localisations, Cursus, Armand Colin, Paris, France, 192 pp., 2010.

Renard, K., Foster, G. R., Weessies, G. A., McCool, D. K., and Yodler, D. C.: Predicting soil erosion by water: a guide to conservation planning with the Revised Universal Soil Loss Equation (RUSLE), U.S. Department of Agriculture, Washington, USA, Agriculture Handbook, 703, 1–384, 1997.

Reulier, R., Delahaye, D., Caillault, S., Viel, V., Douvinet, J., and Bensaid, A.: Mesurer l'impact des entités linéaires paysagères sur les dynamiques spatiales du ruissellement: une approche par simulation multi-agents. Cybergéo. European Journal of Geography [Online], Systèmes, Modélisation, Géostatistiques, document 788, doi:10.4000/cybergeo.27768, 2016.

Rodrigue, J. P.: Geography of transport system, Routledge, New York, USA, 441 pp., 2017.

Schumm, S.: River variability and complexity, Cambridge University Press, Cambridge, UK, 219 pp., 2005.

Shreve, R. L.: Variation of mainstream length with basin area in river networks, Water Resour. Res., 10, 1167–1177, 1974.

Strahler, A. N.: Quantitative analysis of watershed geomorphology, EOS, Trans. Am. Geophys. Union, 38, 913–920, 1957.

Tischendorf, L. and Fahrig, L.: How should we measure landscape connectivity?, Landscape Ecol., 15, 633–641, 2000.

Turnbull, L., Wainwright, J., and Brazier, R. E.: A conceptual framework for understanding semi-arid land degradation: ecohydrological interactions across multiple-space and time scales, Ecohydrology, 1, 23–34, 2008.

Uezu, A., Metzger, J. P., and Vielliard, J. M. E.: Effects of structural and functional connectivity and patch size on the abundance of seven Atlantic forest bird species, Biol. Conserv., 123, 507–519, 2005.

Viel, V.: Analyse spatiale et temporelle des transferts sédimentaires dans les hydrosystèmes normands. Exemple du bassin versant de la Seulles, PhD Thesis, Université de Caen, France, 382 pp., 2012.

Wainwright, J., Turnbull, L., Ibrahim, T. G., Lexartza-Artza, I., Thornton, S. F., and Brazier, R. E.: Linking environmental regimes, space and time: interpretations of structural and functional connectivity, Geomorphology, 126, 387–404, 2011.

Walker, R. G.: Perspective, facies modelling and sequence stratigraphy, J. Sediment. Petrol., 60, 777–786, 1990.

Walling, D. E.: The sediment delivery problem, Hydrology, 69, 209–237, 1983.

Wischmeier, W. H. and Smith, D. D.: Predicting Rainfall Erosion Losses – A Guide to Conservation Planning, U.S. Department of Agriculture, Washington, USA, Agriculture Handbook, 537, 1–58, 1978.

With, K. A. and King, A. W.: The use and misuse of neutral landscape models in ecology, Oikos, 79, 219–229, 1997.

With, K. A., Gardner, R. H., and Turner, M. G.: Landscape connectivity and population distributions in heterogeneous environments, Oikos, 78, 151–169, 1997.

6

Glacial dynamics in pre-Alpine narrow valleys during the Last Glacial Maximum inferred by lowland fluvial records (northeast Italy)

Sandro Rossato[1], Anna Carraro[2], Giovanni Monegato[2], Paolo Mozzi[1], and Fabio Tateo[2]

[1]Department of Geosciences, University of Padova, Padova, 35131, Italy

[2]Institute of Geosciences and Earth Resources (IGG) – National Research Council (CNR), Padova, 35131, Italy

Correspondence: Sandro Rossato (sandro.rossato@unipd.it)

Abstract. During the Last Glacial Maximum (LGM), most of the major glaciated basins of the European Southern Alps had piedmont lobes with large outwash plains; only a few glaciers remained within the valley. Piedmont glaciers have left well-preserved terminal moraines, which allow for investigations to be carried out and inferences to be made regarding their evolution and chronology. Valley glaciers' remnants, on the contrary, are often scantly preserved, and changes can only be detected through correlations with glaciofluvial deposits in downstream alluvial basins. The Brenta glacial system's dynamics in the glacier's terminal tract have been inferred through a wide range of sediment analysis techniques on an alluvial stratigraphic record of the Brenta megafan (northeast Italy), and via the mapping of in-valley glacial/glaciofluvial remnants. Glaciers flowing across narrow gorges could possibly be slowed/blocked by such morphology, and glacial/sediment fluxes may then be diverted to lateral valleys. Moreover, narrow valleys may induce glaciers to bulge and form icefalls at their front, preventing the formation of terminal moraines. The Brenta Glacier was probably slowed/blocked by the narrow Valsugana Gorge downstream of Primolano and was effectively diverted eastwards across a wind gap (Canal La Menor Valley), joining the Cismon/Piave glaciers near Rocca and ending $\sim$ 2 km downstream. The Cismon and Piave catchments started to contribute to the Brenta system just after 27 ka cal BP until at least $\sim$ 19.5 ka cal BP. After the glaciers collapsed, the Piave River once again flowed into its main valley, whilst the Cismon continued to merge with the Brenta.

This investigation shows that glacial catchments may vary significantly over time during a single glaciation in rugged Alpine terrains. Sand petrography and the chemical/mineralogical composition of sediments are powerful proxies for tracing such variations, as they propagate through the glacial and glaciofluvial systems and can be recognized in the alluvial stratigraphic record far downstream from the glacier front.

1 Introduction

Mountain glaciers are complex systems, and the evolution of these systems affects both their mountain basins and the alluvial plains that receive the glaciofluvial water and sediment flux (e.g. Russell et al., 2006). Few minor valley glaciers are currently present in the highest areas of the European Alps (Evans, 2006), whereas during Pleistocene glaciations large ice streams flowed along most of the Alpine valleys, leading to deep landscape modifications (e.g. Koppes and Montgomery, 2009; Preusser et al., 2010; Wirsig et al., 2016). Most of the major glaciated basins had piedmont lobes with large outwash plains, where stratigraphic reconstruction and available chronology has allowed researchers to infer the evolution of the glacial system (Monegato et al., 2007; Preusser et al., 2011; Ravazzi et al., 2012; Fontana et al., 2014). However, there are catchments where glaciers remained within the valley, and their evolution could only be detected through correlations with the related glaciofluvial deposition in the piedmont area (van Husen and Reitner, 2011; Rossato et al., 2013).

The Last Glacial Maximum (LGM) was the last cold extreme on Earth, and it provided the most well-preserved sedimentary and geomorphic record among all Pleistocene glaciations (Bowen, 2009; Clark et al., 2009; Hughes and Gibbard, 2015). In Europe, the Alps and their foreland constitute a key region for LGM studies, as evidence of this event is widespread and chronologically well framed between 30 and 17.5 ka cal BP (e.g. Ivy-Ochs et al., 2008; Ivy-Ochs, 2015; Rossato and Mozzi, 2016; Monegato et al., 2017). Moreover, the Alps constitute an effective barrier for wind circulation (Florineth and Schlüchter, 2000; Luetscher et al., 2015) and offer the unique opportunity to validate climate models (e.g. Smiatek et al., 2009; Torma et al., 2015) and study how glacial processes affect mountain chains (e.g. Norton et al., 2010). LGM glaciers filled the valleys fed by the major accumulation zones located in the axial and highest part of the Alps (e.g. van Husen, 1987; Kelly et al., 2004; Wirsig et al., 2016). In the southern Alpine sector, topography greatly influenced glacier evolution (Wirsig et al., 2016) due to different wind circulation regimes (Luetscher et al., 2015). Alpine glaciers and their sedimentary outputs proved to react differently to climatic signals, in both the north–south (e.g. Luetscher et al., 2015) and west–east directions (Becker et al., 2016, Monegato et al., 2017, Seguinot et al., 2017). Differences are also found when dealing with neighbouring glacial systems, due to their size, catchment topography and possible glaciers confluences and transfluences (e.g. Kelly et al., 2004; Monegato et al., 2007; Rossato et al., 2013).

The present availability of a wide range of proxies, coupled with the increased accuracy of geochronological methods (Brauer et al., 2014), allows for a better assessment of glaciers' behaviour during the LGM compared to earlier studies undertaken in the Alps, which were mostly based on landform and deposit characterization (e.g. Penck and Brückner, 1909; Sacco, 1937; Venzo et al., 1977; Schlüchter, 1986; van Husen, 1987). Furthermore, the analysis of loess and palaeosols has been coupled with biological proxies, such as pollen, chironomids, charcoal and many more (e.g. Heiri et al., 2014; Samartin et al., 2016). Petrographic/mineralogical studies of sediments have also supported palaeoenvironmental reconstructions and allowed researchers to infer variations in the sedimentary systems (Garzanti et al., 2011).

In this paper we investigate the interaction between the LGM glaciers in the middle Brenta Valley (also known as "Valsugana", in the Italian Fore-Alps) and the related glaciofluvial system in the piedmont plain (Fig. 1). During the LGM, the Brenta glacial system received contributions from the major Adige (Etsch) Glacier (Trevisan, 1939; Tessari, 1973; Avanzini et al., 2010; Rossato et al., 2013); the Brenta system also fed a fluvial megafan, which is one of the most prominent sedimentary alluvial systems on the southern side of the Alpine chain (Mozzi, 2005; Fontana et al., 2014). The main aim of this paper is to explore and define the possibility of correlating glacial advances and transfluences in mountain areas with sedimentation pulses in the lowlands using integrated geomorphic, sedimentary, petrographic, mineralogical, geochemical and geochronological evidence.

2 Study site

The Venetian Fore-Alps are geologically detached from the Dolomites by the Valsugana fault (Castellarin et al., 2006), and are characterized by a belt of carbonate plateaus deeply carved by the lower reaches of the Astico, Brenta, Cismon and Piave valleys (Fig. 1). Except for the Astico River, which has a catchment that is restricted to the western part of the Venetian Fore-Alps, the other rivers have their upper catchments in the Dolomites. The Dolomites region includes the Permo-Triassic sedimentary successions (sandstones, dolostones, limestones and volcanic rocks), the low-grade metamorphic basement and the Permian porphyries (e.g. Bartolomei et al., 1969; Avanzini et al., 2010; Fig. 1). The Brenta and Cismon catchments also include the Permian plutonic rocks of Cima d'Asta (Fig. 1). In the footwall of the Valsugana fault the Jurassic–Tertiary sedimentary succession crops out, including different types of limestones, turbiditic sequences and the terrigenous units of the Southalpine foredeep (e.g. Massari et al., 1986; Barbieri and Grandesso, 2007; Stefani et al., 2007). In particular, in the Cismon catchment, all around Lamon (Fig. 1), micritic limestones crop out extensively (Tessari, 1973). Here, a remarkable fluvial aggradation took place during/after deglaciation, later followed by river incision and the formation of a terrace staircase. The highest and largest of these terraces is located about 600 m a.s.l., 200 m above the Cismon Valley bottom. The deposition of such a large amount of sediment was probably due to the damming of the Cismon Valley south of Lamon, which possibly occurred due to a large dead-ice mass or a landslide event, although no direct evidence of this has been found (Tessari, 1973).

The plateaus characterizing the Venetian Fore-Alps (Castiglioni et al., 1988) have mean elevations ranging from 900 to 2000 m a.s.l. and hosted small cirque or plateau glaciers during the last glaciation (Trevisan, 1939; Carraro and Sauro, 1979; Baratto et al., 2003; Barbieri and Grandesso, 2007). The major Astico (Cucato, 2001; Rossato et al., 2013) and Piave (Venzo, 1977; Carton et al., 2009) valleys were carved by ice streams that extended to the lower valley reaches and left well-preserved terminal moraines (Fig. 1). Despite the large megafan ascribed to the LGM (Mozzi, 2005; Fontana et al., 2008; Rossato and Mozzi, 2016), the Brenta Glacier is the only major system in the southeastern Alps that left no clear evidence of terminal moraines (Castiglioni, 2004). The reason that the Brenta system did not follow this evolutionary path remains unresolved, although some early authors have already broached this challenging topic (e.g.

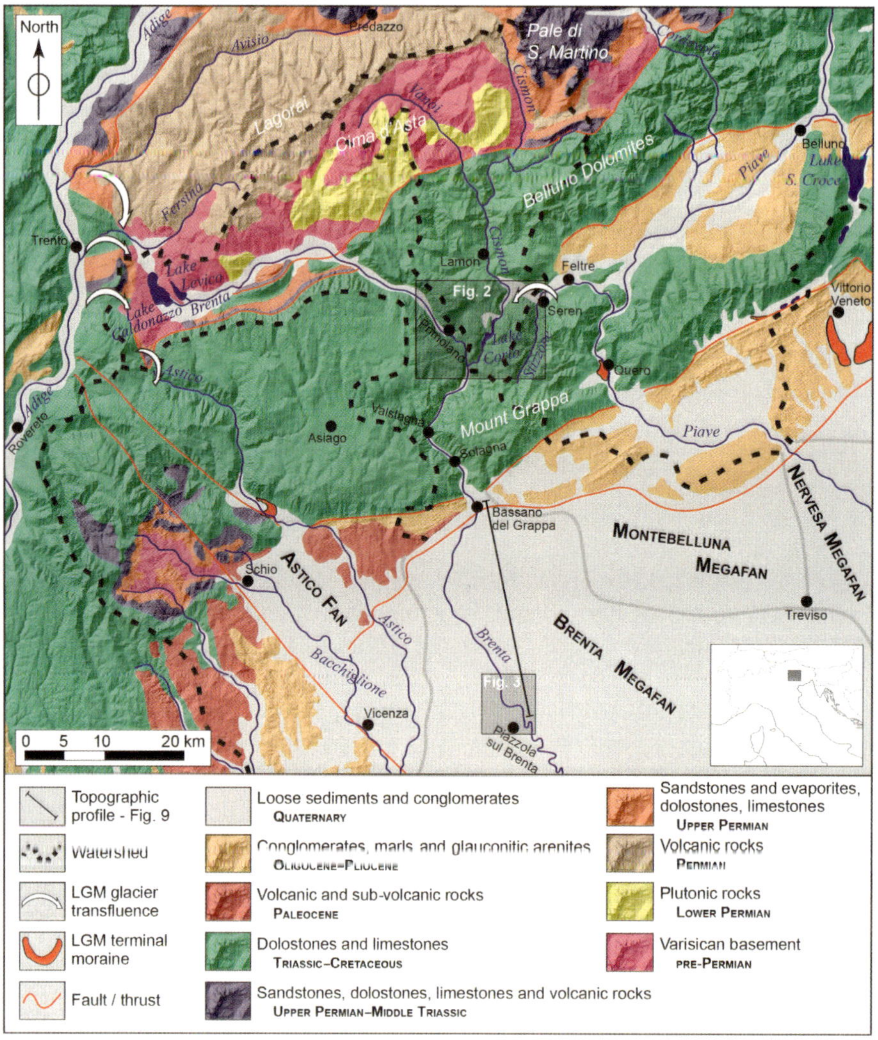

Figure 1. A geological sketch of the study area. The map is based on the "Structural model of Italy" (Bigi et al., 1990) and local geological maps (Bartolomei et al., 1969; Dal Piaz et al., 1946; Barbieri and Grandesso, 2007; Avanzini et al., 2010). It also overlies a Shuttle Radar Topography Mission (SRTM) derived digital elevation model (DTM; 30 m cell size) (source: http://viewfinderpanoramas.org/, last access: 12 April 2018).

Taramelli, 1882; Penck and Brückner, 1909; Castiglioni, 1940). Taramelli (1882) and Castiglioni (1940) placed the Brenta Glacier front near the town of Valstagna, about 10.5 km south of the confluence between the Brenta and Cismon valleys (Fig. 1), while Penck and Brückner (1909) located it about 6 km further to the south, near the town of Solagna. Due to the lack of direct geomorphic or sedimentary evidence of the glacier's front, both interpretations were based on speculation that the glacier gradually thinned downstream of the last preserved lateral moraine at Enego (elevation 760–790 m a.s.l.). The terminal tract of the Cismon Valley, called Corlo Valley, was closed by a dam in 1953 and filled by the artificial Corlo Lake (Fig. 1), so no further survey is presently possible.

The Brenta fluvial megafan is part of the foreland basin of the uplifting eastern Southalpine chain (e.g. Castellarin et al., 2006; Stefani et al., 2007). This megafan extends from the Brenta Valley mouth to the present Venetian coastline. It can be roughly separated into two sectors: (i) piedmont (∼ 20 km from the apex), made of gravel and with an average topographic gradient of 3–5 and (ii) low plain, made of sandy fluvial ridges and silty-clayey floodplain, with a topographic gradient that decreases to less than 1 ‰ (Mozzi, 2005; Mozzi et al., 2013; Fontana et al., 2008).

3 Methods

This investigation was carried out via a field survey, coring, petrographic and mineralogical analyses, and remote sens-

Table 1. Conventional, calibrated and median probability ^{14}C ages obtained from samples collected from RB1 and PM1 cores. The calibration was carried out with OxCal (version 4.2, Ramsey, 2009), based on the IntCal13 calibration curve (Reimer et al., 2013).

					Uncalibrated age [years BP]		Calibrated age (IntCal13 – 2σ) [years BP]		
Sample name	Lab. code	Material	Depth [m]	Thickness [cm]	Lab. age	Uncertainty (±)	Min	Max	Median prob.
PM1-1	UZ-6073	Peat	9.00	1.5	16 700	60	19 945	20 362	20 147
PM1-2	UZ-6074	Peat	13.45	1	19 500	75	23 167	23 754	23 495
RB1-1	UZ-6075	Peat	9.30	1.5	19 050	70	22 648	23 221	22 938
RB1-2	UZ-6076	Peat	21.80	1	21 610	90	25 715	26 058	25 888
RB1-3	UZ-6077	Peat	27.50	1	22 660	90	26 634	27 296	27 013

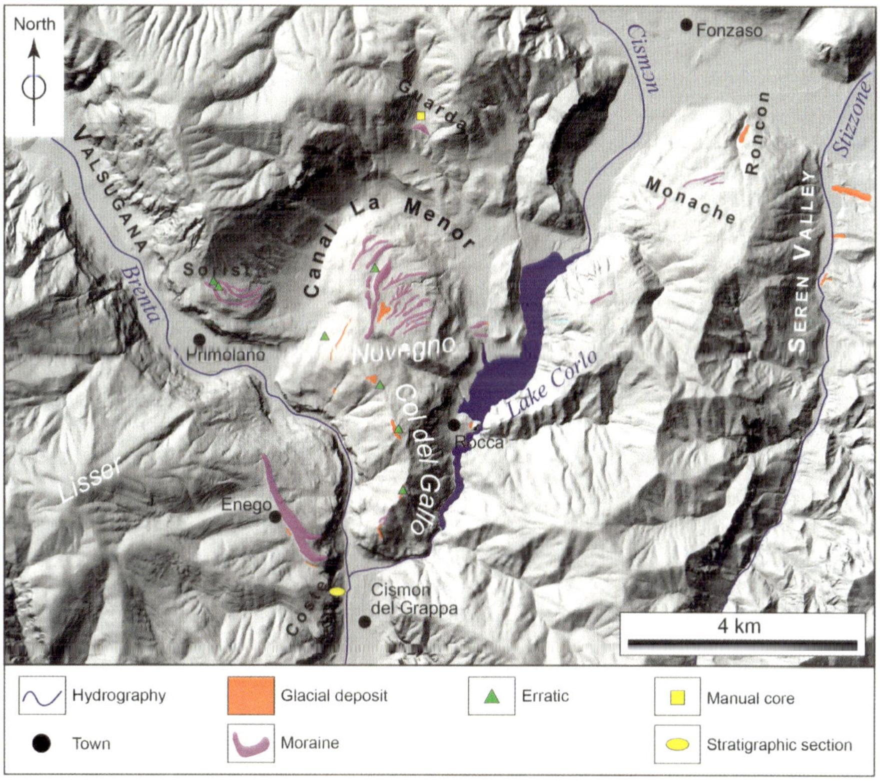

Figure 2. Outcrops map of the middle Valsugana sector derived from field surveys and remote sensing data. Polygons/symbols overlie a 5 m cell DTM (modified from data distributed by Regione Veneto; source: http://idt.regione.veneto.it/app/metacatalog/, last access: 13 April 2018).

ing. All methods and results are presented independently, whilst their meaning, implications and interconnections are discussed thereafter.

3.1 Field survey

Extensive mapping of the Quaternary sediments cropping out in the middle Valsugana and at the junctions with the Canal La Menor and lower Cismon (Corlo) valleys, was performed (Fig. 2). Sedimentary facies were observed and described in the outcrops to identify their original depositional setting; particular attention was devoted to the lithology of the clasts, which provided information on the provenance of the sediments. In this perspective, the geology of the eastern South Alps (Fig. 1) allowed for a clear distinction of each drainage basin with different petrographic signatures. Porphyries, granites and metamorphic rocks (i.e. Permian volcanic rocks, Lower Permian plutonic rocks and pre-Permian Varisican basement in Fig. 1, respectively) are indicators of the Brenta and Cismon valleys and, thanks to their high resistance to weathering, can help identify glacial deposits upon a carbonate bedrock.

Relevant landforms were accurately surveyed, with particular attention being paid to the LGM moraines, which are generally well preserved and quite easily identifiable in these areas (as little anthropogenic activity has taken place).

3.2 Cores

Two 30 m-long cores were drilled near the city of Piazzola sul Brenta, in the upper part of the Brenta megafan. The cores were drilled about 5 km apart, at a topographic elevation of approximately 30 m a.s.l. (Fig. 3) and were part of a pilot study on groundwater geochemistry (Carraro et al., 2013, 2015). They were described based on the lithofacies of the sediments and sampled for sand petrography and mineralogical analyses. Five organic samples were collected from the inner part of the cores to minimize contamination and were dated using the ^{14}C AMS method at the radiocarbon laboratory of the University of Zurich (Switzerland) (Table 1). OxCal software (version 4.2, Bronk Ramsey, 2009; IntCal13 calibration curve, Reimer et al., 2013) was used to calibrate laboratory ages.

A single hand core was drilled in the Guarda area, on the northern flank of the Canal La Menor Valley, uphill from an elongated ridge that has been identified as a LGM lateral moraine (see Section 4 and Fig. 2 for location), at about 660 m a.s.l. This borehole was drilled with a hand auger (Edelman combination type, Ejikelkamp™), which allowed for a semi-disturbed sequence of fine sediments to be obtained (with the limitation that the maximum grain size of the sediments that can be sampled is coarse sand to fine gravel).

3.3 Sand petrography

Ten samples were collected from the two long cores for provenance analysis. The entire sand fraction was isolated and impregnated in an epoxy resin according the methodology from Gazzi et al. (1973), in order to obtain samples for thin-section analysis comparable with river endmembers as per Garzanti et al. (2006) and Monegato et al. (2010). These samples were subsequently stained with Alizarin Red solution for the determination of the carbonate phases. Following the Gazzi–Dickinson method, 400 points per thin section were counted using a 0.5 mm grid spacing (Ingersoll et al., 1984). Data and parameters are reported in Table 2 and plotted in ternary diagrams.

Principal component analysis (PCA) was performed using CoDaPack (Comas and Henestrosa, 2011), and a compositional biplot (Aitchinson and Greenacre, 2002) was utilized to better visualize the dataset and the variables. The ternary diagrams with the most representative variables are presented in order to discuss the data structure. Centring (e.g. Buccianti et al., 1999; von Eynatten et al., 2002) was used to avoiding disturbances related to the dominance of one or two components. In the statistical analysis, petrofacies of modern rivers (Garzanti et al., 2006; Monegato et al., 2010) were included for comparison in order to assess the sediment provenance in the different stratigraphic intervals.

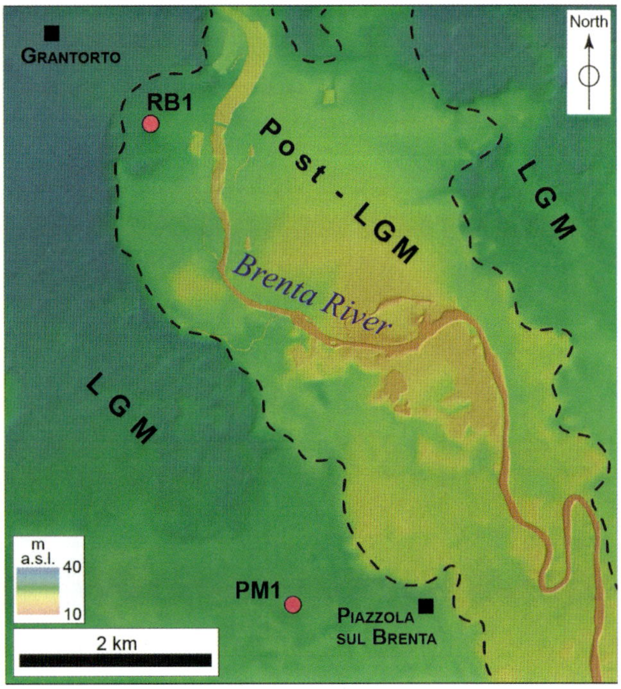

Figure 3. Location of the PM1 and RB1 cores (purple circles). The background is a 5 m cell DTM (modified from data provided by Regione Veneto, 2011), stretched to highlight elevation changes. Scarps bounding the post-glacial incision of the Brenta megafan are denoted by black dashed lines.

In addition, the statistical technique from Vezzoli and Garzanti (2009) was also adopted to assess the contribution of the major drainage systems of the area related to the Brenta, Cismon and Piave rivers (Fig. 1) as endmembers. The parameters were then compared to evaluate possible matches between these catchments, and results with $R^2 > 0.7$ were considered robust. Except for one sample (RB1-8), all the samples show a good interpretation confidence (Table 2).

3.4 X-ray diffraction (XRD)

In order to minimize preferred orientation, samples were pulverized and backloaded. A Philips X'Pert Pro diffractometer (Cu tube and secondary monochromator) was utilized. The mineral constituents were quantified using High-Score plus software following Rietveld refinement (Young, 1993); zincite (ZnO) was added as an internal standard (sample : standard = 10 : 1 weight ratio). Rietveld refinement also allowed for the estimation of the amorphous components (labelled Am.XRD, in Table S1).

To check the reliability of the mineral quantification, the chemical composition of samples was calculated us-

Table 2. Detrital modes of the sand fraction collected on RB1 and PM1 cores. List of acronyms: Q represents quartz; F represents feldspars; Lvf represents felsic volcanic and sub-volcanic lithic fragments; Lvi represents intermediate and mafic lithic fragments; Lvp represents plutonic lithic fragments; Lcc represents limestone grains; Lcd represents dolostone grains; Lp represents shale and siltstone lithic fragments; Lch represents chert grains; Lms represents low-grade metamorphic lithic fragments; and Lmi represents medium-grade metamorphic lithic fragments.

Sample	Depth [m]	Q	F	Lvf	Lvi	Lvp	Lcc	Lcd	Lp	Lch	Lms	Lmi	tot	Contribution of endmembers (R^2)
PM1-12	12.1	30.5	4.7	6		0.5	14.4	20.8	3.5	0.2	19.4		100	50 % Brenta, 20 % Cismon, 30 % Piave (0.76)
PM1-17	16.6	18.2	10.6	9.1	0.3	2.6	23.9	14.8	1.4	1.1	17.9	0.3	100	30 % Brenta, 20 % Cismon, 50 % Piave (0.80)
PM1-19	19.4	19.5	9.9	13.1		2.1	15.5	19.3	0.6	1.6	18.4		100	50 % Brenta, 20 % Cismon, 30 % Piave (0.93)
RB1-8	7.9	10.3	2.1	8		0.3	47.3	15.3	1.6	10.1	5.2		100	All combinations (<0.7)
RB1-16	15.7	25.1	5.5	7.5			18.9	22.5	4.6	0.5	15.5		100	40 % Brenta, 10 % Cismon, 50 % Piave (0.79)
RB1-21	21.4	26.8	10.4	9.7		1.7	5.6	18.9	1.6	0.7	24.4	0.2	100	50 % Brenta, 30 % Cismon, 20 % Piave (0.88)
RB1-24	24.45	19.3	11.1	11.1		1.1	12	21.6	2.2	1.8	19.7	0.2	100	50 % Brenta, 20 % Cismon, 30 % Piave (0.93)
RB1-27	27.3	25.1	8.3	18.8	0.3		3.3	24.3	2.8	0.5	16		100	80 % Brenta, 10 % Cismon, 10 % Piave (0.94)
RB1-29	28.95	28.3	20.2	19.2		1.5	2.5	6.6	1.1	1	19.2		100	100 % Brenta (0.83)
RB1-30	29.95	27.9	18.5	21.1		2.6	7.8	5.3	0.3	1	14.1		100	100 % Brenta (0.83)

ing stoichiometric mineral composition (for quartz, calcite, dolomite, feldspars, illite and kaolinite); a Fe-rich, Mg-poor dolomite was also included in the calculation (according to the XRD evidence of non-stoichiometric dolomite in almost all the samples), in addition to two Al-chlorites (Fe- or Mg-enriched) and organic matter. The comparison between the calculated and the measured chemical composition was considered satisfactory (total least square within 10 for 29 samples over 32).

3.5 Chemical composition

The major elements, some trace elements (including N) and the organic carbon (C_{org}) were determined as reported by Carraro et al. (2015). Samples were digested with concentrated $HNO_3 + HCl + HF$ at 140 °C for 90 min, then rinsed with H_3BO_3 and heated at 140 °C for 60 min. Afterwards, the solutions were analysed by inductively coupled plasma (ICP-OES) and atomic absorption spectroscopy (AAS). The loss on ignition (LOI) was measured after heating the sample to 860 °C for 20 min and to 980 °C for 2 h.

An automatic elemental analyzer was used to obtain C_{org} and total N for the carbonate-free residues. The measurements were repeated at least twice using Ag sample containers.

All of the mineralogical and chemical data were used for hierarchical cluster analysis (carried out using SPSS 6.0), in order to point out sample similarities and/or anomalies. Five dark sediments were grouped in the same cluster; this group was characterized by a percentage of organic matter higher than 12 %. This threshold value was used for peat identification (see Carraro et al., 2015 for analytical details). Macroscopic features (black colour and fine-grained) were used for peat identification when chemical data were not available.

3.6 Remote sensing

Publicly available aerial photographs and satellite images (Google Earth and Bing databases) were processed and analysed for the mapping of relict landforms. Panoramic and detail photographs were acquired during field surveys and subsequently used in conjunction with the other abovementioned images.

A digital elevation model (DEM) provided by the Regione Veneto, based on topographic data derived from 1 : 5000 topographic maps (5 m cell size and an XY accuracy of 2 m – http://idt.regione.veneto.it/app/metacatalog/; last access: 13 April 2018) was used. It covered the entire study area, assuring the uniform accuracy of the investigation. The various DEM tiles were assembled using ArcGIS (10.4.1 version) to allow for better processing and a more uniform visualization. All topographic profiles presented here are based on this DEM.

4 Results

4.1 Field survey

The field survey allowed for the identification of some key stratigraphic sections and outcrops that were regarded as being representative of specific sectors of the study area (see Fig. 2 for locations). Their descriptions are presented here in geographical order, from east to west and from north to south.

4.1.1 The Seren Valley and the Roncon outcrops

On the bottom of the Seren Valley (390 m a.s.l.) widespread outcrops of matrix-supported diamicton can be found (Fig. 2). These deposits are overconsolidated and characterized by centimetric clasts with sparse boulders, sub-rounded to angular in shape and embedded in an abundant silty matrix. In the thalweg of the Stizzone Creek, which runs along the valley, silty matrix-supported diamicton crops out, with striated clasts. Different sedimentary, volcanic and metamorphic lithologies are present in the deposit, all of which show no evident weathering. The Seren Valley deposits can be ascribed to melt-out tills, whilst those located along the thalweg are lodgment till; both of these tills belong to the Piave Glacier. In the nearby Roncon section (Fig. 2), a deposit similar to these melt-out tills was observed in a few decametric outcrops at an elevation of about 450 m a.s.l.

4.1.2 The Monache ridges

On the Monache Plateau a series of ridges, which is more than 0.5 km long, is present at an elevation ranging from 750 to 650 m a.s.l. (Fig. 2). These ridges are elongated and asymmetric in shape. This is due to the fact that they present a very steep, high downhill flank (3 to 20 m), whilst the uphill side is lower and consists of a low-angle slope that rests on the bedrock. These landforms mainly consist of matrix-supported diamicton characterized by white/grey limestone clasts, which are up to 50 cm in size, with dark flint nodules (black and brown). All of the observed clasts are made of "Maiolica" limestone, a Lower Cretaceous formation that crops out extensively along the whole Grappa Massif (Dal Piaz et al., 1946; Carraro et al., 1989), where the ridges are located. Therefore, these ridges are interpreted as lateral moraines of the Piave Glacier that are fed by local limestones debris.

4.1.3 The Guarda Ridge and core

The Guarda Ridge is located on the northern side of the Canal La Menor Valley, in the Casere alla Guarda locality, at an elevation ranging from 660 to 700 m a.s.l. (Fig. 2). It is about 400 m long, up to 30 m high and has very steep flanks. This ridge consists of a polygenic diamicton with clasts of porphyry, limestone and siltstone, up to 50 cm in size, which are sub-rounded to angular in shape. Weathering of the clasts is minimal, and the material is not overconsolidated. The ridge is interpreted as a lateral moraine of the tongue of the Brenta Glacier that was flowing through the Canal La Menor Valley.

A 4.7 m long core (Guarda1 core) was manually drilled in the Guarda locality, at an elevation of 664 m a.s.l., on a wide valley bottom that is closed downstream by the Guarda moraine (Figs. 3 and 6). The cored sequence consists of clayey silt layers, which are grey in colour, with the sporadic presence of centimetric sandy intervals. A single 30 cm-thick gravelly layer is present between 3.2 and 3.5 m b.s. (metres below surface), with sparse angular gravel clasts (3–4 mm in size) immersed in a sandy matrix. The Guarda1 core shows a phase of low-energy sedimentation in a confined environment, with a single high-energy event. According to the lack of deep weathering of the deposit and the presence of a poorly developed soil on top of the succession, the whole sequence is likely to have deposited in a fluvio-lacustrine basin during the late stages of LGM and/or during the Lateglacial and early Holocene; during this period the Guarda moraine was blocking the run-off from this lateral valley to the main Canal La Menor Valley.

4.1.4 The Novegno and Col del Gallo deposits

The top of the Novegno Plateau is characterized by a low-gradient plateau at elevations ranging from about 600 to 720 m a.s.l. In this area, many elongated ridges are found, which locally merge into one another, and have a general southwest–northeast direction (Fig. 2). They are up to 2 km long, 50 m wide and 40 m high, usually with steep slopes. The observed outcrops show that these ridges have polygenic clast compositions, are constituted by matrix-supported diamicton with clasts of porphyry, limestone and siltstone, up to 40 cm in size, and are sub-rounded to angular in shape (Fig. 4b). Between the ridges, as well as on the northern and

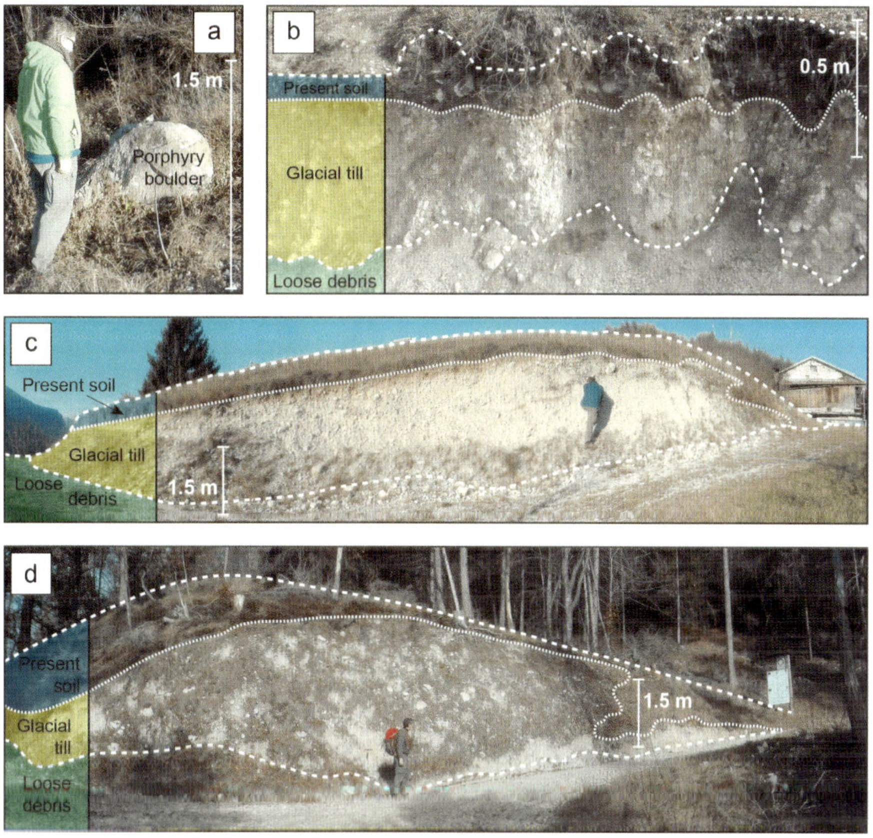

Figure 4. Photos taken in the middle sector of the Valsugana Valley. When present, stratigraphic layers are separated by dotted/dashed white lines. **(a)** A porphyry boulder, located on top of Col del Gallo mount; **(b)** a section of a lateral moraine of the Brenta Glacier, located on top of the Novegno mount; **(c)** a moraine, located on the southern side of the Col del Gallo; and **(d)** a moraine, located in the Sorist area.

southern sides of Col del Gallo (600–780 m a.s.l.), scattered patches of similar deposits can be found, which are characterized by a slightly higher matrix content (Figs. 2, 4c). The deposit is not overconsolidated and the weathering is minimal. Locally, some big boulders, up to 1.5 m in size, were found; these consisted mainly of porphyry (Fig. 4a) and granite. These sediments can be classified as glacial in origin, or, more specifically, as melt-out and flow tills.

On top of Col del Gallo (870 m a.s.l.), no proper outcrops are present, but centimetric clasts of volcanic and metamorphic phyllites were found in the surface soil.

All evidence shows that the Novegno Plateau is occupied by many lateral moraines formed by a glacier which was collecting material from an area located at least 25 km to the north, where the nearest outcrops of porphyry are found (Fig. 1). Similar glacial sediments are found on the northern and southern sides of Col del Gallo (Fig. 4c). It is likely that they relate to the same glacier. The deposits located on top of Col del Gallo are about 100 m higher than the other deposits and may possibly be related to a previous glacial advance.

4.1.5 The Sorist ridges

The Sorist mount area is a series of at least three ridges, which is more than 0.8 km long, at an elevation ranging from 650 to 760 m a.s.l. (Fig. 2). These ridges are elongated and present steep flanks, up to 10 m high. Locally the deposits forming these landforms crop out, showing a polygenic nature, with clasts of porphyry, limestone and siltstone, up to 60 cm in size; these deposits are also sub-rounded to angular in shape and embedded in an abundant matrix (Fig. 4d). Weathering of the clasts is minimal, and the material is not overconsolidated. Some large porphyry boulders, up to 1 m in size, were also found.

These ridges are interpreted as left moraines of the LGM Brenta Glacier, which flowed through the Valsugana Valley.

4.1.6 The Enego Ridge

Enego Ridge is 1 km long, more than 70 m high and is located on the western side of the Brenta Valley at an elevation of 760–790 m a.s.l., where Enego village is also situated (Fig. 2). This ridge is a lateral moraine of the Brenta Glacier, which has already been described by former scholars

(Trevisan, 1939; Dal Piaz et al., 1946), and is considered to have formed during the LGM. These deposits are the southernmost known Brenta glacial deposits. During this particular survey, no large boulders (> 1 m) were discovered; however, a second, smaller ridge was found at lower elevation (~ 650 m a.s.l.).

These ridges are interpreted as the right-lateral moraines of the Brenta Glacier during the LGM.

4.1.7 The Coste and Valstagna sections

In the Coste area, on the western side of the Valsugana Valley bottom, about 600 m south of the Enego Ridge, a stratigraphic section was observed at the excavation front of a gravel pit (Fig. 5), with the base at about 210 m a.s.l. (~ 10 m above the present valley bottom) (Fig. 2). The section is about 50 m high, 100 m long and is composed of two main units:

- The lower unit consists of a 10–15 m thick gravel body rich in sandy matrix, which presents cross-to-planar stratification. Clasts are centimetric in size (maximum diameter is about 30 cm), sub-rounded to sub-angular and mainly consist of carbonate rocks (limestones/marlstones), with abundant siltstones and some granites and porphyries. The lower boundary of this unit is not visible due to a covering of loose debris. The upper boundary with the "upper unit" shows the interfingering of the two units in the western part of the section, near the foot of the rock wall that constitutes the side of the valley.

- The upper unit begins from the top of the lower unit and stretches up to the topographic surface. It is a sedimentary body made of angular clasts, which are centimetric-to-decimetric in size, with a sandy matrix and scattered boulders (maximum diameter of about 1 m). This unit is crudely stratified, with bedding dipping 25–30° towards the valley axis. Clasts are lithologically homogeneous, and only consist of the local carbonate rocks of the overlying rock walls.

The lower unit is attributed to fluvial deposition by the Brenta River, interfingered and superimposed by scree deposits falling from the overlying rock walls.

A similar succession was found on the eastern side of the Brenta Valley, in front of the town of Valstagna, about 9 km south of Coste section. Here, a 15 m-high outcrop has been exposed by quarry activity and testifies to the occurrence of Brenta River fluvial deposits at elevations of ~ 165 to 170 m a.s.l., covered by 10 m of slope deposits.

4.1.8 The Rocca deposits

About 1 km north of the village of Rocca, two 20 m-high ridges are present; the top of these ridges is 310 m a.s.l.

Figure 5. The Coste section. It is possible to appreciate how inclined-bedding scree deposits overlie and interfinger with horizontal-bedding glaciofluvial sediments.

(Fig. 2). The northernmost ridge is made of a sub-rounded polygenic matrix-supported diamicton with clasts up to 20 cm, which are sub-rounded to angular in shape. These ridges are interpreted as frontal moraines of a glacier tongue flowing into the Canal La Menor Valley.

The town of Rocca is built upon carbonate bedrock, but a small patch of sediments crops out next to the local graveyard (Fig. 2). This deposit consists of a matrix-supported diamicton with sub-rounded clasts, up to 30 cm in size, which are sub-rounded to angular in shape. The clasts are minimally weathered and mainly carbonate, although porphyry clasts are also present. The abundant matrix is mainly constituted by sandy particles.

These deposits are interpreted as LGM till deposited by a glacier tongue flowing into the Corlo Valley, where the artificial lake is currently located.

4.2 Remote sensing

Whilst investigation of aerial and satellite images was not very profitable due to the dense vegetation coverage, the DEM provided very valuable data. The DEM allowed for the lateral tracing of those scattered landforms already recognized in the field, such as the Enego moraine, and for the mapping of other new landforms, based on morphological similarity. In particular, many moraines belonging to the Novegno group were mapped using this approach.

4.3 Alluvial plain cores

Cores are described here as lithofacies assemblages, from the bottom up. The depth of the various layers is referred to from the top of the core (indicated by the acronym "b.s." – below surface). Detailed logs are presented in Fig. 6. Data concerning the samples collected for radiocarbon dating are summarized in Table 1; the reader can refer to the specific core descriptions for details on the positions of the samples in the stratigraphy.

4.3.1 RB1 core

The RB1 borehole was drilled in the Brenta megafan near the town of Carturo, about 5 km north of Piazzola sul Brenta (see Fig. 3 for location and Fig. 6 for the stratigraphic log), at a topographic elevation of 30 m a.s.l.

The basal part of the core is made of an alternation of sand bodies and silty and clayey layers with varying sand content. The sand is normally well sorted, and is made up of grains that are a maximum of 0.5 mm wide. Six thin (6–18 cm thick) very fine-grained intervals characterized by high organic content are present. Three almost equally spaced layers (27.5, 21.8 and 9.3 m b.s.) were radiocarbon dated to 26.6–27.3, 25.7–26.1 and 22.6–23.2 ka cal BP, respectively.

At about 9 m b.s. a clear erosional surface is present, marked by an abrupt transition from clayey silt to a coarsening upward sequence from medium sand to coarse gravel, which is sub-angular to sub-rounded (largest clasts are about 2 cm). A single, 15 cm-thick silty layer interrupts this succession at 7.85 m b.s. The uppermost 2.5 m of the core is constituted of a fining upward sequence of sandy and silty layers. No clear evidence of the modern soil was found at the top.

The core is interpreted as sandy proximal overbank deposits, intercalated with more distal overbank fines indicative of low-energy floodplain deposition. Locally, some intra-ridge swampy areas developed, inducing the formation of peat layers where organic deposition prevailed over minerogenic contribution, as was common in the southeastern Alpine piedmont during the LGM (Miola et al., 2006; Rossato and Mozzi, 2016). The coarsening upward sequence is ascribable to sandy-gravelly channel sediments that eroded the older deposits at the end of the LGM and formed fluvial incised valleys through the whole Brenta megafan (Mozzi et al., 2013). Finally, the top of the core consists of lower energy channel deposits or proximal overbank sandy and silty fines.

4.3.2 PM1 core

The PM1 core was drilled about 1 km west of the town of Piazzola sul Brenta, at an elevation of 27 m a.s.l. (see Fig. 3 for location and Fig. 6 for stratigraphic log).

The entire core is quite homogeneous, and is composed of an alternation of silty layers, with a variable content of clay and sand, interbedded with fine sand intervals. The sand is normally well sorted, with a maximum grain size of 0.5 mm. Two layers of medium sand located at 19–20 and 16.3–17.2 m b.s. are the only exception to this rather monotonous sequence. Three thin (10 cm maximum) layers with a very high organic content are present in the uppermost 15 m of the succession, two of which have been radiocarbon dated to 23.2–23.7 (13.45 m b.s.) and 19.9–20.4 (9 m b.s.) ka cal BP, respectively. The uppermost 2 m of the core shows evidence of a well-developed soil, constituted of pedogenic horizons C, Bk and Bw, from the bottom up. The entire sequence is topped by a 0.5 m thick anthropogenic landfill made of gravels with a silty matrix.

As in core RB1, the whole sequence can be interpreted as sandy proximal overbank sediments intercalated with finer floodplain deposits. Locally, peat deposition took place when minerogenic contribution was very low, as was also found in RB1 core. The topmost soil has characteristic calcic horizons (Calcisol after FAO, 1998), which allow its correlation to the "caranto paleosol" that developed on top of the LGM deposits in the whole Venetian area (Mozzi et al., 2003; ARPAV, 2005; Donnici et al., 2011).

4.4 Sand petrography

The results of the petrographic analysis of the sand fraction (and the related statistics) of cores RB1 and PM1 are reported in Table 2 and Fig. 7. According to the distribution in the biplot diagram (Fig. 7a) samples clustered in different sectors allow two petrofacies (A and B) to be distinguished, with a single sample (RB1-8) clearly differentiated by a carbonaticlastic composition.

Samples related to petrofacies A belong to the RB1 core below the peat layer at 27.5 m b.s. Sandy grains are mainly quartz, feldspar and lithic fragments (> 30 % of which are volcanic), whereas the carbonate fragments are scarce and account for around 10 % of the sandy grains (Fig. 7b). No sediments belonging to this petrofacies have been found in the PM1 core.

Petrofacies B is clustered towards the centre of the biplot. Although the spectrum of lithic fragments contained in this petrofacies is similar to that of petrofacies A, petrofacies B contains more carbonate clasts (Fig. 7c), generally above 35 %, and sedimentary lithic fragments (Fig. 7d). Micritic limestone fragments are also particularly common. Petrofacies B in the RB1 core starts above the 27.5 m b.s. peat, beginning with the RB1-27 sample (Fig. 6). All samples from the PM1 core belong to petrofacies B.

The outlier sample was ascribed to petrofacies C. This sample displayed a composition dominated by carbonate (> 60 % of the total amount) and cherts (10 %), which are normally embedded in the micritic limestones as nodules (Barbieri and Grandesso, 2007). However, this sample also had the lowest amount of quartz (10 %) and all of the other parameters, which were below 10 %.

These results, compared with the present-day sands of the Brenta, Cismon and Piave rivers (Garzanti et al., 2006; Monegato et al., 2010), show that petrofacies A (RB1-29 and 30) is shifted towards the felsic volcanic component. No modern river shows a petrofacies such as this. Conversely, petrofacies B matches well with the modern Brenta River, which includes the Cismon catchment. Ternary diagrams show a

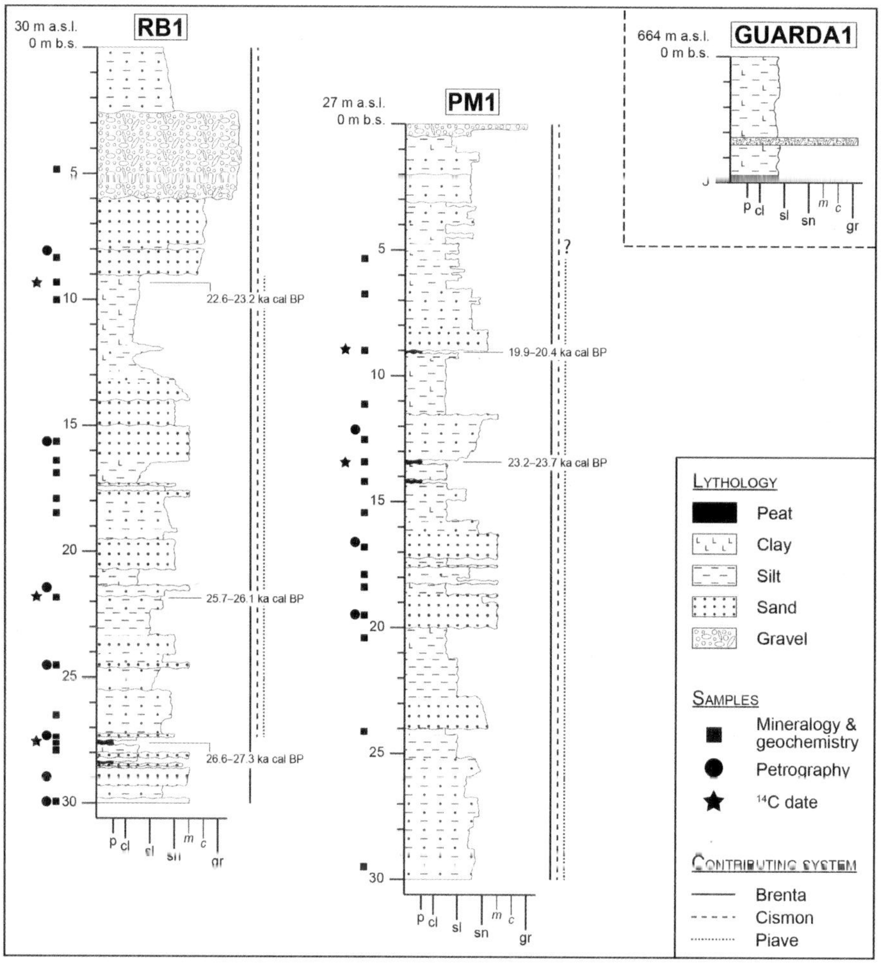

Figure 6. Stratigraphic logs of the RB1, PM1 and Guarda1 cores. Different catchments contributing to the sedimentation are marked by lateral solid/dashed lines. Samples are shown using different symbols according to the technique that was adopted.

slight shift of petrofacies B towards carbonate components (Fig. 7).

Using the mixing technique of Vezzoli and Garzanti (2009), petrofacies A was confirmed to be the most similar to the Brenta River sediments upstream of the junction with the Cismon River. Petrofacies B is quite similar to the present Brenta River, with an enrichment of carbonate rock fragments that suggests the contribution of a catchment rich in these components. This input could be from the Piave drainage basin; however, input from the Astico–Bacchiglione system in the lowlands can be discarded as the river was pushed to the west by the development of the Brenta megafan during the LGM (Rossato et al., 2013; Fontana et al., 2014), which also managed to dam Lake Fimon south of Vicenza (Monegato et al., 2011). Finally, petrofacies C shows high values of carbonate and chert fragments (Table 2) and no similarity to the endmembers. The mixing technique used rules out any possible scenario involving modern endmembers ($R^2 < 0.7$; Table 2). Petrofacies C belongs to the filling of a post-glacial incision of the Brenta megafan (Mozzi et al., 2013) when only the Brenta and Cismon catchments were contributing. Most of the carbonate clasts are micritic limestones that, coupled with the abundance of cherts, suggest an erosion event/phase in the lower Cismon (Corlo) Valley north of Rocca (Figs. 1, 3) or in the upper Cismon Valley close to Lamon, where these rocks are dominant (Tessari, 1973).

4.5 Mineralogy and geochemistry

Mineralogical analyses of the bulk sediments related to the petrofacies of the cores are reported in terms of main minerals, such as phyllosilicates (mainly micas and chlorites), dolomites (two different crystal chemical terms) and feldspars (plagioclase and k-feldspar), as shown in Fig. 8a. The complete dataset, including mineralogical and geochemical data, is reported in the Supplement (Table S1).

Basal sediments in the RB1 core (i.e. the three deepest samples and a peat sample) are characterized by an enrichment of feldspars and correspond to petrofacies A. The over-

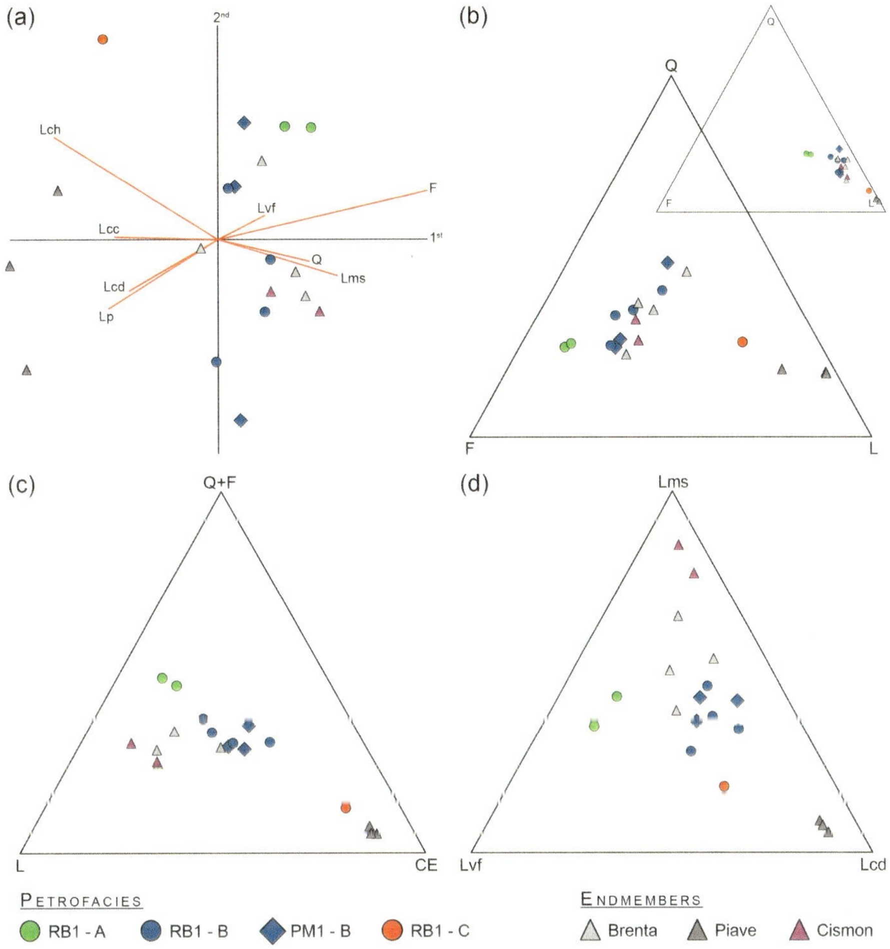

Figure 7. **(a)** Compositional biplot of the principal components. **(b)** The original (small) and centred (large) QFL diagrams. **(c)** The Q+F, L (non-carbonate lithic) and CE (carbonate lithic and cherts), and **(d)** Lms, Lvf and Lcd ternary diagrams showing the results of the sand petrography analyses performed on the RB1 and PM1 cores. Endmembers following Garzanti et al. (2006) and Monegato et al. (2010). See Table 2 for component acronyms.

head sediments, below the main erosional surface along the core, are characterized by an abundance of phyllosilicates, whereas dolomites are depleted; the stratigraphic position of these sediments corresponds to petrofacies B. The two topmost samples (above the erosional surface) are distinctive due to their higher dolomite content, and correspond to petrofacies C.

Samples from the PM1 core mainly plot as those recognized for petrofacies B in the RB1 core, without significant stratigraphic differences along the PM1 core. Whilst calcite is one of the main minerals observed in the cores, it is not considered to be a discriminating variable because it is strongly depleted or even absent in peat sediments (i.e. > 12 wt % organic matter); therefore, this mineral is deemed to be mainly influenced by the depositional environment and poorly indicative of the sediment provenance.

The distinctive mineralogical features of samples corresponding to petrofacies A, B and C are also evidenced by sediment chemistry, as reported in Fig. 8b in terms of MgO, Na_2O and Fe_2O_3, because of their affinity for dolomite, plagioclases and fine-grained minerals (clay minerals, oxides and hydroxides). In this case, petrofacies A, B and C are even more clearly discriminated for both cores.

A more detailed evaluation of the bulk mineralogical composition along the RB1 core highlights the fact that both samples corresponding to petrofacies A, B and C are characterized by a peculiar feldspar/quartz ratio (Fig. S1 in the Supplement), and that the transition between each group is rather sharp. Conversely, the feldspar/quartz ratio throughout the PM1 core does not show abrupt variations.

5 Discussion

Data gathered in the mountainous areas of this study and in the piedmont megafan are discussed here in order to reconstruct the evolution of the LGM Brenta Glacier and the tributary glacial systems.

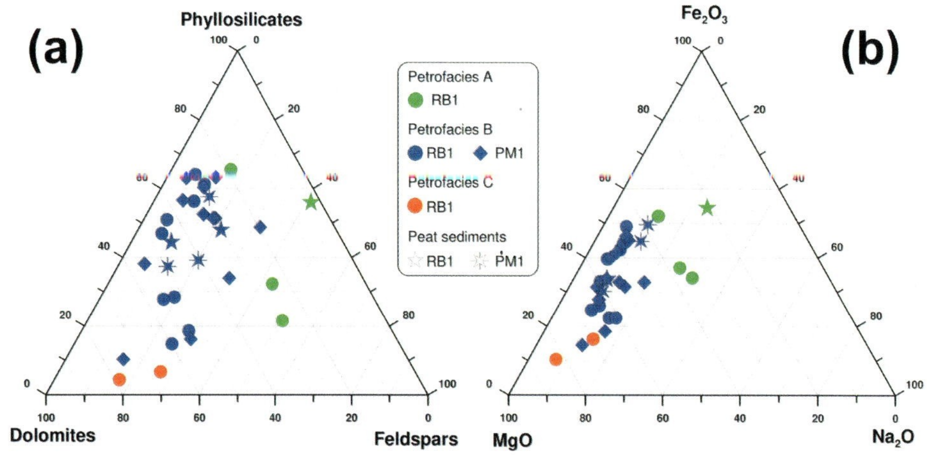

Figure 8. Ternary diagram reporting the main selected mineral components **(a)** and geochemical components **(b)** of the bulk sediments in the RB1 and PM1 cores. Peat samples are represented by the two star-like symbols.

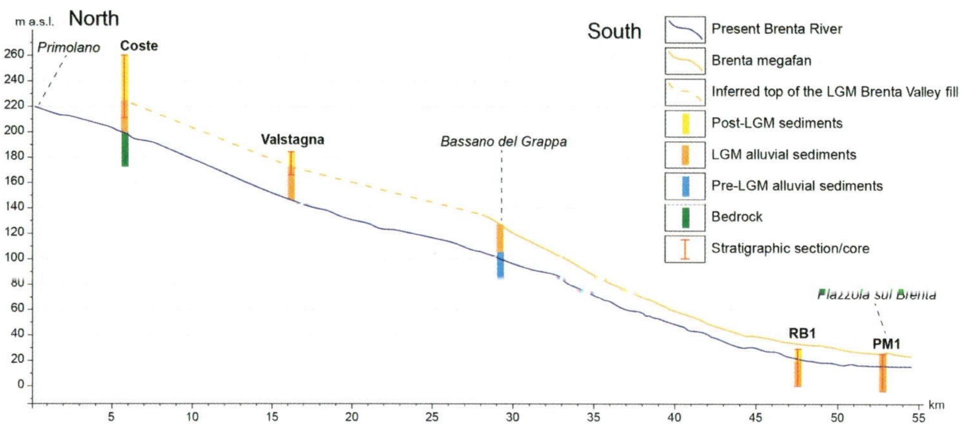

Figure 9. Longitudinal profiles of the present Brenta River (blue solid line), from Primolano to Piazzola sul Brenta, including the Brenta megafan (orange solid line), from the apex to Piazzola sul Brenta, and the possible profile of the Brenta Valley bottom during the LGM (orange dashed line), inferred from stratigraphic sections. The age of sediments/bedrock is shown using different colours.

5.1 Glaciers in the mountainous study area

While all the major LGM valley glaciers of the southeastern Alps preserved all or parts of their end-moraine systems in the terminal valley tracts and/or in the piedmont plain (Venzo et al., 1977; Monegato et al., 2007; Carton et al., 2009; Rossato et al., 2013), in the Brenta Valley there is no evidence of the LGM (nor older) terminal moraines (Castiglioni, 2004). The Brenta Glacier that used to flow through the Valsugana during the LGM was mainly fed by the Adige Glacier, the largest glacier on the southern side of the Alps (Bassetti and Borsato, 2005; Monegato et al., 2017), and some local tributary glaciers from the eastern side of the valley. The transfluence of Adige Glacier into the Valsugana was through the Fersina saddle (550 m a.s.l.), the Vigolo Vattaro wind gap (lowermost altitude: 680 m a.s.l.) and, more significantly, above the Calisio Plateau (ca 1000 m a.s.l.) during the maximum glacier expansion. Flowing for about 50 km along the Valsugana, the glacier reached the Primolano sector where the Valsugana narrows from about 1 km to 100 m and the Canal La Menor wind gap opens eastwards at about 350 m a.s.l. (Fig. 2).

Based on the location of marginal glacial deposits, the sudden narrowing of the Brenta Valley may have caused the glacier front to stabilize, which is known to have occurred in the northwestern Himalayas during the late Pleistocene, when narrow steep-walled canyons constituted an effective barrier to glacial advance (Burbank and Fort, 1985). The subsequent growth of the glacier forced it to rise up to the Canal La Menor wind gap and split into two lobes (Fig. 2). The eastern lobe flowed across the wind gap and formed the Sorist (650 to 760 m a.s.l.), Novegno (600 to 720 m a.s.l.) and Guarda (660 to 700 m a.s.l.) lateral moraines, whilst the western lobe built the Enego and Col del Gallo (both ~ 780 m a.s.l.) lateral moraines. The abundance of porphyry clasts and boulders in the deposits related to both the eastern and western lobes indicates sediment provenance from the upper Valsugana and the Adige valleys. The differences in elevation of the various moraines suggest that the western lobe grew more than the eastern lobe, possibly bulging due to the narrowing of the Valsugana. A secondary effect of the glacier bulging would have been the formation of transverse crevasses, which would have caused the fall of supraglacial debris into the ice mass; both of these processes would have hindered the transport of sediments to the glacier front and increased the hydraulic conductivity of the otherwise effectively impermeable glacier ice (Gulley and Benn, 2007). At the end of the gorge the Valsugana Valley widens again, a morphology that likely induced the formation of splaying/radial crevasses and icefalls in the frontal glacier mass (Nye, 1952; Harper et al., 1998; Colgan et al., 2016).

The large stratigraphic section in the Coste quarry (and the minor Valstagna section) displays no evidence of glacial deposits, while it indicates the presence of important LGM glaciofluvial aggradation in front of the western glacier's fronts. Henceforth, the front of the Brenta Glacier flowing in the main valley (western lobe) should have been located between the southernmost end of the Enego and Col del Gallo moraines and the Coste quarry. Glaciers confined in narrow valleys normally experience rapid advances under positive mass balance conditions: the narrower the valley, the faster the glacier's speed (e.g. Egholm et al., 2011). However, it has been proved that narrowing valleys promoted glacier blockage in the Himalayas during late Pleistocene (Burbank and Fort, 1985). This is believed to have been the case for the Brenta tongue, with the Coste section only being about 250 m downstream of the Enego moraine. The presence of the high-elevation lateral moraines hanging above the valley (the Enego moraine is about 550 m higher than the present valley floor) at such a short distance from proglacial sediments suggests that the glacier's front probably consisted of an icefall when the glacier was at its maximum size. The tongue of the Brenta Glacier, being a debris-free glacier, has been characterized by a more effective ablation due to solar energy compared to glaciers covered by several centimetres or more of debris (Lardeux et al., 2015; Wei et al., 2010). The resulting abundant meltwaters fed a well-developed proglacial stream, inducing aggradation along the whole Brenta Valley, as evidenced by the Coste and Valstagna sections. The development of a subglacial/englacial drainage system related to debris-filled crevasses probably controlled the seasonal variation of the glacier and resulted in fewer fluctuations at the glacier's front (van der Veen, 2007). Such stability would have resulted in the Enego and Col del Gallo lateral moraines growing. This Brenta Glacier tongue may have brought debris at its front both during stability, through supra- and/or englacial transport (Barr and Lovell, 2014 and reference therein), and while advancing, bulldozing pre-existing valley-floor sediments (e.g. Winkler and Matthews, 2010). Such debris may have formed end moraines, which were later eroded by post-glacial fluvial and slope processes. However, considering the narrow gorge that hosted the glacier front, it seems more probable that high-discharge proglacial streams occupied the entire gorge and continuously removed the incoming debris and precluded the formation of terminal moraines.

When the Brenta Glacier reached the elevation of the Canal La Menor wind gap, the eastern path became an effective glacial flux, which is known to occur when stabilized valley glaciers can extend laterally (Barr and Lovell, 2014). The higher the glacier, the more effective the glacial flux through the eastern path would have been, which would have even allowed the glacier to exceed the western side of the Novegno Plateau. As glacial sediment transport is likely to follow the main glacial flux, the eastern flow probably subtracted increasingly higher portions of the glacially transported debris to the western flow, depleting the sedimentary flux through the Valsugana Gorge and further hindering the formation of an end moraine at the western front. The eastern glacial lobe flowed along the Canal La Menor wind gap down to the confluence with the Corlo Valley. Here, it merged with the glacier coming from the northeast, formed by contributions from both the Cismon Glacier and the westernmost lobe of the Piave Glacier, as evidenced by the Roncon till. The Roncon till is polygenic, with pebbles of dolostones and Triassic volcanic rocks belonging to the Piave catchment. This till also crops out extensively in the nearby Seren Valley, which had a lateral tongue of the Piave Glacier flowing through it. Geomorphic evidence of the westward flow of Piave Glacier is provided by the lateral moraines at Monache (650 to 750 m a.s.l.), even though they are mostly made of limestone clasts; thus, they reflect a local glacigenic sedimentary input.

The glacier, which was derived from the merging of the eastern Brenta lobe and the Cismon/Piave glaciers at the confluence of Canal La Menor and the Corlo valleys, left no traces of frontal moraines. Glacial till crops out close to Rocca (Fig. 2) and till patches have been described about 0.5 km south of Rocca (Dal Piaz et al., 1946), suggesting that the front of this glacier was located at the beginning of the narrow gorge now occupied by the southern end of the artificial Lake Corlo. The geomorphological setting is very similar to the western tongue, suggesting that the glacier's front may also have consisted of an icefall with deep crevasses without terminal moraines at this location.

The Valsugana Glacier left a remarkably small amount of erratics. During the survey for this study, a total number of seven boulders made of porphyry and granite, up to 1 m in size, were found (in the Novegno and Sorist areas). Other authors have mentioned erratics on top of the Novegno Plateau (up to 2 m in size and made of crystalline rocks), next to the Enego moraine ("very large" boulders made of carbonate rocks; Secco, 1883; Venzo, 1940) and on the Canal La Menor Valley bottom ("extremely large" porphyry boulders; Taramelli, 1882). However, no erratics have been found, nor mentioned, downstream along the Valsugana Valley bottom.

5.2 The fluvial record of the glaciers' changes

The fluvial sediments cropping out in the Coste and Valstagna quarries are the only remnants of the LGM glaciofluvial aggradation that took place downstream of both the eastern and western glaciers' fronts. This glaciofluvial sedimentation led to the infilling of the Valsugana Valley bottom up to some tens of metres above the present Brenta River. The elevation of the top of the LGM valley fill at Coste (about 225 m a.s.l.) is consistent with that of the Valstagna section (about 175 m a.s.l.) as well as with the top LGM surface of the Brenta megafan southeast of Bassano del Grappa (about 130 m a.s.l.). This allows for the correlation of these depositional top surfaces and the related sediments, as well as the reconstruction of the longitudinal profile of the LGM Brenta Valley bottom (Fig. 9). The LGM fluvial aggradation was followed by the incision of the valley bottom and the piedmont megafan at around 17.5 ka cal BP, as the fluvial system reacted to the downwasting of the glacial system (Mozzi, 2005; Fontana et al., 2014; Rossato and Mozzi, 2016).

The growth/collapse of glacier tongues and modifications to the fluvial networks can be detected from changes in the mountain catchments and in the alluvial plain. Sedimentary systems developing at the mouth of major valleys are highly valuable databases of sedimentary, climatic and tectonic data (e.g. Mozzi, 2005; Carton et al., 2009; Pini et al., 2009; Piovan et al., 2012; Rossato and Mozzi, 2016). The paucity of radiocarbon datable material usually available in glacial deposits can be balanced with the abundant organic samples that can be collected in fluvial sedimentary sequences of this area. In this investigation, the petrographic, mineralogical and geochemical analyses of the RB1 and PM1 cores in the glaciofluvially fed Brenta megafan, chronologically framed through radiocarbon dating, integrate the evidence obtained in the mountainous area of this study. This allows for specific evolutionary phases in the drainage network feeding the Brenta megafan to be distinguished (Fig. 10). They are described in the following from oldest to youngest:

The first phase, seen in the RB1 core, corresponds to petrofacies A and lasted until just after 27 ka cal BP. At that time glaciers were growing at the onset of the LGM (Monegato et al., 2007, 2017; Ivy-Ochs et al., 2008; Preusser et al., 2011), and the mountain drainage systems began to modify. The Brenta megafan sedimentation rates were still comparable with pre-LGM rates (Rossato and Mozzi, 2016). Sediments indicate that this megafan was fed by a river with a drainage system limited to the Valsugana, as shown by the petrographic samples (RB1-29 and 30). The Cismon drainage system, which is currently merging with the Brenta system, was not contributing, as it was probably flowing eastwards into the Piave system (Fig. 10).

The beginning of the second phase coincided with a gradually increasing change in the drainage system, as evidenced by mineralogical and petrographic analyses of petrofacies B (samples RB1-27, 24, 21 and 16 and PM1-19, 17 and 12)

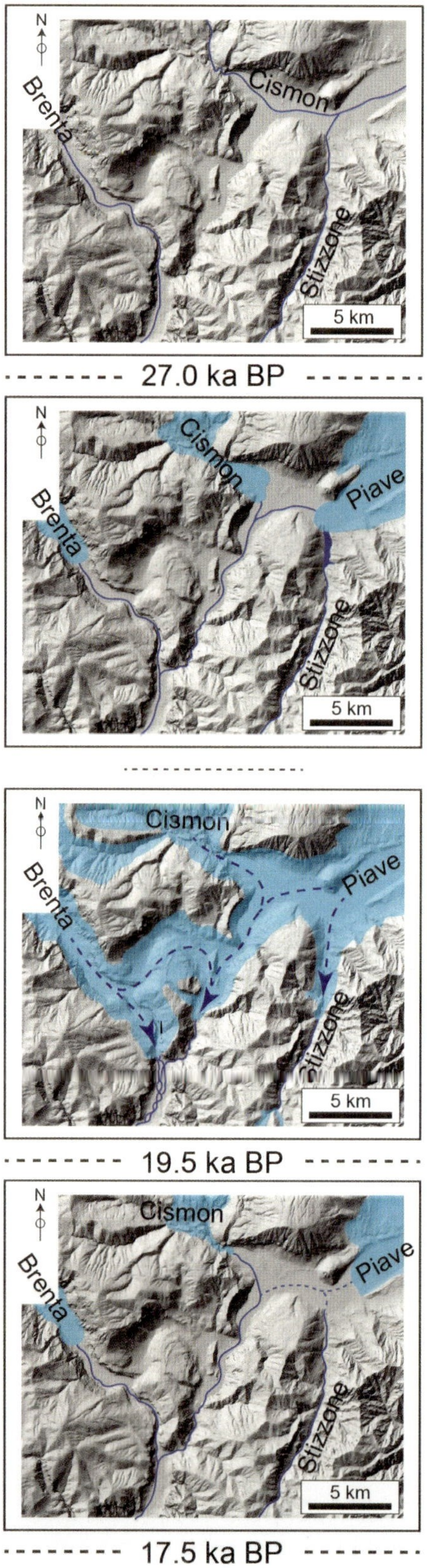

Figure 10. Evolution of the middle sector of the Valsugana Valley since the onset of the LGM; see text for details.

(Figs. 7–8). Radiocarbon dates indicate that this phase started just after 27 ka cal BP and continued until after 20.1 ka cal BP (Fig. 6). The lack of major unconformities in the succession of the PM1 core suggests that the top of this phase corresponds to the LGM surface of the Brenta megafan, which implies that sedimentation lasted until 17.5 ka cal BP (Rossato and Mozzi, 2016). Soon after 27 ka cal BP, the Piave and Cismon systems began to contribute to the Brenta system through the Corlo Valley. Whilst the Cismon River diversion does not require a direct connection with glacier dynamics, the Piave Glacier must have grown enough to overcome the Seren saddle (~ 330 m a.s.l.; Figs. 1, 10), which is about 100 m above the current Piave Valley bottom. At the acme of the LGM, the contribution of these systems is likely to have greatly increased the sedimentation rates in the Brenta megafan, which nearly doubled in the 26.7–23.8 ka cal BP period (Rossato and Mozzi, 2016; Fig. 10). The Piave Glacier tongue, having overcome the Seren saddle, probably survived the ~ 19.5 ka cal BP glacial retreat, as the main glacier was still 500 m thick near Belluno ("Val Piana" stage: 16210 ± 50 years BP, Pellegrini et al., 2005, recalibrated to 19386–19772 years cal BP with IntCal13 calibration curve, Reimer et al., 2013).

The youngest phase is only recorded in the topmost sediments of the RB1 core, and constitutes petrofacies C. It is ascribable to the infilling of a fluvial entrenchment (Fig. 2) developed during the late stages of deglaciation (Mozzi et al., 2010, 2013), which has also been seen elsewhere throughout the central and eastern Po Plain (Fontana et al., 2014). Mineralogy confirms the present configuration of the drainage system, with the Cismon River flowing into the Brenta River through the Corlo Valley and the Piave River already flowing along its modern valley. However, a remarkable enrichment of carbonates (micritic limestones) with respect to modern Brenta sediments, evidenced by petrography, mineralogy and geochemistry, highlights an anomalous setting. Two possible scenarios may be proposed to explain a signal such as this: (i) a significant bedrock erosion of these lithologies in the mountainous area of this study (i.e. especially in the Cismon catchment), or (ii) a connection with the dismantling of the postulated landslide that induced the aggradation of the Lamon terraces (if the "landslide scenario" is assumed; see the Section 2 of this paper for details). The first hypothesis could relate to the carving of the Corlo Valley by the Cismon River during the Lateglacial when high meltwater discharges could easily induce such enhanced erosion stages, which normally occurs during glacier recessional phases (Herman et al., 2011).

6 Conclusions

The acquired dataset casts new light on the dynamics of the LGM glaciers in the canyon-like middle tract of a major Alpine valley, the Valsugana. The knot of the Valsugana Glacier has been disentangled for the first time, indicating a singular configuration of the glacier snouts crossing narrow and deep valley reaches.

Our data indicate that the Brenta Glacier split at Primolano at the LGM acme. One tongue flowed southwards along the Valsugana Valley through a narrow gorge which prevented an effective glacier flux and caused the glacier to bulge. This bulging forced the right side of the glacier tongue to reach high elevation at Enego, while its front was probably constituted by an icefall located just upstream of Coste. Meltwaters flowed in subglacial and englacial streams removing debris from the glacier front and filling the Valsugana Valley bottom. The other tongue collected most of the glacial and debris fluxes, and flowed eastwards along the Canal La Menor Valley before joining the Cismon/Piave glaciers near Rocca.

The coupling of data gathered in the mountainous area of the study site with those collected in the piedmont alluvial plain allows a coherent reconstruction of the dynamics of the LGM glacier tongues. Prior to the arrival of the glacier fronts in the study area, the Brenta megafan only received sediments from the upper Valsugana catchment. A major alteration of the drainage system occurred just before 27 ka cal BP, when the Cismon and Piave systems joined the Brenta River in the Valsugana Valley, through the Seren saddle and the Corlo Valley. Between ~ 27 and, at least, ~ 19.5 ka cal BP, the Brenta, Cismon and Piave glaciers merged in the surroundings of the Novegno mount. Their meltwaters built up the largest alluvial landform on the Po Plain at that time: the Brenta megafan. At the end of the LGM, the waning of glaciers induced the fluvial incision of the Brenta megafan. The abundance of micritic carbonates in the sedimentary fill of incisions such as that near Piazzola sul Brenta suggests concomitant remarkable bedrock erosion in the Corlo Valley and/or the reworking of sediments from the Cismon catchment at Lamon.

As general conclusive remarks, this study highlights the following:

- The narrowing of a main glaciated valley may result in the blockage/slowing of the glacier flux. A larger lateral valley may easily represent an alternative path for the glacier, even if its bottom lies at higher elevation, subtracting a large part of the glacial and debris flux from the main valley.
- Valley glaciers flowing across narrow gorges may be subject to bulging and likely have icefalls at their front, while proglacial meltwater streams prevent the formation of end moraines.
- In rugged Alpine terrains, glacial catchments may significantly vary over time during a single glaciation. Such changes affect both the glacial and glaciofluvial systems and can be recognized in the alluvial stratigraphic record far downstream from the glacier front.

Sand petrography and chemical/mineralogical composition of sediments are good tracers of glacial catchment variations.

Author contributions. All authors contributed to interpreting the results and improving the text, which was primarily written by SR. Each author contributed to different sections of the text: the field survey was undertaken by SR, GM and PM; the sand petrography was carried out by GM; the X-ray diffraction was undertaken by AC and FT; the chemical composition was carried out by AC and FT; the core descriptions were carried out by SR and PM; and the remote sensing was carried out by SR.

Competing interests. The authors declare that they have no conflict of interest.

Acknowledgements. We sincerely thank Stefan Winkler and Laura Stutenbecker for thorough critical reviews that greatly improved the quality of the paper.

References

Aitchinson, J. and Greenacre, M.: Biplots of compositional data, J. Roy. Stat. Soc. C-App., 51, 375–392, https://doi.org/10.1111/1467-9876.00275, 2002.

ARPAV: Carta dei suoli del Veneto in scala 1: 250.000, Grafiche Vianello, Treviso, 2005.

Avanzini, M., Bargossi, G. M., Borsato, A., and Selli, L.: Note Illustrative della Carta Geologica d'Italia alla scala 1: 50.000, foglio 060- Trento, ISPRA-Servizio Geologico d'Italia, Trento, 2010.

Baratto, A., Ferrarese, F., Meneghel, M., and Sauro, U.: La ricostruzione della glaciazione wurmiana nel Gruppo del Monte Grappa (Prealpi Venete), in: Risposta dei processi geomorfologici alle variazioni ambientali, edited by: Biancotti, A. and Motta, M., G. Brigati, Genova, Italy, 67–77, 2003.

Barbieri, G. and Grandesso, P.: Note illustrative della Carta Geologica d'Italia alla scala 1: 50.000, foglio 082-Asiago, APAT, S.EL.CA., Firenze, 135, 2007.

Barr, I. D. and Lovell, H.: A review of topographic controls on moraine distribution, Geomorphology, 226, 44–64, https://doi.org/10.1016/j.geomorph.2014.07.030, 2014.

Bartolomei, G.: Nuovi elementi su un morenico antico a Bassano del Grappa, Vicenza, in: Studi geografici e geologici in onore di Severino Belloni, edited by: Orombelli, G., 9–18, 1999.

Bartolomei, G., Corsi, M., Dal Cin, R., D'Amico, C., Gatto, G. O., Gatto, P., Nardin, M., Rossi, D., Sacerdoti, M., and Semenza, E.: Note illustrative della carta geologica d'Italia alla scala 1: 100.000, foglio 021-Trento, Poligrafica e Cartevalori, Ercolano, 1969

Bassetti, M. and Borsato, A.: Evoluzione geomorfologica della Bassa Valle dell'Adige dall'Ultimo Massimo Glaciale: sintesi delle conoscenze e riferimenti ad aree limitrofe, Studi Trent. Sci. Nat.-Acta Geol., 82, 31–42, 2005.

Becker, P., Seguinot, J., Jouvet, G., and Funk, M.: Last Glacial Maximum precipitation pattern in the Alps inferred from glacier modelling, Geogr. Helv., 71, 173–187, https://doi.org/10.5194/gh-71-173-2016, 2016.

Bigi, G., Castellarin, A., Coli, M., Dal Piaz, G. V., Sartori, R., Scandone, P., and Vai, G. B.: Structural Model of Italy, sheets 1, CNR, S.EL.CA., Firenze, 1990.

Bowen, D. Q.: Last glacial maximum, in: Encyclopedia of Paleoclimatology and Ancient Environments, Springer Netherlands, 493–495, 2009.

Brauer, A., Hajdas, I., Blockley, S. P., Ramsey, C. B., Christl, M., Ivy-Ochs, S., Moseley, G. E., Nowaczyk, N. N., Rasmussen, S. O., Roberts, H. M., Spötl, C., Staff, R. A., and Svensson, A.: The importance of independent chronology in integrating records of past climate change for the 60-8 ka INTIMATE time interval, Quaternary Sci. Rev., 106, 47–66, https://doi.org/10.1016/j.quascirev.2014.07.006, 2014.

Buccianti, A., Pawlowsky-Glahn, V., Barceló-Vidal, C., and Jarauta-Bragulat, E.: Visualization and modeling of natural trends in ternary diagrams: a geochemical case study, in: Proceedings of IAMG'99 – The fifth annual conference of the International Association for Mathematical Geology, edited by: Lippard, S. J., Naess, A., and Sinding-Larsen, R., Trondheim, 139–144, 1999.

Burbank, D. W. and Fort, M. B.: Bedrock control on glacial limits: examples from the Ladakh and Zanskar ranges, north-western Himalaya, India, J. Glaciol., 31, 143–149, 1985.

Carraro, A., Fabbri, P., Giaretta, A., Peruzzo, L., Tateo, F., and Tellini, F.: Arsenic anomalies in shallow Venetian Plain (Northeast Italy) groundwater, Environ. Earth Sci., 70, 3067–3084, https://doi.org/10.1007/s12665-013-2367-2, 2013.

Carraro, A., Fabbri, P., Giaretta, A., Peruzzo, L., Tateo, F., and Tellini, F.: Effects of redox conditions on the control of arsenic mobility in shallow alluvial aquifers on the Venetian Plain (Italy), Sci. Total Environ., 532, 581–594, https://doi.org/10.1016/j.scitotenv.2015.06.003, 2015.

Carraro, F. and Sauro, U.: Il Glacialismo "locale" Wurmiano del Massiccio del Grappa (Provincie di Treviso e di Vicenza), Geogr. Fis. Din. Quat., 2, 6–16, 1979.

Carraro, F., Grandesso, P., Sauro, U., and Paoletti, G.: Incontri con il Grappa: i segreti della geologia, Moro Edizioni, 1989.

Carton, A., Bondesan, A., Fontana, A., Meneghel, M., Miola, A., Mozzi, P., Primon, S., and Surian, N.: Geomorphological evolution and sediment transfer in the Piave River system (northeastern Italy) since the Last Glacial Maximum, Géomorphologie, 15, 155–174, https://doi.org/10.4000/geomorphologie.7639, 2009.

Castellarin, A., Nicolich, R., Fantoni, R., Cantelli, L., Sella, M., and Selli, L.: Structure of the lithosphere beneath the Eastern Alps (southern sector of the TRANSALP transect), Tectonophysics, 414, 259–282, https://doi.org/10.1016/j.tecto.2005.10.013, 2006.

Castiglioni, B.: L'Italia nell'età quaternaria, Atlante Fisico-economico d'Italia, Consociazione Turistica Italiana, Milano, Italy, 1940.

Castiglioni, G. B.: Quaternary glaciations in the eastern sector of the Italian Alps, in: Quaternary Glaciations-Extent and Chronology, edited by: Elhers, J. and Gibbard, P., 209–215, https://doi.org/10.1016/S1571-0866(04)80072-1, 2004.

Castiglioni, G. B., Meneghel, M., and Sauro, U.: Elementi per una ricostruzione dell'evoluzione morfotettonica delle Prealpi Venete, Geogr. Fis. Din. Quat., 1, 31–44, 1988.

Clark, P. U., Dyke, A. S., Shakun, J. D., Carlson, A. E., Clark, J., Wohlfarth, B., Mitrovica, J. X., Hostetler, S. W., and McCabe, A. M.: The last glacial maximum, Science, 325, 710–714, https://doi.org/10.1126/science.1172873, 2009.

Colgan, W., Rajaram, H., Abdalati, W., McCutchan, C., Mottram, R., Moussavi, M. S., and Grigsby, S.: Glacier crevasses: Observations, models and mass balance implications, Rev. Geophys., 54, 119–161, https://doi.org/10.1002/2015RG000504, 2016.

Comas, M. and Thió- Henestrosa, S.: CoDaPack 2.0: a stand-alone multi-platform compositional software, in: CoDaWork'11: 4th International Workshop on Compositional Data Analysis, edited by: Egozcue, J. J., Tolosana-Delgado, R., and Ortego, M. I., Saint Feliu de Guixols, Girona, Spain, 2011.

Cucato, M.: Rilevamento della media Val d'Astico (Provincia di Vicenza): saggio per l 'applicazione della normativa sulla cartografia geologica del Quaternario continentale, B. Serv. Geol. Ital., 99–130, 2001.

Dal Piaz, G., Fabiani, R., Trevisan, L., and Venzo, S.: Carta geologica delle tre Venezie al 100.000, foglio 37-Bassano del Grappa, Ufficio Idrografico Magistrato delle Acque, Venezia, 1946.

Donnici, S., Serandrei-Barbero, R., Bini, C., Bonardi, M., and Lezziero, A.: The caranto paleosol and its role in the early urbanization of Venice, Geoarchaeology, 26, 514–543, 2011.

Egholm, D. L., Knudsen, M. F., Clark, C. D., and Lesemann, J. E.: Modeling the flow of glaciers in steep terrains: The integrated second-order shallow ice approximation (iSOSIA), J. Geophys. Res.-Earth, 116, F02012, https://doi.org/10.1029/2010JF001900, 2011.

Evans, I. S.: Glacier distribution in the Alps: statistical modelling of altitude and aspect, Geogr. Ann. A, 88, 115–133, https://doi.org/10.1111/j.0435-3676.2006.00289.x, 2006.

FAO: World Reference Base for Soil Resources, Food and Agriculture Organization of the United Nations, Rome, Italy, 1998.

Florineth, D. and Schlüchter, C.: Alpine evidence for atmospheric circulation patterns in Europe during the Last Glacial Maximum, Quaternary Res., 54, 295–308, https://doi.org/10.1006/qres.2000.2169, 2000.

Fontana, A., Mozzi, P., and Bondesan, A.: Alluvial megafans in the Venetian-Friulian Plain (north-eastern Italy): Evidence of sedimentary and erosive phases during Late Pleistocene and Holocene, Quaternary Int., 189, 71–90, https://doi.org/10.1016/j.quaint.2007.08.044, 2008.

Fontana, A., Monegato, G., Zavagno, E., Devoto, S., Burla, I., and Cucchi, F.: Evolution of an Alpine fluvioglacial system at the LGM decay: the Cormor megafan (NE Italy), Geomorphology, 204, 136–153, https://doi.org/10.1016/j.geomorph.2013.07.034, 2014.

Garzanti, E., Andò, S., and Vezzoli, G.: The continental crust as a source of sand (southern Alps cross section, northern Italy), J. Geol., 114, 533–554, https://doi.org/10.1086/506159, 2006.

Garzanti, E. Andó, S., France-Lanord, C., Censi, P., Vignola, P., Galy, V., and Lupker, M.: Mineralogical and chemical variability of fluvial sediments 2. Suspended-load silt (Ganga-Brahmaputra, Bangladesh), Earth Planet. Sc. Lett., 302, 107–120, https://doi.org/10.1016/j.epsl.2010.11.043, 2011.

Gazzi, P., Zuffa, G. G., Gandolfi, G., and Paganelli, L.: Provenienza e dispersione litoranea delle sabbie delle spiagge adriatiche fra le foci dell'Isonzo e del Foglia: inquadramento regionale, Mem. Soc. Geol. Ital., 12, 1–37, 1973.

Gulley, J. and Benn, D. I.: Structural control of englacial drainage systems in Himalayan debris-covered glaciers, J. Glaciol., 53, 399–412, 2007.

Harper, J. T., Humphrey, N. F., and Pfeffer, W. T.: Crevasse patterns and the strain-rate tensor: a high-resolution comparison, J. Glaciol., 44, 68–76, https://doi.org/10.3189/S0022143000002367, 1998.

Heiri, O., Koinig, K. A., Spötl, C., Barrett, S., Brauer, A., Drescher-Schneider, R., Gaar, D., Ivy-Ochs, S., Kerschner, H., Luetscher, M., Moran, A., Nicolussi, K., Preusser, F., Schmidt, R., Schoeneich, P., Schwörer, C., Sprafke, T., Terhorst, B., and Tinner, W.: Palaeoclimate records 60-8 ka in the Austrian and Swiss Alps and their forelands, Quaternary Sci. Rev., 106, 186–205, https://doi.org/10.1016/j.quascirev.2014.05.021, 2014.

Herman, F., Anderson, B., and Leprince, S.: Mountain glacier velocity variation during a retreat/advance cycle quantified using sub-pixel analysis of ASTER images, J. Glaciol., 57, 197–207, https://doi.org/10.3189/002214311796405942, 2011.

Hughes, P. D. and Gibbard, P. L.: A stratigraphical basis for the Last Glacial Maximum (LGM), Quatern. Int., 383, 174–185, https://doi.org/10.1016/j.quaint.2014.06.006, 2015.

Ingersoll, R. V., Bullard, T. F., Ford, R. L., Grimm, J. P., Pickle, J. D., and Sares, S. W.: The effect of grain size on detrital modes: a test of the Gazzi-Dickinson point-counting method, J. Sediment. Res., 54, 103–116, 1984.

Ivy-Ochs, S.: Glacier variations in the European Alps at the end of the last glaciation, Cuadernos Invest. Geogr., 41, 295–315, https://doi.org/10.18172/cig.2750, 2015.

Ivy-Ochs, S., Kerschner, H., Reuther, A., Preusser, F., Heine, K., Maisch, M., Kubik, P. W., and Schlüchter, C.: Chronology of the last glacial cycle in the European Alps, J. Quaternary Sci., 23, 559–573, https://doi.org/10.1002/jqs.1202, 2008.

Kelly, M. A., Buoncristiani, J. F., and Schlüchter, C.: A reconstruction of the last glacial maximum (LGM) ice-surface geometry in the western Swiss Alps and contiguous Alpine regions in Italy and France, Eclogae Geol. Helv., 97, 57–75, https://doi.org/10.1007/s00015-004-1109-6, 2004.

Koppes, M. N. and Montgomery, D. R.: The relative efficacy of fluvial and glacial erosion over modern to orogenic timescales, Nat. Geosci., 2, 644–647, https://doi.org/10.1038/ngeo616, 2009.

Lardeux, P., Glasser, N. F., Holt, T., Irvine-Fynn, T. D., and Hubbard, B. P.: Area and Elevation Changes of a Debris-Covered Glacier and a Clean-Ice Glacier Between 1952–2013 Using Aerial Images and Structure-from-Motion, AGU Fall Meeting Abstracts, https://doi.org/10.13140/RG.2.1.5059.4329, 2015.

Luetscher, M., Boch, R., Sodemann, H., Spötl, C., Cheng, H., Edwards, R. L., Frisia, S., Hof, F., and Müller, W.: North Atlantic storm track changes during the Last Glacial Maximum recorded by Alpine speleothems, Nat. Commun., 6, 6344, https://doi.org/10.1038/ncomms7344, 2015.

Massari, F., Grandesso, P., Stefani, C., and Zanferrari, A.: The Oligo-Miocene Molasse of the Veneto-Friuli region, Southern Alps, Giorn. Geol., 48, 235–255, 1986.

Miola, A., Bondesan, A., Corain, L., Favaretto, S., Mozzi, P., Piovan, S., and Sostizzo, I.: Wetlands in the Venetian Po Plain (northeastern Italy) during the Last Glacial Maximum: Interplay between vegetation, hydrology and sedimentary environment, Rev. Palaeobot. Palyno., 141, 53–81, https://doi.org/10.1016/j.revpalbo.2006.03.016, 2006.

Monegato, G., Ravazzi, C., Donegana, M., Pini, R., Calderoni, G., and Wick, L.: Evidence of a two-fold glacial advance during the last glacial maximum in the Tagliamento end moraine system (eastern Alps), Quaternary Res., 68, 284–302, https://doi.org/10.1016/j.yqres.2007.07.002, 2007.

Monegato, G., Stefani, C., and Zattin, M.: From present rivers to old terrigenous sediments: the evolution of the drainage system in the eastern Southern Alps, Terra Nova, 22, 218–226, https://doi.org/10.1111/j.1365-3121.2010.00937.x, 2010.

Monegato, G., Pini, R., Ravazzi, C., Reimer, P. J., and Wick, L.: Correlating Alpine glaciation with Adriatic sea-level changes through lake and alluvial stratigraphy, J. Quaternary Sci., 26, 791–804, https://doi.org/10.1002/jqs.1502, 2011.

Monegato, G., Scardia, G., Hajdas, I., Rizzini, F., and Piccin, A.: The Alpine LGM in the boreal ice-sheets game, Sci. Rep.-UK, 7, 2078, https://doi.org/10.1038/s41598-017-02148-7, 2017.

Mozzi, P., Bini, C., Becattini, R., and Mariotti Lippi, M.: Stratigraphy, palaeopedology and palynology of Late Pleistocene and Holocene deposits in the landward sector of the Lagoon of Venice (Italy), in relation to the "Caranto" level, Il Quaternario-Ital, J. Quaternary Sci., 16, 193–210, 2003.

Mozzi, P.: Alluvial plain formation during the Late Quaternary between the southern Alpine margin and the Lagoon of Venice (northern Italy), Geogr. Fis. Din. Quat., 7, 219–230, 2005

Mozzi, P., Piovan, S., Rossato, S., Cucato, M., Abbà, T., and Fontana, A.: Palaeohydrography and early settlements in Padua (Italy), Il Quaternario-Ital. J. Quaternary Sci., 23, 387–400, 2010.

Mozzi, P., Ferrarese, F., and Fontana, A.: Integrating digital elevation models and stratigraphic data for the reconstruction of the post-LGM unconformity in the Brenta alluvial megafan (North-Eastern Italy), Alp. Mediterr. Quaternary, 26, 41–54, 2013.

Norton, K. P., Abbühl, L. M., and Schlunegger, F.: Glacial conditioning as an erosional driving force in the Central Alps, Geology, 38, 655–658, https://doi.org/10.1130/G31102.1, 2010.

Nye, J. F.: The mechanics of glacier flow, J. Glaciol., 2, 82–93, https://doi.org/10.3189/S0022143000033967, 1952.

Pellegrini, G. B., Albanese, D., Bertoldi, R., and Surian, N.: La deglaciazione alpina nel Vallone Bellunese, Alpi meridionali orientali, Geogr. Fis. Din. Quat., 7, 271–280, 2005.

Penck, A. and Brückner, E.: Die Alpen im Eiszeitalter, Tauchnitz, Leipzig, 1909.

Piovan, S., Mozzi, P., and Zecchin, M.: The interplay between adjacent Adige and Po alluvial systems and deltas in the late Holocene (Northern Italy), Géomorphologie, 18, 427–440, https://doi.org/10.4000/geomorphologie.10034, 2012.

Pini, R., Ravazzi, C., and Donegana, M.: Pollen stratigraphy, vegetation and climate history of the last 215 ka in the Azzano Decimo core (plain of Friuli, north-eastern Italy), Quaternary Sci. Rev., 28, 1268–1290, https://doi.org/10.1016/j.quascirev.2008.12.017, 2009.

Preusser, F., Reitner, J. M., and Schlüchter, C.: Distribution, geometry, age and origin of overdeepened valleys and basins in the Alps and their foreland, Swiss J. Geosci., 103, 407–426, https://doi.org/10.1007/s00015-010-0044-y, 2010.

Preusser, F., Graf, H. R., Keller, O., Krayss, E., and Schlüchter, C.: Quaternary glaciation history of northern Switzerland, E&G Quaternary Sci. J., 60, 282–305, https://doi.org/10.3285/eg.60.2-3.06, 2011.

Ramsey, C. B.: Bayesian analysis of radiocarbon dates, Radiocarbon, 51, 337–360, https://doi.org/10.1017/S0033822200033865, 2009.

Ravazzi, C., Badino, F., Marsetti, D., Patera, G., and Reimer, P. J.: Glacial to paraglacial history and forest recovery in the Oglio glacier system (Italian Alps) between 26 and 15 ka cal BP, Quaternary Sci. Rev., 58, 146–161, https://doi.org/10.1016/j.quascirev.2012.10.017, 2012.

Reimer, P. J., Bard, E., Bayliss, A., Beck, J. W., Blackwell, P. G., Bronk Ramsey, C., Buck, C. E., Cheng, H., Edwards, R. L., Friedrich, M., Grootes, P. M., Guilderson, T. P., Haflidason, H., Hajdas, I., Hatté, C., Heaton, T. J., Hoffmann, D. L., Hogg, A. G., Hughen, K. A., Kaiser, K. F., Kromer, B., Manning, S. W., Niu, M., Reimer, R. W., Richards, D. A., Scott, E. M., Southon, J. R., Staff, R. A., Turney, C. S. M., and van der Plicht, J.: IntCal13 and Marine13 radiocarbon age calibration curves 0–50,000 years cal BP, Radiocarbon, 55, 1869–1887, https://doi.org/10.2458/azu_js_rc.55.16947, 2013.

Rossato, S. and Mozzi, P.: Inferring LGM sedimentary and climatic changes in the southern Eastern Alps foreland through the analysis of a 14C ages database (Brenta megafan, Italy), Quaternary Sci. Rev., 148, 115–127, https://doi.org/10.1016/j.quascirev.2016.07.013, 2016.

Rossato, S., Monegato, G., Mozzi, P., Cucato, M., Gaudioso, B., and Miola, A.: Late Quaternary glaciations and connections to the piedmont plain in the prealpine environment: the middle and lower Astico Valley (NE Italy), Quaternary Int., 288, 8–24, https://doi.org/10.1016/j.quaint.2012.03.005, 2013.

Russell, A. J., Roberts, M. J., Fay, H., Marren, P. M., Cassidy, N. J., Tweed, F. S., and Harris, T.: Icelandic jökulhlaup impacts: implications for ice-sheet hydrology, sediment transfer and geomorphology, Geomorphology, 75, 33–64, https://doi.org/10.1016/j.geomorph.2005.05.018, 2006.

Sacco, F.: Il Glacialismo veneto, L'Universo, 7, 1–40, 1937.

Samartin, S., Heiri, O., Kaltenrieder, P., Kühl, N., and Tinner, W.: Reconstruction of full glacial environments and summer temperatures from Lago della Costa, a refugial site in Northern Italy, Quaternary Sci. Rev., 143, 107–119, https://doi.org/10.1016/j.quascirev.2016.04.005, 2016.

Schlüchter, C.: The Quaternary glaciations of Switzerland, with special reference to the Northern Alpine Foreland, Quaternary Sci. Rev., 5, 413–419, https://doi.org/10.1016/0277-3791(86)90206-4, 1986.

Secco, A.: Note geologiche sul Bassanese, Stabilimento Tipografico Sante Pozzato, 1883.

Seguinot, J., Jouvet, G., Huss, M., Funk, M., and Preusser, F.: Modelling last glacial cycle ice dynamics in the Alps, EGU Abstracts, 19, 8982, 2017.

Smiatek, G., Kunstmann, H., Knoche, R., and Marx, A.: Precipitation and temperature statistics in high-resolution regional climate models: Evaluation for the European Alps, J. Geophys. Res.-Atmos., 114, D19107, https://doi.org/10.1029/2008JD011353, 2009.

Stefani, C., Fellin, M. G., Zattin, M., Zuffa, G. G., Dalmonte, C., Mancin, N., and Zanferrari, A.: Provenance and paleogeographic evolution in a multi-source foreland: the Cenozoic

Venetian-Friulian Basin (NE Italy), J. Sediment. Res., 77, 867–887, https://doi.org/10.2110/jsr.2007.083, 2007.

Taramelli, T.: Geologia delle provincie venete, Atti della reale Accademia dei Lincei, 8, 303–541, 1882.

Tessari, F.: Geomorfologia del bacino di Lamon, Val Cismon, Alpi Dolomitiche, Museo Tridentino di Scienze Naturali, 1973.

Torma, C., Giorgi, F., and Coppola, E.: Added value of regional climate modeling over areas characterized by complex terrain – Precipitation over the Alps, J. Geophys. Res.-Atmos., 120, 3957–3972, https://doi.org/10.1002/2014JD022781, 2015.

Trevisan, L.: Il glacialismo quaternario nell'Altipiano dei Sette Comuni (Vicenza), Stabilimento tipografico Villarboito F. e figli, 1939.

van der Veen, C. J.: Fracture propagation as means of rapidly transferring surface meltwater to the base of glaciers, Geophys. Res. Lett., 34, L01501, https://doi.org/10.1029/2006GL028385, 2007.

van Husen, D.: Die Ostalpen in den Eiszeiten, Geol. Bundes., 1987.

Van Husen, D. and Reitner, J. M.: An Outline of the Quaternary Stratigraphy of Austria, E&G Quaternary Sci. J., 60, 366–387, https://doi.org/10.3285/eg.60.2-3.09, 2011.

Venzo, S.: Studio geotettonico del Trentino meridionale-orientale tra Borgo Valsugana e il Monte Coppolo, Poligrafico dello Stato, Roma, 86 pp., 1940.

Venzo, S.: I depositi quaternari e del neogene superiore nella bassa valle del Piave da Quero al Montello e del Paleopiave nella valle del Soligo (Treviso), Memorie degli Istituti Mineralogia e Geologia dell'Università di Padova, 30, 1–64, 1977.

Vezzoli, G. and Garzanti, E.: Tracking paleodrainage in Pleistocene foreland basins, J. Geol., 117, 445–454, https://doi.org/10.1086/598946, 2009.

von Eynatten, H., Pawlowsky-Glahn, V., and Egozcue, J. J.: Understanding perturbation on the simplex: A simple method to better visualize and interpret compositional data in ternary diagrams, Math. Geol., 34, 249–257, 2002.

Wei, Y., Tandong, Y., Baiqing, X., and Hang, Z.: Influence of supraglacial debris on summer ablation and mass balance in the 24K Glacier, southeast Tibetan Plateau, Geogr. Ann. A, 92, 353–360, https://doi.org/10.1111/j.1468-0459.2010.00400.x, 2010.

Winkler, S. and Matthews, J. A.: Observations on terminal moraine-ridge formation during recent advances of southern Norwegian glaciers, Geomorphology, 116, 87–106, https://doi.org/10.1016/j.geomorph.2009.10.011, 2010.

Wirsig, C., Zasadni, J., Christl, M., Akçar, N., and Ivy-Ochs, S.: Dating the onset of LGM ice surface lowering in the High Alps, Quaternary Sci. Rev., 143, 37–50, https://doi.org/10.1016/j.quascirev.2016.05.001, 2016.

Young, R. A.: The Rietveld method, Oxford University Press, 298 pp., 1993.

7

Impact of grain size and rock composition on simulated rock weathering

Yoni Israeli and Simon Emmanuel

Institute of Earth Sciences, The Hebrew University of Jerusalem, Jerusalem, Israel

Correspondence: Simon Emmanuel (swemmanuel@gmail.com)

Abstract. Both chemical and mechanical processes act together to control the weathering rate of rocks. In rocks with micrometer size grains, enhanced dissolution at grain boundaries has been observed to cause the mechanical detachment of particles. However, it remains unclear how important this effect is in rocks with larger grains, and how the overall weathering rate is influenced by the proportion of high- and low-reactivity mineral phases. Here, we use a numerical model to assess the effect of grain size on chemical weathering and chemo-mechanical grain detachment. Our model shows that as grain size increases, the weathering rate initially decreases; however, beyond a critical size no significant decrease in the rate is observed. This transition occurs when the density of reactive boundaries is less than $\sim 20\,\%$ of the entire domain. In addition, we examined the weathering rates of rocks containing different proportions of high- and low-reactivity minerals. We found that as the proportion of low-reactivity minerals increases, the weathering rate decreases nonlinearly. These simulations indicate that for all compositions, grain detachment contributes more than $36\,\%$ to the overall weathering rate, with a maximum of $\sim 50\,\%$ when high- and low-reactivity minerals are equally abundant in the rock. This occurs because selective dissolution of the high-reactivity minerals creates large clusters of low-reactivity minerals, which then become detached. Our results demonstrate that the balance between chemical and mechanical processes can create complex and nonlinear relationships between the weathering rate and lithology.

1 Introduction

The rate at which rocks weather influences the development of landscapes, the formation of soil, and the preservation of buildings and monuments (Wilson, 2004; Brantley, 2008; Price and Doehne, 2011; Putnis and Ruiz-Agudo, 2013). Rock weathering is affected by both chemical and physical processes (Fletcher et al., 2006; Buss et al., 2008; Røyne et al., 2008) that operate from the outcrop scale down to the atomic scale. Although mechanical weathering is typically thought to be important mainly at large scales, enhanced chemical dissolution at grain boundaries allows tiny grains to become mechanically detached from rock surfaces (Silveira and Aarão Reis, 2013; Emmanuel and Levenson, 2014; di Caprio et al., 2016; Krklec et al., 2016; Fischer and Luttge, 2017). This effect was estimated to represent around $40\,\%$ of the total weathering in fine-grained limestone and is likely to be important in other rock types as well (Levenson and Emmanuel, 2016).

Although grain detachment has been observed in rocks with micrometer size grains, it is unclear how important this effect is in rocks comprising larger grains. From a chemical weathering perspective, a fine-grained rock should dissolve more rapidly than a coarse-grained rock due to its higher reactive surface area (Beckingham et al., 2016; Kirstein et al., 2016). Moreover, at the nanoscale, tiny grains become intrinsically more reactive than large grains due to thermodynamic and kinetic effects (Arvidson et al., 2003; Emmanuel and Ague, 2011; Briese et al., 2017). Importantly for mechanical detachment, a rock comprising large grains will have a low density of grain boundaries relative to a rock with small grains. Since the grain detachment mechanism involves preferential dissolution along intergranular boundaries, it might be expected that fine-grained rocks will also be more susceptible to mechanical detachment (Emmanuel and Levenson, 2014). Thus, both the chemical and mechanical effects could make the overall weathering rate strongly dependent on grain size. However, in a study examining experimental weathering rates in carbonate rocks with different textures and crystal sizes, no significant difference was found (Levenson et al., 2015). This apparent lack of dependence on grain size could be due to the inherent difficulty associated with comparing the reaction rates in rocks with varying chemistry and crystal form. A complete understanding of grain size effects during weathering has yet to be achieved.

Another factor expected to affect weathering rates is the presence of mineral phases with different reactivities (Kump et al., 2000; Critelli et al., 2014; White and Buss, 2014). Beckingham et al. (2016) showed that the presence of a highly reactive volcanic glass phase indeed enhanced the chemical weathering rate of a volcanogenic sandstone. Deng et al. (2017) indicated that preferential dissolution of high-reactivity phases enhances fracture evolution in carbonate-rich shales. However, these experiments did not examine mechanical processes, such as grain detachment. Less reactive grains should become detached if their surrounding mineral matrix dissolves. Such complexity was in fact observed in bimineralic rocks comprising calcite and dolomite, when relatively inert dolomite crystals detached as the surrounding calcite matrix became dissolved (Kirklec et al., 2013; Levenson et al., 2017).

As well as having implications for weathering rates at the Earth's surface, detachment of low-reactivity particles could also occur in subsurface environments. Enhanced oil recovery, hydraulic fracturing, and CO_2 sequestration all involve the injection of reactive fluids into rock formations (Al-Mjeni et al., 2010; Barati and Liang, 2014; Lions et al., 2014). The selective dissolution of high-reactivity minerals during such operations could result in the detachment of mineral particles, potentially impacting rock porosity and permeability. Furthermore, if these mineral phases contain toxic elements or compounds, grain detachment could represent an important mode of mobilizing contaminants in groundwater. An example of such a system is carbonate-rich shales. Such shales often contain pyrite, which is a potential source of arsenic contamination. The pyrite in these rocks, however, is much less reactive than the calcite. Kreisserman and Emmanuel (2018) showed that pyrite particles become detached from shales during rock–fluid interactions as a result of the dissolution of surrounding carbonate minerals. However, in both surface and subsurface environments, it remains unclear how the overall weathering rates, which include mechanical processes, are influenced by the proportion of high- and low-reactivity mineral phases.

In this study, we investigate how grain size and rock composition affect the chemical and mechanical weathering rate of rocks. To achieve this, we constructed a numerical model to simulate the weathering of rocks with a range of grain sizes and different proportions of high-reactivity and low-reactivity minerals. Our model incorporates both chemical dissolution and grain detachment whilst exploring the impact of each of these processes on weathering rates.

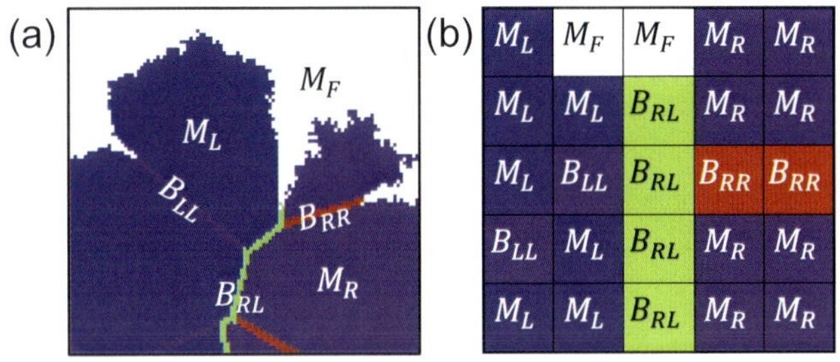

Figure 1. Part of the domain simulating a rock in the model at **(a)** the macro scale and **(b)** the pixel scale. The rock is simulated as a 2-D cross section using Voronoï tessellation. Grains are represented as Voronoï cells containing a mineral phase, and cell edges represent the grain boundaries. Every element is given a characteristic value of either a fluid state (M_F, white), a high-reactivity grain (M_R, blue), a low-reactivity grain (M_L, dark blue), a grain boundary between two high-reactivity grains (B_{RR}, red), a boundary between two low-reactivity grains (B_{LL}, blue), or a boundary between one high-reactivity and one low-reactivity grain (B_{RL}, green).

2 Methods

2.1 Modeling rock weathering

We developed a numerical model in MATLAB to simulate the combined effects of chemical and mechanical weathering. A 2-D cross section of a polycrystalline rock was represented by a domain (560 elements wide and 420 elements in height) constructed with Voronoï tessellation (Taleb and Stafiej, 2011; di Caprio et al., 2016). Every Voronoï cell represents a rock grain comprising a single mineral phase, and the cell edges represent the intergranular boundaries (Fig. 1a). The range of grain size was varied in our simulations by dividing the domain into a different number of grains (100, 200, 400, 800, 1600, 3200, 6400) while keeping the number of elements constant. The grain sizes here are arbitrarily normalized to the size of the grains in the simulation with the largest number of grains (i.e., 6400), so that the relative grain size for the simulation with 6400 grains is 1, while the grain size for the simulation with 100 grains is 64. The impact of grain size on weathering was examined by comparing simulations of a coarse-grained rock with that of a rock comprising small grains (relative grain size equals 64 and 2, respectively).

To account for the heterogeneity of surface reactivity (Fischer et al., 2014), every element in the domain is assigned a characteristic value (M_i; Fig. 1b) that corresponds to a fluid state (M_F), a high-reactivity grain (M_R), a low-reactivity grain (M_L), or a grain boundary (B_{ii}). There are three possible kinds of grain boundaries: boundaries between two high-reactivity grains (B_{RR}), boundaries between two low-reactivity grains (B_{LL}), and boundaries between one high-reactivity and low-reactivity grain (B_{RL}). These values are reflective of the kinetic rate of dissolution, so that 1 represents the maximum rate of dissolution and 0 represents an inert state. In our formulation, grain boundaries dissolve 1 order of magnitude more rapidly than the mineral core, which is consistent with experimental work (Lüttge et al., 2013; Emmanuel, 2014; Bray et al., 2015).

To determine the impact of rock composition on weathering rate, we carried out simulations in which the proportion of low-reactivity minerals within the rock is varied from 0 to 100 %. Over 600 simulations of rocks with different compositions were carried out.

In the initial state in our simulations, no elements are designated as fluid, although the rock is exposed to a reactive solution at the top boundary of the domain. We chose to simulate a system that is far from chemical equilibrium, which is representative of weathering rocks exposed at Earth's surface. To model the time-dependent evolution of the rock, we used a cellular automaton approach, similar to that used by Taleb and Stafiej (2011) to model metal corrosion. Each element in our domain was subject to well-defined rules, although we used a stochastic approach to simulate the random nature of dissolution (Fischer et al., 2014). At each time step, the only elements that dissolved were those in direct contact with the solution. The probability for any given element to undergo dissolution was calculated from the product of two parameters: the characteristic value for each element (M_i) and the probability that the element would dissolve in that time step (P_i). This probability increased linearly with the number of fluid elements by which it was surrounded, so that a solid element in contact with a single fluid element was assigned a P value of 12.5 %, determined from a random Gaussian distribution, while an element surrounded by eight fluid elements was given a P value of 100 %. The number of surrounding elements was defined using the Moore neighborhood method (range = 1; Lishchuk et al., 2011). Any element that underwent dissolution became a fluid and was assigned a characteristic value of M_F. At the end of each time step, the rock was scanned for regions that were completely surrounded by fluid. Such regions were then detached from the surface and were replaced by a characteristic value of M_F. The number of elements that underwent mechanical grain detachment and the number of detachment events is updated after each time step.

To minimize boundary affects, a bounding box was defined in the center of the domain. In all the simulations, the distance of the bounding box from the sides and the base of the entire domain corresponded to the maximal grain diameter in the simulation with the largest grains. The simulations were terminated when all the elements within the calculation bounding box were assigned a fluid state (M_F sites).

For each simulation, the main parameters that were recorded were (i) the number of time steps until complete dissolution, (ii) the ratio of mechanical detachment to the overall weathering, and (iii) the total number of grain detachment events. The dimensionless weathering rate, R^*, was calculated by dividing the number of elements inside the bounding box by the total number of time steps.

3 Results and discussion

3.1 Impact of grain size on weathering rates

During the time-dependent simulations, grain size was found to strongly impact the weathering rate of rocks. In a comparison between the weathering rates in fine-grained rocks and coarse-grained rocks, we found that fine-grained rocks weathered more rapidly (Fig. 2a). Increasing the relative grain size from 2 to 64 slowed the weathering rate by 37 %. This was due mainly to the higher rate of chemical weathering in the small-grained rocks (Fig. 2b), although in both cases the mechanical weathering component (Fig. 2c) contributed at least 33 % to the overall weathering rate. Even though the rate of mechanical weathering was similar for both grain sizes, we observed a higher frequency of detachments in fine-grained rocks, with tiny grains being detached during most time steps. By contrast, in the coarse-grained rock, mechanical detachments were less frequent, although much larger grains were removed during each event as is evident from the step-like function in Fig. 2c and from simulation snapshots (Fig. 2d–e). The snapshot from the fine-grained rock simulation (Fig. 2d) shows a higher density of grain boundaries (39 %) than the coarse-grained rock (7 %). This high density of reactive grain boundaries is the factor responsible for the increased rate of dissolution.

Although we found that fine-grained rocks weathered faster than coarse-grained rocks, the relationship between grain size and weathering rate was nonlinear and appeared to show an exponential decline with increasing grain size. This relationship was observed in all the rock compositions that we examined (Fig. 3a). In our simulations, above a certain size, the decrease in weathering rate slows significantly, suggesting that in real rocks the influence of grain size may be limited to relatively small scales. The transition we observed occurs when the density of reactive boundaries is less than ~ 20 % of the entire domain. This seemingly high density of grain boundaries could be expected to occur in rocks with sub-micrometer grains and may therefore be relatively rare in natural systems.

Even though the chemical weathering rate is slower for rocks with large grains, the contribution of grain detachment to the overall weathering rate increases with grain size (Fig. 3b). In coarse-grained rocks, detachment can contribute as much as 45 % to the total weathering. This trend in mechanical weathering with increasing grain size (Fig. 3b) shows an opposite trend to that of the overall rate (Fig. 3a). The main reason for this is that in large-grained rocks, the

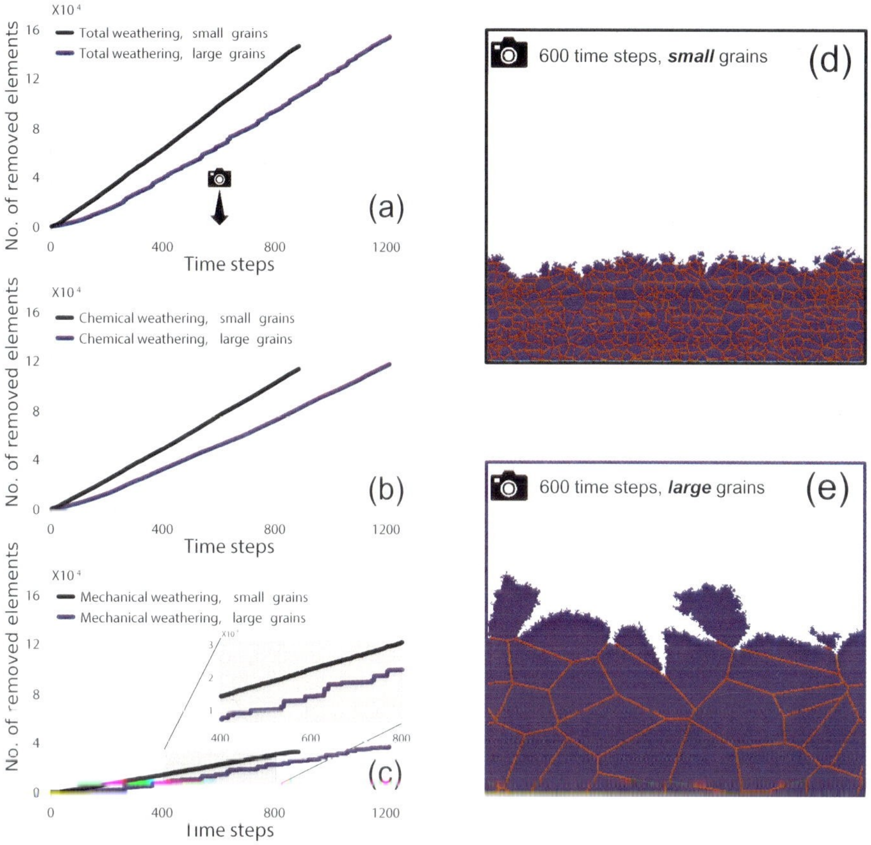

Figure 2. Number of removed elements as a function of time steps for two different simulations. Elements removed by **(a)** total weathering, **(b)** chemical weathering, and **(c)** mechanical weathering. The black lines represent the simulation with small grains (relative grain size = 2), while the blue lines represent the number of removed elements in the simulation with the large grains (relative grain size = 64). The camera icon in **(a)** indicates the timing of the snapshots for the **(d)** small-grain simulation and **(e)** the large grain simulation. The domain is comprised entirely of high-reactivity grains. For both grain sizes, chemical weathering contributes more to the total weathering than mechanical weathering. In the fine-grained rock **(d)**, 67 % of the material has been removed; by contrast, in the snapshot from the coarse-grained rock **(e)**, taken after the same number of time steps, only 42 % of the rock has been removed. The coarse-grained rock has a 7 % density of grain boundaries, while in the fine-grained rock the boundaries cover 39 % of the simulated rock area. The higher density of reactive grain boundaries seen in the fine-grained rock promotes the increased rate of dissolution.

grains undergo less dissolution before detachment because of the lower density of grain boundaries.

There is both experimental and field-based evidence to support the trends we observed in our simulations. Experimental observations by Kirstein et al. (2016) showed a preferential dissolution of small grains in micrite compared to larger-grained rocks. This was explained by the micritic structure that promotes the exposure of new, highly reactive grain boundaries. In addition, using vertical scanning interferometry (Fischer et al., 2012; Lüttge et al., 2013), micrite was found to dissolve 1 order of magnitude faster than single calcite crystals. Field-based measurements presented by Emmanuel and Levenson (2014) also showed that micritic limestone weathered significantly faster than coarse grainstone.

However, not all studies show that grain size affects the weathering rate significantly. In experiments on rocks with different textures (ranging from micrite to marble), Levenson et al. (2015) found no significant difference in the rates of calcite dissolution. In addition, simulations of electrochemical particle detachments by Taleb and Stafiej (2011) showed that the effect of grain size is insignificant at long reaction times. It may be that in these experiments and simulations, the grain sizes were in the region of the size-independent weathering regime we identified in Fig. 3a. To determine if this is indeed the case, future experiments and field-based work on a wider range of grain sizes are necessary.

3.2 Impact of rock composition on weathering rate

Our simulations show a nonlinear relationship between rock composition and weathering rate (Figs. 4, 5a). We found that the presence of a small amount of a low-reactivity phase in a high-reactivity rock produced a disproportionally large reduction in the overall rate. This reduction in rate is primar-

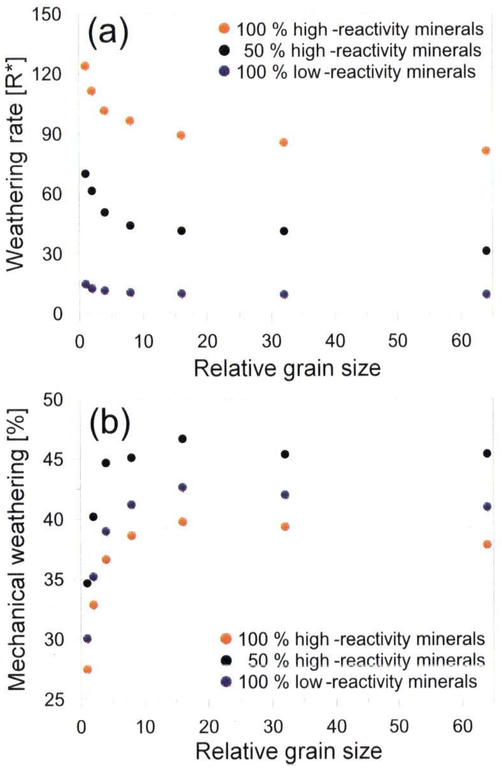

Figure 3. Effect of grain size on weathering for three different compositions: **(a)** weathering rate versus relative grain size in units of R^* and **(b)** the ratio of mechanical weathering to overall weathering versus relative grain size. Red circles represent simulations for a rock comprising 100 % high-reactivity mineral phase. Black circles represent an equal mix of high- and low-reactivity mineral phases. Blue circles represent a rock comprising 100 % low-reactivity mineral phases. Weathering rate decreases strongly with relative grain size until a plateau is reached at larger grain sizes. Similarly, the mechanical contribution also increases with increasing grain size until a plateau is reached.

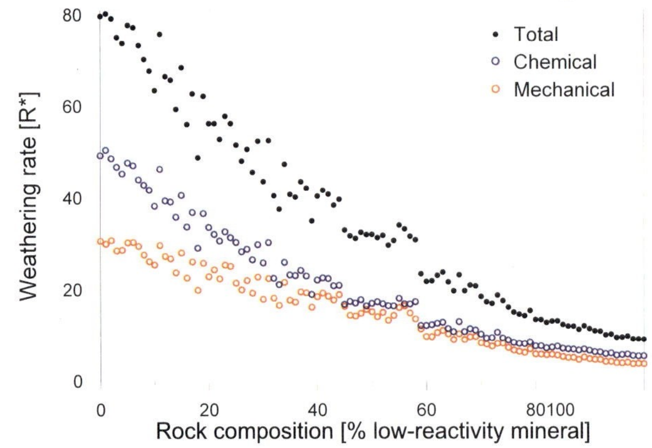

Figure 4. Impact of rock composition on weathering rate. The overall weathering rate (full black circles) decreases with the addition of low-reactivity minerals to the rock. Chemical weathering (hollow blue circles) shows a disproportionally large decrease with a small addition of low reactive phases. Mechanical weathering (hollow red circles) shows a more uniform decrease with the change in the rock composition, indicating that the main factor controlling the overall weathering rate is the chemical weathering rate.

ily due to a decline in chemical weathering, although mechanical weathering also slows over the same range of compositions (Fig. 4). However, when the low-reactivity phase becomes dominant, the further addition of low-reactivity grains has no significant impact on both the chemical and mechanical rate. We suggest that this trend is a result of the increasing density of low-reactivity grain boundaries: as more low-reactivity grains are added, the slow dissolution of their boundaries retards the rate at which new reactive grain boundaries are exposed to the fluid.

In contrast to the mechanical rate, which decreases as the amount of low-reactivity phase increases (Fig. 5a), the relative contribution of mechanical grain detachment to the overall weathering shows a more complex trend (Fig. 5b); as the proportion of low-reactivity grains increases from 0 to 40 %, the contribution of grain detachment increases from 37 to 46 %. When low-reactivity grains make up 40–60 % of the rock, the mechanical contribution reaches a maximum with values as high as 50 %. However, as the proportion of low-reactivity grains increases beyond 60 %, the mechanical contribution drops linearly to a value of about 41 %.

The peak in mechanical weathering between 40 and 60 % is not a result of an increasing number of detachment events in this range. This can clearly be seen at Fig. 5c, which shows that in fact, a minimum number of events occurs in this range. Instead, the observed trend can be explained by the size of detached clusters. The number of detachment events is greater than 2000, while the number of grains in the simulation is only 100, demonstrating that most of the detachments are sub-grains. In the snapshots of rocks with different compositions, it can be seen that when the rock comprises mainly high-reactivity minerals, numerous detachment events occur, and the removed clusters are relatively small (Fig. 6a). When the proportion of low-reactivity minerals increases to 40–60 % (Fig. 6b), the rapid dissolution of the high-reactivity phase leaves behind a skeleton of large clusters of low-reactivity grains. It is the removal of these large clusters that increases the contribution of mechanical weathering. The detachment of small clusters and individual grains again becomes the dominant mode when the rock comprises mainly low-reactivity minerals (Fig. 6c).

Although previous studies have shown that the chemical dissolution rate is determined by the abundance of lower-reactivity phases in a rock (White and Brantley, 2003; Wigley et al., 2013; Critelli et al., 2014), we show that the combined effects of chemical and small-scale physical processes control the overall weathering rate. Importantly, there is evidence from both field and laboratory experiments to support this.

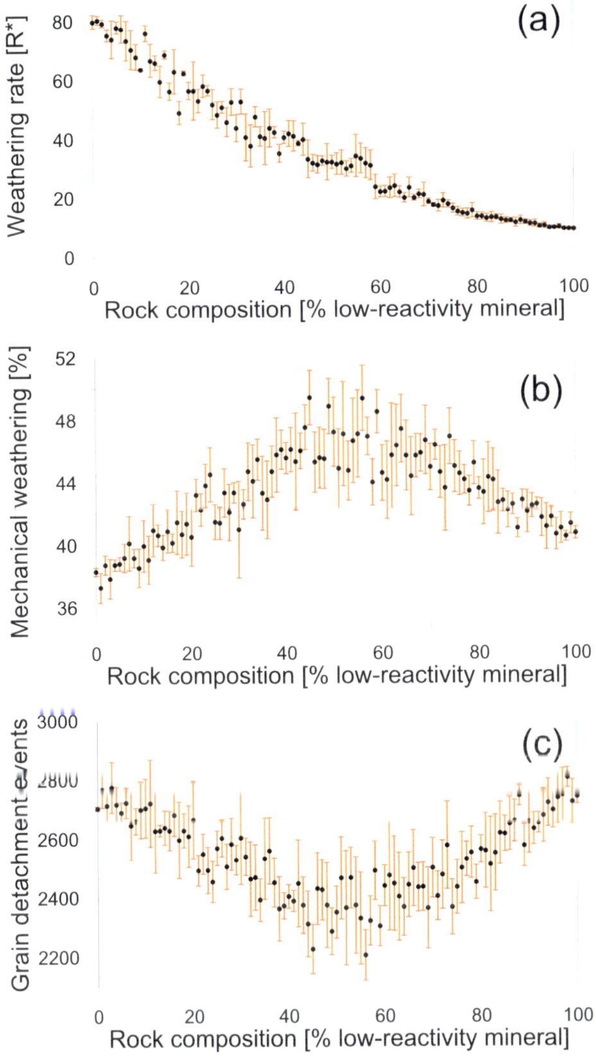

Figure 5. Impact of rock composition on weathering. **(a)** Weathering rate in R^* units versus composition; **(b)** ratio of mechanical weathering to total weathering as a function of composition; and **(c)** the number of detachment events versus composition. Black circles show the average of six different realizations for each composition, while the red bars show the standard deviation. The stochastic nature of dissolution and the random position of low-reactivity grains in the simulated rocks causes variation in the weathering rate between rocks with the same composition. The weathering rate decreases nonlinearly with the addition of low-reactivity minerals. When there are approximately equal proportions of high- and low-reactivity minerals, mechanical weathering reaches a maximum, although the actual number of detachment events is at a minimum.

For example, Deng et al. (2017) showed that the presence of a high-reactivity phase (such as calcite) strongly affected the stability of fractured shale surfaces, reporting that erosion was initiated only when the high-reactivity phase exceeded 35 %. Krklec et al. (2016) invoked a combination of chemical and micron-scale physical processes to explain the results

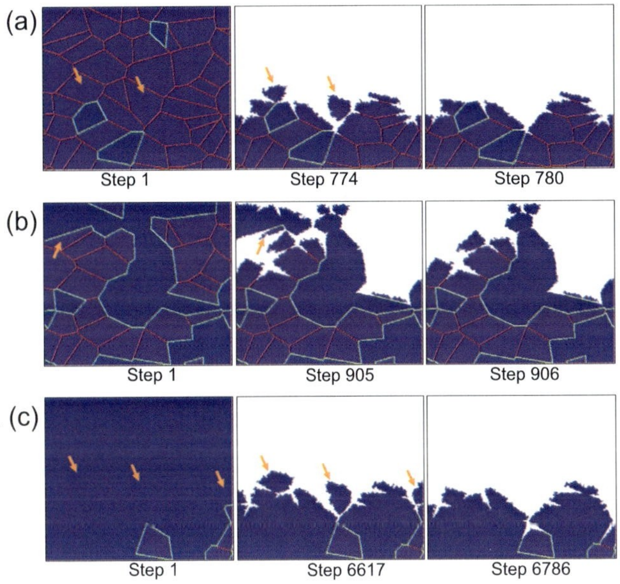

Figure 6. Simulation snapshots for three representative compositions. Rocks comprising **(a)** 10 % low-reactivity minerals, **(b)** an equal mix of high and low-reactivity minerals, and **(c)** 90 % low-reactivity minerals. In all simulations, the grain size is constant (relative grain size = 64). Orange arrows indicate clusters that are removed via mechanical grain detachment. Small sized grains are detached in **(a)** and **(c)**; in contrast, large clusters detach in **(b)**. Note that although **(a)** and **(c)** have similar granular structures, the overall weathering rate in **(c)** is almost 1 order of magnitude slower.

of weathering experiments on carbonate tablets. They suggested that most of the material loss was due to grain detachment that was triggered by dissolution along crystal edges. In a laboratory study of fine-grained limestone, Levenson and Emmanuel (2016) reported that grain detachment accounted for approximately 38 % of the material removed from the rock surface, which is within the range (37–50 %) predicted by our model. However, the precise proportion of mechanical weathering for any given rock is likely to be strongly dependent on the rate of dissolution along grain edges.

3.3 Effect of the rate of dissolution along grain boundaries on weathering

The last set of simulations was aimed at examining how the rate of dissolution along grain boundaries impacts the weathering process. In addition to the existing model in which the grain boundaries reacted at a rate that was dependent on its mineral phase (which we define here as the heterogeneous model), we constructed an alternative model in which all the grain boundaries reacted at the same rate (defined as the homogeneous model). A comparison of the two kinds of simulations shows that the weathering rate of rocks with homogeneous grain boundaries is almost constant, regardless of the mineral composition (Fig. 7a). This is due to a higher

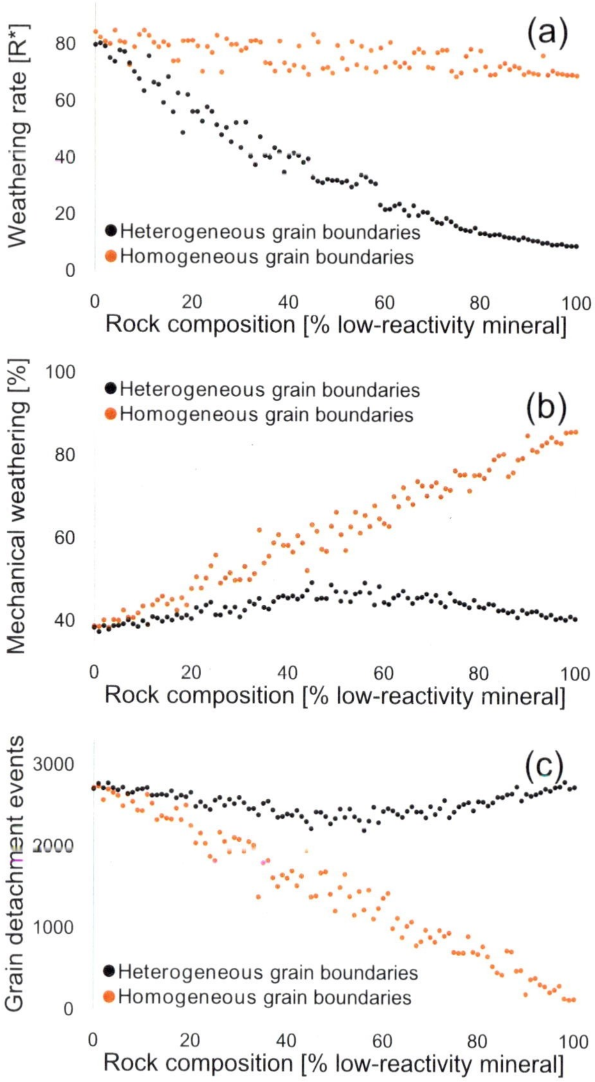

Figure 7. Impact of rock composition on weathering in heterogeneous and homogeneous grain boundary models. **(a)** Weathering rate in R^* units versus composition; **(b)** ratio of mechanical weathering to total weathering as a function of composition; and **(c)** the number of detachment events versus composition. The heterogeneous model assumes that the reactivity for each grain boundary is dependent on the type of the adjacent minerals, while in the homogeneous model all boundaries have the same reactivity.

density of reactive grain boundaries (relative to the heterogeneous model), which enhances the contribution of mechanical weathering to the overall rate (Fig. 7b). The proportion of grain detachment in the homogeneous model shows a linear increase with an increasing proportion of low-reactivity minerals, reaching a maximum value of 86 %. This result shows that the reactivity of the grain boundaries is the principal factor in determining whether or not mechanical detachment is the dominant mode of weathering. This is a result of the rapid dissolution of highly reactive grain boundaries which promotes grain detachment and effectively compensates for the increasing presence of low-reactivity minerals. In addition, the homogeneous model also shows a dramatic decrease in the number of detachment events with an increasing proportion of the low-reactivity phase (Fig. 7c). This reflects the transition from the detachment of tiny sub-grains to the detachment of large complete grains.

4 Conclusions

In this study, we used 2-D cellular automaton simulations to examine the impact of grain size and rock composition on chemical weathering and chemo-mechanical grain detachment. While our results indicate that the weathering rate decreases with increasing grain size, we also found that the weathering rate shows a nonlinear dependence on the proportion of low-reactivity minerals. This is due to the influence of chemical weathering, which is mainly controlled by the density of grain boundaries. In addition, we found that the contribution of mechanical weathering to the overall weathering is determined by the rate of dissolution along grain boundaries.

Our model is the first step towards accurately simulating the coupling between physical and chemical processes during weathering in rocks. As such, there are clearly a number of limitations. Our simulations do not include standard kinetic formulations or physical processes, such as diffusion and fluid flow. In addition, little is known about the rates of dissolution along grain boundaries for different mineral systems, so that it cannot yet be determined whether the heterogeneous model or the homogeneous model is more realistic. Moreover, crystals often contain defects so that their reactivity may in fact be heterogeneous, which could affect both reaction rates and dissolution patterns (Fischer et al., 2018). Future work focusing on the development of more sophisticated models and the experimental measurement of dissolution rates along grain boundaries, may provide a way to resolve these issues.

In addition to shedding light on the behavior of rocks comprising different minerals, our modeling approach could also be used to simulate rocks with different crystallographic orientations. Different surfaces of the same crystal often react at different rates (Godinho et al., 2014), so that the two different mineral reactivities in our model could represent a rock containing one mineral phase with two different orientations. In that sense, an advantage of using such a model is that it can be modified to represent various kinds of rocks and provide insight into effects of both mineral orientation and spatial distribution on mechanical and chemical processes during weathering. In addition, the model could provide a way to simulate the mobilization of particles and contaminants in a way that cannot be achieved using existing geochemical models.

Competing interests. The authors declare that they have no conflict of interest.

Acknowledgements. This research was supported by student scholarships from the Israeli Ministry of National Infrastructures, Energy, and Water Resources and The Hebrew University Center for Nanoscience and Nanotechnology. The Israel Science Foundation is thanked for additional financial support. We also thank Kristina Krklec and Cornelius Fischer for their constructive reviews.

References

Al-Mjeni, R., Arora, S., Cherukupalli, P., Van Wunnik, J., Edwards, J., Felber, B. J., Gurpinar, O., Hirasaki, G. J., Miller, C. A., and Jackson, C.: Has the time come for EOR?, Oilfield Rev., 22, 16–35, 2010.

Arvidson, R. S., Ertan, I. E., Amonette, J. E., and Luttge, A.: Variation in calcite dissolution rates, Geochim. Cosmochim. Ac., 67, 1623–1634, https://doi.org/10.1016/S0016-7037(02)01177-8, 2003.

Barati, R. and Liang, J.-T.: A review of fracturing fluid systems used for hydraulic fracturing of oil and gas wells, J. Appl. Polym. Sci., 131, 40735, https://doi.org/10.1002/app.40735, 2014.

Beckingham, L. E., Mitnick, E. H., Steefel, C. I., Zhang, S., Voltolini, M., Swift, A. M., Yang, L., Cole, D. R., Sheets, J. M., Ajo-Franklin, J. B., DePaolo, D. J., Mito, S., and Xue, Z.: Evaluation of mineral reactive surface area estimates for prediction of reactivity of a multi-mineral sediment, Geochim. Cosmochim. Ac., 188, 310–329, https://doi.org/10.1016/j.gca.2016.05.040, 2016.

Brantley, S. L.: Kinetics of Mineral Dissolution, in: Kinetics of Water-Rock Interaction, edited by: Brantley, S. L., Kubicki, J. D., and White, A. F., Springer New York, New York, NY, 151–210, 2008.

Bray, A. W., Oelkers, E. H., Bonneville, S., Wolff-Boenisch, D., Potts, N. J., Fones, G., and Benning, L. G.: The effect of pH, grain size, and organic ligands on biotite weathering rates, Geochim. Cosmochim. Ac., 164, 127–145, https://doi.org/10.1016/j.gca.2015.04.048, 2015.

Briese, L., Arvidson, R. S., and Luttge, A.: The effect of crystal size variation on the rate of dissolution – A kinetic Monte Carlo study, Geochim. Cosmochim. Ac., 212, 167–175, https://doi.org/10.1016/j.gca.2017.06.010, 2017.

Buss, H. L., Sak, P. B., Webb, S. M., and Brantley, S. L.: Weathering of the Rio Blanco quartz diorite, Luquillo Mountains, Puerto Rico: Coupling oxidation, dissolution, and fracturing, Geochim. Cosmochim. Ac., 72, 4488–4507, https://doi.org/10.1016/j.gca.2008.06.020, 2008.

Critelli, T., Marini, L., Schott, J., Mavromatis, V., Apollaro, C., Rinder, T., De Rosa, R., and Oelkers, E. H.: Can the dissolution rates of individual minerals be used to describe whole rock dissolution?, EGU General Assembly, Vienna, Austria, 27 April–2 May, EGU2014-11285, 2014.

Deng, H., Voltolini, M., Molins, S., Steefel, C., DePaolo, D., Ajo-Franklin, J., and Yang, L.: Alteration and Erosion of Rock Matrix Bordering a Carbonate-Rich Shale Fracture, Environ. Sci. Technol., 51, 8861–8868, https://doi.org/10.1021/acs.est.7b02063, 2017.

di Caprio, D., Stafiej, J., Luciano, G., and Arurault, L.: 3D cellular automata simulations of intra and intergranular corrosion, Corros. Sci., 112, 438–450, https://doi.org/10.1016/j.corsci.2016.07.028, 2016.

Emmanuel, S.: Mechanisms influencing micron and nanometer-scale reaction rate patterns during dolostone dissolution, Chem. Geol., 363, 262–269, https://doi.org/10.1016/j.chemgeo.2013.11.002, 2014.

Emmanuel, S. and Ague, J. J.: Impact of nano-size weathering products on the dissolution rates of primary minerals, Chem. Geol., 282, 11–18, 2011.

Emmanuel, S. and Levenson, Y.: Limestone weathering rates accelerated by micron-scale grain detachment, Geology, 42, 751–754, 2014.

Fischer, C. and Luttge, A.: Beyond the conventional understanding of water–rock reactivity, Earth Planet. Sc. Lett., 457, 100–105, https://doi.org/10.1016/j.epsl.2016.10.019, 2017.

Fischer, C., Arvidson, R. S., and Lüttge, A.: How predictable are dissolution rates of crystalline material?, Geochim. Cosmochim. Ac., 98, 177–185, https://doi.org/10.1016/j.gca.2012.09.011, 2012.

Fischer, C., Kurganskaya, I., Schäfer, T., and Lüttge, A.: Variability of crystal surface reactivity: What do we know?, Appl. Geochem., 43, 132–157, https://doi.org/10.1016/j.apgeochem.2014.02.002, 2014.

Fischer, C., Kurganskaya, I., and Luttge, A.: Inherited control of crystal surface reactivity, Appl. Geochem., 91, 140–148, https://doi.org/10.1016/j.apgeochem.2018.02.003, 2018.

Fletcher, R., Buss, H., and Brantley, S.: A spheroidal weathering model coupling porewater chemistry to soil thicknesses during steady-state denudation, Earth Planet. Sc. Lett., 244, 444–457, https://doi.org/10.1016/j.epsl.2006.01.055, 2006.

Godinho, J. R. A., Putnis, C. V., and Piazolo, S.: Direct Observations of the Dissolution of Fluorite Surfaces with Different Orientations, Cryst. Growth. Des., 14, 69–77, https://doi.org/10.1021/cg401119p, 2014.

Kirstein, J., Hellevang, H., Haile, B. G., Gleixner, G., and Gaupp, R.: Experimental determination of natural carbonate rock dissolution rates with a focus on temperature dependency, Geomorphology, 261, 30–40, https://doi.org/10.1016/j.geomorph.2016.02.019, 2016.

Kreisserman, Y. and Emmanuel, S.: Release of Particulate Iron Sulfide during Shale-Fluid Interaction, Environ. Sci. Technol., 52, 638–643, 10.1021/acs.est.7b05350, 2018.

Krklec, K., Marjanac, T., and Perica, D.: Analysis of "standard" (Lipica) limestone tablets and their weathering by carbonate staining and SEM imaging, a case study on the Vis Island, Croatia, Acta Carsologica, 42, 135–142, 2013.

Krklec, K., Domínguez-Villar, D., Carrasco, R. M., and Pedraza, J.: Current denudation rates in dolostone karst from central Spain: Implications for the formation of unroofed caves, Geomorphology, 264, 1–11, https://doi.org/10.1016/j.geomorph.2016.04.007, 2016.

Kump, L. R., Brantley, S. L., and Arthur, M. A.: Chemical Weathering, Atmospheric CO_2, and Climate, Annu. Rev. Earth Pl. Sc., 28, 611–667, https://doi.org/10.1146/annurev.earth.28.1.611, 2000.

Levenson, Y. and Emmanuel, S.: Quantifying micron-scale grain detachment during weathering experiments on limestone, Geochim. Cosmochim. Ac., 173, 86–96, 2016.

Levenson, Y., Schiller, M., Kreisserman, Y., and Emmanuel, S.: Calcite dissolution rates in texturally diverse calcareous rocks, Geological Society, London, Special Publications, 406, 81–94, 2015.

Levenson, Y., Ryb, U., and Emmanuel, S.: Comparison of field and laboratory weathering rates in carbonate rocks from an Eastern Mediterranean drainage basin, Earth Planet. Sc. Lett., 465, 176–183, https://doi.org/10.1016/j.epsl.2017.02.031, 2017.

Lions, J., Devau, N., de Lary, L., Dupraz, S., Parmentier, M., Gombert, P., and Dictor, M.-C.: Potential impacts of leakage from CO2geological storage on geochemical processes controlling fresh groundwater quality: A review, Int. J. Greenh. Gas Con., 22, 165–175, https://doi.org/10.1016/j.ijggc.2013.12.019, 2014.

Lishchuk, S. V., Akid, R., Worden, K., and Michalski, J.: A cellular automaton model for predicting intergranular corrosion, Corros. Sci., 53, 2518–2526, https://doi.org/10.1016/j.corsci.2011.04.027, 2011.

Lüttge, A., Arvidson, R. S., and Fischer, C.: A stochastic treatment of crystal dissolution kinetics, Elements, 9, 183–188, 2013.

Price, C. A. and Doehne, E.: Stone conservation: an overview of current research, Getty Publications, Los Angeles, 2011.

Putnis, C. V. and Ruiz-Agudo, E.: The Mineral–Water Interface: Where Minerals React with the Environment, Elements, 9, 177–182, https://doi.org/10.2113/gselements.9.3.177, 2013.

Røyne, A., Jamtveit, B., Mathiesen, J., and Malthe-Sørenssen, A.: Controls on rock weathering rates by reaction-induced hierarchical fracturing, Earth Planet. Sc. Lett., 275, 364–369, https://doi.org/10.1016/j.epsl.2008.08.035, 2008.

Silveira, F. A. and Aarão Reis, F. D. A.: Detachment of non-dissolved clusters and surface roughening in solid dissolution, Electrochim. Acta, 111, 1–8, https://doi.org/10.1016/j.electacta.2013.08.007, 2013.

Taleb, A. and Stafiej, J.: Numerical simulation of the effect of grain size on corrosion processes: Surface roughness oscillation and cluster detachment, Corros. Sci., 53, 2508–2513, https://doi.org/10.1016/j.corsci.2011.04.008, 2011.

White, A. F. and Brantley, S. L.: The effect of time on the weathering of silicate minerals: why do weathering rates differ in the laboratory and field?, Chem. Geol., 202, 479–506, https://doi.org/10.1016/j.chemgeo.2003.03.001, 2003.

White, A. F. and Buss, H. L.: Natural Weathering Rates of Silicate Minerals, Treatise on Geochemistry, 2nd Edn., Elsevier, Oxford, 115–155, https://doi.org/10.1016/b978-0-08-095975-7.00504-0, 2014.

Wigley, M., Dubacq, B., Kampman, N., and Bickle, M.: Controls of sluggish, CO_2-promoted, hematite and K-feldspar dissolution kinetics in sandstones, Earth Planet. Sc. Lett., 362, 76–87, https://doi.org/10.1016/j.epsl.2012.11.045, 2013.

Wilson, M. J.: Weathering of the primary rock-forming minerals: processes, products and rates, Clay Miner., 39, 233–266, https://doi.org/10.1180/0009855043930133, 2004.

8

Fluvial response to changes in the magnitude and frequency of sediment supply in a 1-D model

Tobias Müller and Marwan A. Hassan

Department of Geography, the University of British Columbia, Vancouver, BC, Canada

Correspondence: Tobias Müller (tobias.mueller@geog.ubc.ca)

Abstract. In steep headwater reaches, episodic mass movements can deliver large volumes of sediment to fluvial channels. If these inputs of sediment occur with a high frequency and magnitude, the capacity of the stream to rework the supplied material can be exceeded for a significant amount of time. To study the equilibrium conditions in a channel following different episodic sediment supply regimes (defined by grain size distribution, frequency, and magnitude of events), we simulate sediment transport through an idealized reach with our numerical 1-D model "BESMo" (Bedload Scenario Model). The model performs well in replicating flume experiments of a similar scope (where sediment was fed constantly, in one, two, or four pulses) and allowed the exploration of alternative event sequences. We show that in these experiments, the order of events is not important in the long term, as the channel quickly recovers even from high magnitude events. In longer equilibrium simulations, we imposed different supply regimes on a channel, which after some time leads to an adjustment of slope, grain size, and sediment transport that is in equilibrium with the respective forcing conditions. We observe two modes of channel adjustment to episodic sediment supply. (1) High-frequency supply regimes lead to equilibrium slopes and armouring ratios that are like conditions in constant-feed simulations. In these cases, the period between pulses is shorter than a "fluvial evacuation time", which we approximate as the time it takes to export a pulse of sediment under average transport conditions. (2) In low-frequency regimes the pulse period (i.e., recurrence interval) exceeds the "fluvial evacuation time", leading to higher armouring ratios due to the longer exposure of the bed surface to flow. If the grain size distribution of the bed is fine and armouring weak, the model predicts a decrease in the average channel slope. The ratio between the "fluvial evacuation time" and the pulse period constitutes a threshold that can help to quantify how a system responds to episodic disturbances.

1 Introduction

Mass movements in mountainous regions often deliver sediment directly to the stream network, resulting in coupled conditions that can trigger immediate channel responses during relatively large delivery events. Notably, delivery events can reset the local channel profile and govern construction and maintenance of channel bed architecture downstream of delivery points. The local response rate and trajectory following a delivery event is a function of the prevailing watershed flow regime, the magnitude of the delivery event, gradients in channel width (Ferrer-Boix et al., 2016), and in some instances the concentrated activity of aquatic species such as salmon (Hassan et al., 2008a). While lowland river systems have been the focus of a substantial body of research, less work has been carried out within steep mountain streams, particularly concerning mountain channel responses to changes in flow or sediment supply regimes.

In mountain streams, large, episodic inputs may temporarily dominate channel processes and morphology, significantly altering sediment transport and storage within the stream channel (Hassan et al., 2005, 2008b). Lisle et al. (1997) conducted flume experiments and numerical modelling that showed that sediment pulses are mainly reworked in situ, in contrast to a downstream translation in the form of a sediment wave. This finding is supported by Lisle et al. (2001), where little evidence of sediment waves was found in the field. Lisle and Church (2002) suggested that a stream channel responds to changes in the sediment supply by altering both storage and sediment transport rates. They describe that after a sediment pulse occurs, a first phase with low armouring rates allows for high transport rates in reworking the introduced material, corresponding to supply limited conditions. This is followed by a second phase in which armouring develops and transport rates decrease, corresponding to transport limited conditions. If the fluvial capacity is too low to evacuate the mass or grain size of material in the current hydrological regime, lag sediment can remain in the channel and dominate the local morphology for a long time (e.g., Benda et al., 2005; Brummer and Montgomery, 2006). Patterns of cyclic behaviour, associated with the rapid input of sediment from external sources, have been described in a number of field observations (e.g., Roberts and Church, 1986; Madej and Ozaki, 1996; Madej, 1999, 2001; Miller and Benda, 2000; Hoffman and Gabet, 2007; Hassan et al., 2008b) and some experimental studies (e.g., Cui et al., 2003; Sklar et al., 2009; Venditti et al., 2010; von Flotow, 2013; Elgueta, 2014; Ferrer-Boix and Hassan, 2014, 2015; Johnson et al., 2015; Elgueta-Astaburuaga and Hassan, 2017; An et al., 2017a). These studies observed fining of the bed surface, higher mobility and thus increased transport rates following such episodic sediment supply events. A reverse trend for coarsening and stabilizing of the bed was also noted as the supply was exhausted and a decrease in sediment transport rates followed (e.g., Dietrich et al., 1989; Church et al., 1998; Hassan and Church, 2000; Nelson et al., 2009). Cui and Parker (2005) support these findings using numerical modelling and further point out that abrasion can play an important role in the reworking of sediment pulses. Field observations (e.g., Benda, 1990; Pryor et al., 2011), and flume experiments (e.g., Pryor et al., 2011; Luzi, 2014) also document cycles of aggradation and degradation due to changes in the sediment supply. The observations discussed above suggest that changes in the sediment supply rate may lead to significant changes in bed elevation, bed surface texture, channel stability, and bed morphology. An analytical model developed by Blom et al. (2017) showed that the local channel geometry and surface grain size composition is mainly governed by long-term mean sediment supply rates and not by short-term changes in supply conditions. In contrast to cyclic sediment supply, cyclic hydrographs were found to mainly affect sediment transport rates and have a lesser impact on bed surface texture and channel morphology (Parker et al., 2007). Wong and Parker (2006) reported that cyclic hydrographs cause a part of the channel bed to undergo cyclic aggradation and degradation forming a hydrograph boundary layer. Cyclic sediment supply causes a similar effect which is termed the sedimentograph boundary layer (An et al., 2017b).

These findings imply that the morphological impact of the sediment pulse is most prevalent at the point of entrance, while the downstream portion mostly conveys the subsequently eroded material. The time needed for channel adjustments to occur after a large sediment input event depends on the amount and the texture of the delivered material. Brummer and Montgomery (2006) reported that 2 years after the supply event, the most mobile fractions (e.g., the fine fractions) were evacuated by a series of moderate floods while the largest grain sizes remained in the channel as lag deposits because flows were below their flow competence. Further, they showed that selective transport of sediment led to the development of an armour layer after only a few flow events. This armour layer protects the supplied material in the bed subsurface, increases bed stability, and causes lateral erosion and channel widening.

We expect the sediment transport rate in a channel to reach a long-term balance between erosional and depositional forces, even though there can be periodic changes in the short term. This state is defined as a "dynamic equilibrium" following Ahnert (1994). If the external forcing on the system changes, the channel will be in a transient state of adjustment towards a new dynamic equilibrium. Little attention has been directed to the question of what effect a change in the frequency of sediment supply events may have on the response of alluvial streams. Brunsden and Thornes (1979) proposed that if the frequency (i.e., recurrence interval) of disturbing events is shorter than the time necessary for a system to adjust to new boundary conditions ("relaxation time"), then transience will dominate the system and it may never achieve equilibrium (Brunsden, 1980). Wolman and Miller (1960) suggest that mountain channels experiencing direct inputs of sediment are good examples of such systems, where form is defined by extreme events rather than events of intermediate magnitude and frequency. The concept of this so-called "temporal sensitivity" was later elaborated on by Thomas (2001) and Brunsden (2001), although since their studies little attempt has been made in fluvial geomorphology to address this issue in practice. Bull (1991) applies the theory of Brunsden and Thornes (1979) to the impact of a hypothetical temporal succession of disturbance events on sediment storage, which can either increase or decrease the stream-bed elevation. The system processes the disturbance over the relaxation time, which together with a potential reaction time constitutes the total response time. If there is no further disturbance, the system status remains unchanged over a time of persistence. The concept from Bull (1991) is based on a system's trend towards a dynamic equilibrium between the forcing by, and the reworking of, disturbances. Howard (1982) concluded that if episodic inputs occur with a frequency that matches the inverse of the relaxation time, the output of the system will remain in a constant equilibrium with the average value of the forcing. Flume based insights about equilibrium conditions and timescales of adjustment to changes in sediment supply rates are discussed in some studies (e.g., Elgueta-Astaburuaga and Hassan, 2017; Pryor et al., 2011), but response times are only quantified in few cases (e.g., Podolak and Wilcock, 2013).

The paragraphs above illustrate how the frequency at which events occur may be fundamental in defining the response of a fluvial system to a change in boundary conditions. Therefore, it appears that event frequency should be a central aspect in investigations regarding the effect of episodic sediment supply on streams. Consequently, our understanding of involved processes remains incomplete. For example, it is uncertain whether the freshly delivered sediment that buries and is transferred over the bed surface is simply removed by the subsequent floods or whether there is some exchange between armoured and structured bed and the fine and mobile deposits. Furthermore, Hassan and Zimmermann (2012) asserted that it is important to study how quickly internal changes in grain size, channel morphology, and sediment storage occur when the stream shifts between cycles of aggradation and degradation.

Our main research objective is to describe the impact of episodic sediment supply on channel bed evolution in simulations using a 1-D morphodynamic numerical model for a bed of multiple grain sizes. We use the model to recreate conditions from experiments conducted at the Mountain Channel Hydraulic Experimental Laboratory, University of British Columbia, where a set of experiments were carried out to examine the impacts of episodic sediment supply on bed surface evolution and channel adjustment of a gravel bed stream (von Flotow, 2013; Elgueta, 2014; Elgueta-Astaburuaga and Hassan, 2017; Elgueta-Astaburuaga, 2018). Although these experiments provide detailed information on channel adjustment to changes in the sediment supply regime, they are limited in terms of the number of experiments and the range of scenarios that could be conducted. The performance of the model is tested against the experimental results obtained in the laboratory and then used to further explore controls and responses of the fluvial system to changes in flow and sediment supply regimes. The specific research questions addressed in this study are as follows:

1. Can the numerical model recreate the channel response that was observed in flume experiments of similar scope?

2. Does the sequencing of supply events play a role in the reaction of a gravel-bed stream, when several events of specified magnitudes occur in a different order?

3. How will different combinations of episodic sediment supply, obtained by varying their magnitude and frequency, impact channel evolution of a gravel-bed stream?

2 Methods

We applied the 1-D morphodynamic model BESMo (Bedload Scenario Model) to calculate capacity based sediment transport under different sediment supply regimes. We chose values for model parameters to match the flume experiments as closely as possible and used measurements of sediment transport rate, surface grain size distribution, and slope to calibrate the model. Matching our research questions, we then conducted two types of simulations:

1. In "event sequencing simulations", we simulated different permutations of events to understand the role event succession plays in long-term channel response.

2. In "equilibrium simulations", we used the same model setup and imposed different, but within each run regular, supply event frequencies. These simulations were run until we achieved a recurrent pattern in slope and grain size adjustment, allowing us to identify how the channel adjusts to the supply regime in the long term.

2.1 Model setup

The structure of the model is similar to other models designed to reproduce and interpret data from flume experiments (e.g., Cui and Parker, 2005; Wong and Parker, 2006; Ferrer-Boix and Hassan, 2014; An et al., 2017a). Figure 1 gives an overview of the implemented model components and their basic interaction. The model can be subdivided into a "hydraulic part" and a "sediment part", both of which are subject to "external forcing" that varies in according to the modelling scenarios.

We set up the modelling environment to run on a Compute Canada research cluster, which allows us to simulate many different input conditions in parallel and compare results quickly. We use a backwater flow model as suggested by Cui et al. (2006), implementing a threshold Froude number (Fr) to switch conditions between supercritical and subcritical flow:

$$Fr = \sqrt{\frac{Q_\mathrm{w}^2}{g w_\mathrm{r}^2 h^3}}. \tag{1}$$

The Froude number (Fr) is calculated as a function of discharge (Q_w), gravity (g), channel width (w_r) and water depth (h) (Eq. 1). The threshold $Fr = 0.9$ simplifies the calculation of flow conditions, allowing us to spatially iterate through the nodes only once from downstream to upstream. In the case of subcritical flow, the water depth is solved locally as a function of downstream friction slope (S_f) and bed slope (S_0) (Eq. 2a):

$$\frac{\mathrm{d}h}{\mathrm{d}x} = \frac{S_0 - S_\mathrm{f}}{1 - Fr^2} \quad \text{for subcritical: } Fr < 0.9 \tag{2a}$$

$$h = \left(\frac{n^2 Q_\mathrm{w}^2}{S_\mathrm{f}}\right)^{3/10} \quad \text{for supercritical: } Fr \geq 0.9. \tag{2b}$$

Water depth under supercritical flow conditions is calculated locally assuming steady uniform flow using the Manning–Strickler formulation (Eqs. 2b and 3), where α_r is a coeffi-

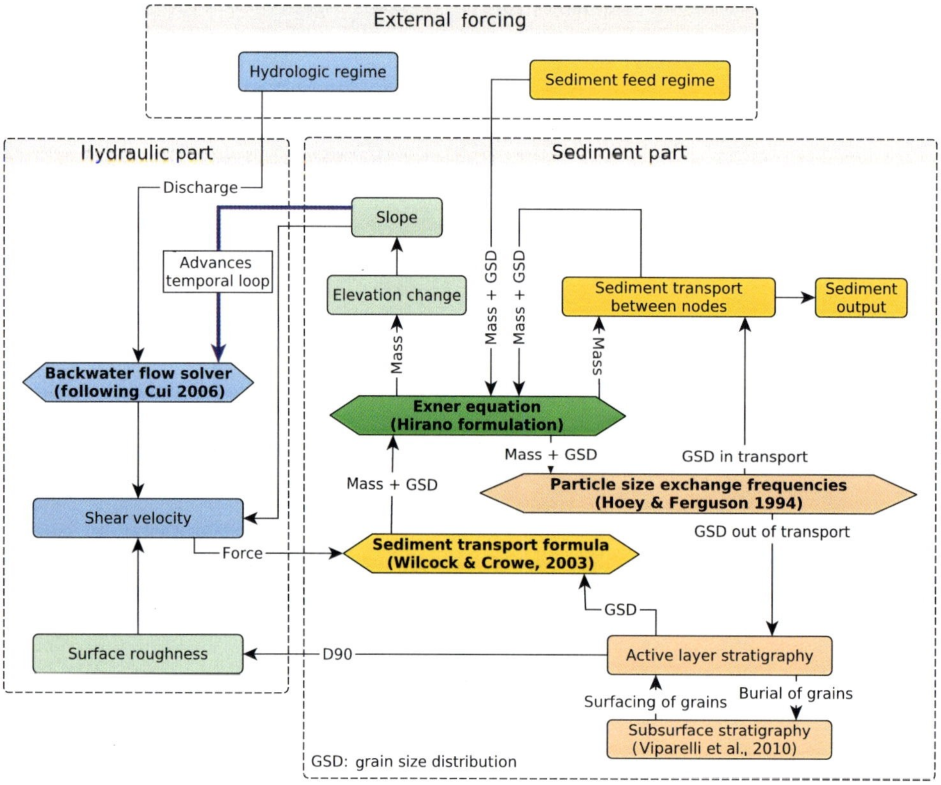

Figure 1. Flowchart stating the main components of the model and the flow of information between them. The temporal loop is advanced as the new elevation affects the slope in the flow model. The components are coloured as follows: blue – flow related, dark yellow – sediment volume related, peach – particle size related, and green – geometry related.

cient of 8.1 (Parker, 1991) and roughness height (k_s):

$$n = \frac{k_s^{1/6}}{\alpha_r g^{1/2}}. \quad (3)$$

k_s is calculated using the constants n_k and D_{s90}, the surface grain size for which 90 % of the surface is finer:

$$k_s = n_k D_{s90}. \quad (4)$$

In the case of steady and uniform flow the bed slope S_0 is equal to the friction slope (S_f) with bed elevation (η_b) at downstream position x:

$$S_f = \frac{-d\eta_b}{dx} = S_0. \quad (5)$$

If the solution of the water depth with Eq. (2a) is numerically unstable on the current node distribution, the model subdivides the channel into more nodes and reiterates the subdivision until a stable backwater curve is found. This approach does not properly represent the location of hydraulic jumps (Cui et al., 2006), which should not be a problem as we average conditions over a node spacing of at least one channel width. Boundary shear stress (τ_b) is then calculated with the depth-slope product:

$$\tau_b = \rho g h S_f, \quad (6)$$

where ρ is the water density. τ_b is then converted to the shear velocity u^*, which is used in the sediment routing component:

$$u^* = \sqrt{\tau_b/\rho}. \quad (7)$$

The volumetric unit bedload transport rate per size class q_{bi} is calculated using the sediment transport function provided by Wilcock and Crowe (2003). The change in bed elevation $\partial\eta_b$ for each node x per time step t follows from the Exner equation of mass conservation:

$$(1-\lambda)\frac{\partial\eta_b}{\partial t} = -\frac{\partial q_b}{\partial x}, \quad (8)$$

where λ is bed porosity. The volumetric bedload transport rate per unit width is given as q_b and is calculated per grain size class i in the sediment mixture of n size classes:

$$q_b = \sum_{i=1}^{n} q_{bi}. \quad (9)$$

The model incorporates subsurface stratigraphy using the active layer concept (Parker, 2008), which gives the Hirano equation:

$$(1-\lambda)\left[\frac{\partial}{\partial t}(L_a F_i) - f_{Ii}\frac{\partial L_a}{\partial t}\right] = -\left(\frac{\partial q_{bi}}{\partial x} - f_{Ii}\frac{\partial q_b}{\partial x}\right), \quad (10)$$

where L_a is the active layer thickness, F_i is the surface frequency of the ith grain size class, f_{Ii} is the ith grain size class proportion exchanged between the surface and the subsurface, and $q_{bi} = p_{bi} q_b$ is the volumetric unit bedload transport rate of the ith grain size class where p_{bi} represents the ith fraction of the bedload transport rate. The active layer thickness is calculated as $L_a = n_a D_{s90}$, with the parameter n_a, representing the scale of bed fluctuations.

The grain size distribution of the sediment flux between the active layer and the substrate is either calculated from the subsurface texture when the bed degrades, or from a linear combination of surface and bedload grain size distributions when the bed aggrades (Hoey and Ferguson, 1994):

$$f_{Ii} = \begin{cases} f_i & \text{for } \frac{\partial \eta_b}{\partial t} < 0 \\ \alpha F_i + (1-\alpha) p_{bi} & \text{for } \frac{\partial \eta_b}{\partial t} > 0, \end{cases} \quad (11)$$

where f_i is the fraction of the ith grain size class in the subsurface, and α is a constant. The vertical stratigraphy is stored in 10 cm high layers following Viparelli et al. (2010). By keeping track of grain size distributions within the surface and subsurface layers, the model can preserve the history of phases of erosion or aggradation. This allows emergent properties such as armouring layers to occur. To study the combined effect that the active layer thickness factor n_a and the active layer exchange ratio α have on the model results, we executed sensitivity runs shown in the Supplement to this paper (Supplement Figs. S1–S4).

The Exner equation (Eq. 8) in combination with the expression for the friction slope in Eq. (2) and the sediment transport function by Wilcock and Crowe (2003) form an non-linear advection–diffusion system that allows the calculation of bed elevation as a function of space and time (An et al., 2017a). An upwind scheme was used for the numerical discretization. The model needs an initial bed profile and an initial value of the surface grain size to be solvable. Sediment boundary conditions are given by the sediment feed rate and grain size distributions on the inlet, and a fixed bed elevation at the outlet of the simulated reach. The flow boundary condition is a water surface height of 0.1 m over normal flow at the outlet. The bedload transport function is used to calculate transport rates for each channel cross section. During the model run, changes of the sediment transport rate are dependent on changes in the sediment supply, channel slope, and surface grain size distribution.

2.2 Model calibration

We used data from flume experiments to calibrate the model. The objective of the flume experiments was to measure the adjustment of an alluvial steep channel to different frequencies of sediment supply. The experiments were carried out in a water recirculating flume which is 18 m long, 1 m wide, and 1 m deep in the Mountain Channel Hydraulic Experimental Laboratory at the University of British Columbia. Here we will provide a brief summary of the flume setup and experimental design, for more details see von Flotow (2013), Elgueta (2014), Ferrer-Boix and Hassan (2015), and Elgueta-Astaburuaga and Hassan (2017).

The experiment consisted of seven 40 h long runs with different sediment supply frequencies, while keeping the total sediment input the same at 300 kg per run. The experiments were run continuously, i.e., the bed surface at the end of run 1 was the starting condition for run 2 and so on. For all runs, flow was held constant so that the sediment feed regime could be studied with no changes in flow regime (Table 1). The difference between the runs was the spreading of the supply over a changing input frequency, which was either constant, in one pulse, in two pulses, or in four pulses. The bed was fixed in the first 1 m downstream section of the flume head box with stones equivalent to about D_{s84} of the experimental bed material. In the remainder of the flume the bed initially consisted of 0.1 m of loose material with particle sizes ranging from 0.5 to 64 mm with a D_{s50} of 5.64 mm, matching downscaled (by a factor of 3) conditions of a study reach in East Creek, British Columbia, Canada. The flume slope was set to 0.022 m m^{-1}. Measurements include water depth, water surface slope, water velocity, bed surface slope, bed surface particle size distribution, bed elevation, sediment transport rate, and bedload texture. Measurements of the water surface elevation were conducted throughout the experiment using a mechanical point gauge with 0.001 m precision. Photos were used to manually sample bed surface grain size distributions and the bed elevation was recorded with a green laser scanner at a 2 mm resolution. The bed surface scans were used to measure the bed surface slope along the thalweg, i.e., the line of lowest elevation along the flume. Flow velocity measurements were conducted using an ADV profiler. The grain size and count of particles exiting the flume were recorded with a camera and a light table at the outlet of flume (Zimmermann et al., 2008). The transport rate measurements were done at 30 Hz and validated after the experiments by total exported weight.

Using the model described, we simulated a 12 m long and 1 m wide channel in 13 downstream nodes each spaced 1 m apart. The model was set to calculate sediment transport for all nodes in time steps of 10 s. All simulations used a constant water discharge of 0.065 m^3 s^{-1} and a geometric mean grain size of 5.64 mm for both initial bed and sediment feed (full distribution shown in Fig. 2a). We chose to use a normally distributed approximation of the flume grain size with a width of $\sigma = 1.6$ to be consistent with distributions used for the equilibrium simulations. This distribution and the original flume GSD are statistically the same. The initial channel slope was 0.022 m m^{-1}, also matching the parameters from the flume experiments.

We calibrated the model by visually reducing the difference between measured and simulated values of bed slope (S), surface grain size parameters (D_{sg}, D_{s90}), and transport rate (q_b). In the calibration runs more importance was given to recreating S, D_{sg}, and q_b than to a good match in D_{s90}. We first increased the reference Shields stress (τ^*_{rm}) in the Wilcock and Crowe formula to roughly match simulated and measured q_b. Afterwards, we varied the grain exchange ratio (α) (Eq. 11) and the coefficient of the active layer thickness (n_a). As we achieved a good visual match between simulation and flume measurements, we did not see the need to calibrate more parameters.

2.3 Event sequencing simulations

We explored the role that the sequencing of the pulse events could have on the flume study by simulating the "original flume" event sequence and comparing the result to alternative sequencing of events (see Table 2). The alternative event sequences are using the same pulse distributions (four pulses, two pulses, or one pulse over 40 h), but the pulse order is either from "few to many" (FtM) (i.e., one pulse, then two pulses, then four pulses) or from "many to few" (MtF) (i.e., four pulses, then two pulses, then one pulse) per 40 h phase. To allow the system more time to recover from pulse events, we simulated two more cases where each pulsed phase is buffered from the next one by a 40 h constant-feed phase. These runs are called "c-buffered: many to few" (cMtF) and "c-buffered: few to many" (cFtM).

2.4 Equilibrium simulations

In our final set of experiments, we kept the frequency and magnitude of pulse events constant to achieve equilibrium slope and grain size conditions. The use of numerical modelling allows for the comparison of many simulations with differing grain size distributions, pulse frequencies, and pulse magnitudes. We expect a channel under episodic sediment supply to adjust synchronously to the frequency of external forcing events. The added sediment volume from a supply event will increase the channel slope at first. After the supply of sediment ends and material is removed from the channel, the slope will decrease, and the surface grain size will begin to reflect the sediment starved conditions. As the long-term sediment input equals the long-term sediment output, the channel will eventually achieve a condition where the capacity to erode material (through increased slope in conjunction with changes in the surface grain size) equals the depositional forcing (i.e., long-term sediment input) of the supply regime. In this state the adjustment of channel slope and grain size to each sediment input event will return to the same values after every pulse. All runs achieved this equilibrium condition within 20 000 simulation hours.

To find the equilibrium slope resulting from different sediment supply regimes, we simulated different combinations of sediment supply frequency (F_{pulse}) and magnitude (M_{pulse}) for nine different grain size distributions (Fig. 2a). All distributions have the same mean grain size of 5.64 mm, but differ in the width of the distribution by the standard deviation (σ), which was chosen for a phi-scaled, normally distributed sample. While $\sigma = 0.05$ represents a nearly uniform sediment mixture, $\sigma = 1.6$ roughly matches the grain size distribution of the flume experiments. We used 11 grain size classes in the simulations and set the initial GSD to the feed GSD. Under the hydraulic conditions applied all grain sizes are initially mobile, which might change during the simulations due to the effect of armouring and changes in bed slope. Figure 2b shows the combinations of frequency and magnitude used in this study. Each model run delivered the same input mass over the simulation time (150 000 kg over 20 000 h). We then distributed this total mass over different pulse frequencies, with four of the combinations matching the flume experiments. Each pulse was 10 min in length. The lowest frequency was chosen to be one pulse every 400 h (pulse period time: $T_{pp} = 400$ h), and the highest frequency was constant feed (one 10 min long pulse every 10 min). We selected a range of 40 pulse frequencies for which the whole number of cycles summed to 20 000 h. In total we executed 360 simulations, 9 different σ with 40 frequencies each.

3 Results

3.1 Model calibration

Elgueta-Astaburuaga and Hassan (2017) describe the flume results we used for model calibration in detail. We will give a short summary of the findings here. An initial run without sediment feed over an unstructured bed showed a high sediment output while the bed armoured. This run was similar in output grain sizes and grain mobility to the run with one sediment pulse over a structured bed. This shows that active restructuring of the bed occurred in both of these runs. On the other end of the spectrum, the constant feed and four pulse runs were similar in their low sediment output and showed different grain mobility. This implies that the system reacts differently at a threshold frequency somewhere between two pulses per 40 h and four pulses per 40 h, which Elgueta-Astaburuaga and Hassan (2017) interpret as the relaxation time of the bed to a pulse event.

We used these experiments to test the ability of the numerical model to recreate the flume results in discharge, pulse frequency, pulse magnitude, slope, and grain sizes. In the experiment, mean grain size (D_{sg}) and slope (S) were measured over a central 2 m long section to avoid a bias of bed surface measurements due to inflow and outflow conditions. The measurement intervals varied between every 1 and every 20 h (depending on the pulse interval). The values for the numerical simulation are averaged over the whole reach and were recorded every 10 min during the simulation time. We achieved a best match in slope S, D_{sg}, D_{s90}, and the transport rate q_b by increasing the reference Shields stress τ^*_{rm} in the Wilcock and Crowe formula by a factor of 2. Wilcock (2001) suggests taking the same approach of increasing the threshold shear stress to match a sediment transport calcula-

Table 1. Overview of runs in the flume experiments. All runs were 40 h long. The texture of the initial bed mixture and the sediment feed were identical. Plots of resulting slope, D_{sg}, D_{s90}, and sediment transport are shown in Fig. 3. The symbols in column two signify the feed regime where 0 represents no feed, C represents constant feed, and 1, 2, and 4 represent the number of pulses in 40 h.

Flume run	Symbol	Feed regime	Mean water depth after 40 h (m)	Bed slope after 40 h ($m\,m^{-1}$)	Water surface slope at 40 h ($m\,m^{-1}$)	Feed rate ($g\,ms^{-1}$)	Feed duration (min)	Pulse magnitude (kg)	Pulse period (h)
R1	0	None	NA	0.017	NA	0	–	–	–
R2	C	Constant	0.073	0.016	0.017	2.0833	–	–	Const.
R3	1	One pulse	0.080	0.018	0.019	83.3	60	300	40 h
R4	4	Four pulses	0.083	0.020	0.020	83.3	30	75	10 h
R5	2	Two pulses	0.072	0.022	0.020	83.3	15	150	20 h
R6	C	Constant	0.075	0.022	0.020	2.0833	–	–	Const.
R7	0	None	0.073	0.022	0.020	0	–	–	–

Initial slope 2.2 %; mixture D_{s50} = 5.64 mm; mixture D_{s90} = 11.2 mm, Q_w = 65 L s^{-1}; duration of each run = 40 h; total feed = 300 kg per run. NA represents not available.

Table 2. Sequencing of events in runs that were simulated as permutations of the original flume experiment. Each of the seven periods was 40 h long and each run lasted 280 h in total. There was no sediment input during the no feed runs (0). Within all other period types, 300 kg of sediment was fed over 40 h, either constantly (C) or in pulses (one, two, or four events). Besides recreating the original flume sequence (OF), we simulated two runs where the pulsed events occur either in order from many pulses to few (MtF) or from few to many (FtM). To explore if the system could rebound from the impact of a certain pulse phase during a constant-feed phase, we created two additional runs where this was the case (cMtF and cFtM), which led to a 600 kg higher total sediment feed.

	Simulation run	Event sequence symbol							Mass fed total (kg)
OF	Original flume	0	C	1	4	2	C	0	1500
MtF	Many to few	0	C	4	2	1	C	0	1500
FtM	Few to many	0	C	1	2	4	C	0	1500
cMtF	C-buffered: many to few	C	4	[illegible]	[illegible]	[illegible]	[illegible]	[illegible]	2100
cFtM	C-buffered: few to many	C	1	C	2	C	4	C	2100

tion to field data. Other researchers use this method to calibrate models to field and flume data (Gary Parker, personal communication, 2018; Chartrand et al., 2015).

Figure 3 shows the comparison between flume measurements and the calibrated model results with the grain exchange ratio $\alpha = 0.45$ (Eq. 11) and the active layer thickness factor $n_a = 2$. The sensitivity of the model to changes in α and n_a is shown in Supplement Figs. S1–S4. Due to the long interval between measurements in the flume experiments, some short-term slope responses to individual sediment pulses might be hidden (e.g., after hours 80 and 180 in Fig. 3a). The model underpredicts both the slope and the mean surface grain size (D_{sg}) (see Fig. 3b) in the first 60 h, while the coarse grain size fractions (D_{s90}) and average transport rate are over predicted for the first 30 h (see Fig. 3c and d). This might be due to imperfect initial conditions or boundary effects in the flume experiments. For the rest of the simulation both slope and D_{sg} show good agreement with the flume results. D_{90} is overpredicted in the model, but this is seen as a minor issue because the simulated transport rates mainly depend on D_{sg} and the slope. The transport rate in the simulation lags behind the light table data, which might be due to our numerical implementation of diffusion. As the model matches the average transport rates well, we did not see the need to improve the temporal agreement.

3.2 Event sequencing simulations

After obtaining a good match between the model and the flume data, we simulated alternative event sequences as described in Table 2. Figure 4a shows the adjustment of slope in runs that preserved the same sediment feed volume and had the same duration as the flume experiments (OF, MtF, and FtM), but the frequency of events is ordered differently. At the end of the simulations, all runs approach the same slope value of 0.022 m m^{-1}, which shows that the main factor determining the long-term slope is the total volume of sediment fed. The sequencing of events seems to play a role in the slope adjustment over the short term, here about 80 h after the events. On this short timescale, large pulses increase the slope quickly, while the smaller, more frequent pulses lead to a more gradual adjustment of slope. Figure 4b shows the runs where pulse phases were buffered by constant-feed phases (cMtF and cFtM), increasing the total sediment feed by 600 kg. This did not change the pattern of adjustment significantly compared to the earlier runs, as the constant-feed phases only prolonged the effect of the previous pulse phase.

The effect of event sequencing on the channel response in D_{sg} is shown in Fig. 4c for all runs in Table 2. The order of events only has a weak impact on patterns of adjustment in D_{sg}, as maximum grain size conditions are reached within 20–30 h. Afterwards no further adjustment occurs un-

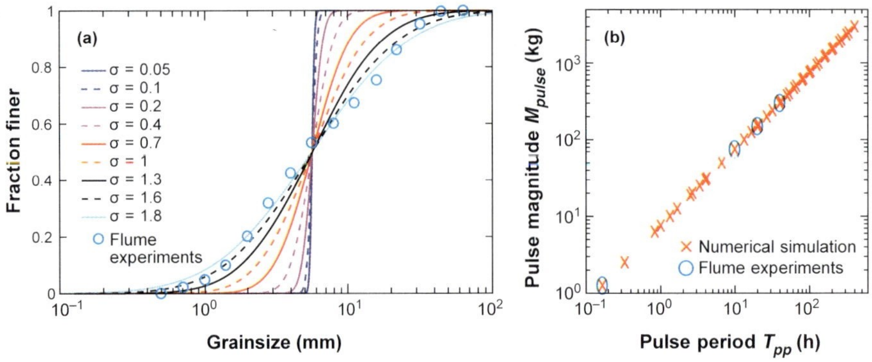

Figure 2. **(a)** Variation of the grain size distribution between model runs (lines) and values from the flume experiments (circles). All distributions have a geometric mean grain size (D_{sg}) of 5.64 mm. The different σ values represent the width of the distribution, calculated for normal distributions in phi-scaled sediment sizes. The data for the flume experiments are roughly matched with $\sigma = 1.6$. **(b)** Combinations of pulse magnitude and pulse period (or recurrence interval) used in the equilibrium model runs. As the total mass of supplied material is the same for all simulations, runs with high pulse frequencies (i.e., low pulse periods T_{pp}) are smaller in magnitude. Four of the combinations match the flume runs (circles).

til the introduction of fine material with the next pulse lowers the surface grain size again. This means that armouring of the channel surface happens quickly in relation to the time between pulse events. The subsurface is made of the same grain size distribution as the sediment feed, so its mean size is 5.56 mm. Therefore, we can infer that an armouring ratio (D_{sg}/D_{subg}) of about 2.2 was reached within 20 h and then increased towards 2.7 over the following 260 h. A finer, less armoured surface at the time of each supply event is followed by a coarsening of the surface as the finer grain sizes are more mobile, and thus more easily evacuated in the time without feed between pulses.

3.3 Equilibrium simulations

Our second set of simulations explored the equilibrium conditions that are reached under different supply regimes. As we will explain in the discussion, the effect of the supply regime can be constrained to changes in slope (S) and the surface grain size distribution (GSD_{fluv}), which in the following will be characterized by the armouring ratio between surface and subsurface mean grain size (D_{sg}/D_{subg}). After different times in the simulations, these parameters reach a time-independent periodic adjustment that is illustrated in Fig. 5 for the last 400 h of two simulations. Figure 6a shows box plots of the distribution of slope values during the last pulse of all 40 runs with $\sigma = 1.6$. The presented normalized slopes (S/S_{const}) indicate how the slope for each pulse frequency compares to the constant-feed slope of runs with the same grain size distribution. Figure 6b shows the change in normalized mean slope (S_{mlp}/S_{const}) with pulse frequency, which corresponds to the red lines in Fig. 6a and is our main indicator for the equilibrium state of the channel slope. Each line represents a different width σ of the GSD over 40 runs with increasing T_{pp}.

4 Discussion

4.1 Extension of flume results with the numerical model

The numerical model shows good agreement with the temporal response of mean surface grain size and slope from the flume experiments. As the initial bed grain size distributions were well mixed, the good match in mean surface grain size also implies a good match in the armouring ratios. A series of runs with an alternative sequencing of events showed that, while the adjustment of mean surface grain size was not sensitive to the order of the pulsed phases, the evolution of slope differed considerably. In the cases where the first pulsed phase consisted of one large magnitude event (FtM and cFtM), the slope increased quickly and the following higher-frequency, lower magnitude pulse phases did not modify the system considerably. In contrast, cases where multiple smaller event phases occurred first (MtF and cMtF), the slope increased more gradually. All runs ended at about the same slope after the 280 h simulation time, which implies that while the low frequency, large magnitude events strongly alter the channel in the short term, the sequencing of events does not play an important role in the long run. The constant-feed-buffered runs (cFtM and cMtF) show similar behaviour to their unbuffered counterparts, as the constant-feed phases preserve the bed state of the previous pulse phase. The main driver of the slope adjustment is the total sediment feed, which is consistent with findings by Blom et al. (2017).

4.2 Development of timescales from the equilibrium simulations

The definition of a sediment supply regime can be based on different aspects of sediment input into a stream, either from outside or within the channel. For simplicity, we restrict the definition of a sediment supply regime to the input of mate-

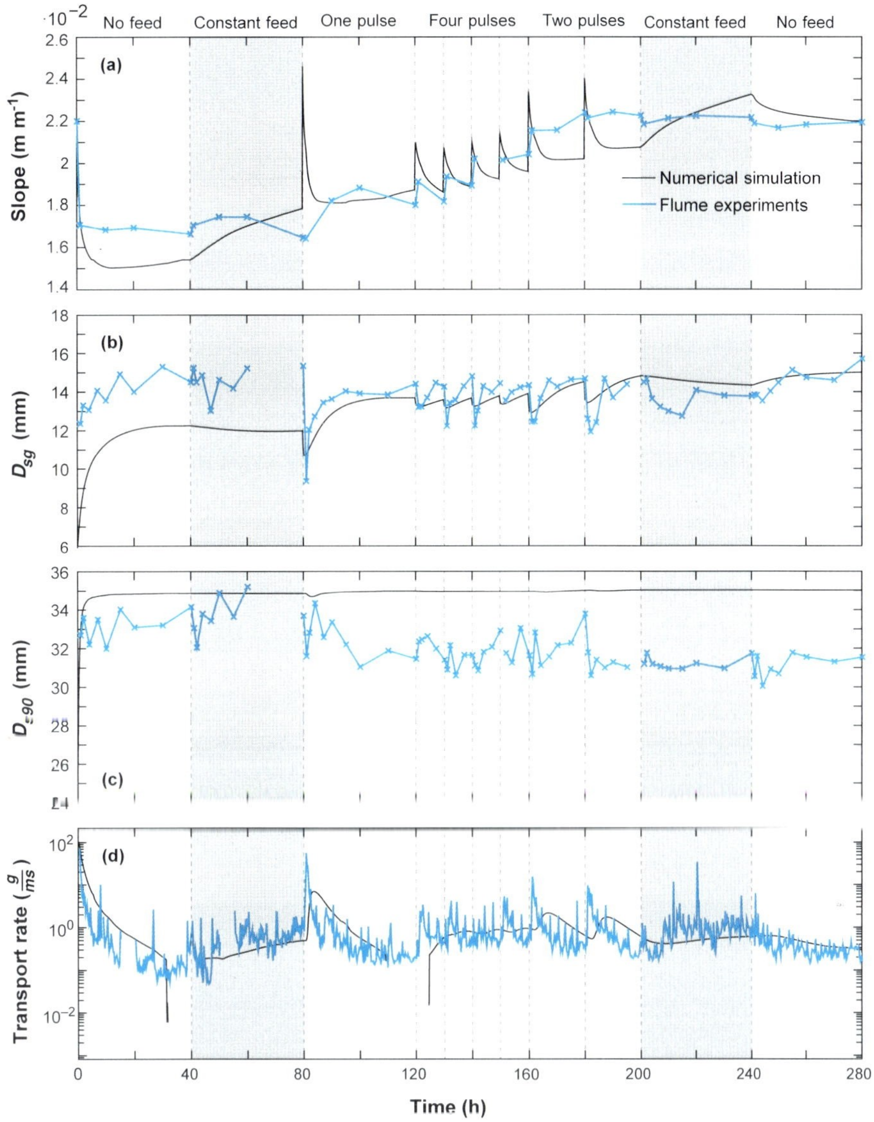

Figure 3. Comparison of **(a)** slope, **(b)** mean surface D_{sg}, **(c)** surface D_{s90}, and **(d)** sediment transport rate between the numerical simulation and the flume experiments.

rial into the fluvial system from outside the active channel. Sediment supply from storage close to the channel can be viewed as external supply if it only occurs episodically (e.g., less than yearly) in large flooding events. This view allows us to describe the sediment supply regime by a combination of frequency, magnitude, and grain size distribution of sediment supply events over a multi-event time frame (as in Benda and Dunne, 1997a). The time frame must be long enough to contain enough sediment supply events to allow the stream bed to adjust to the external forcing by changing its internal configuration of sediment storage and bed structuring. Even though a natural channel might never reach an equilibrium to a certain sediment supply regime, it might produce regular patterns of transient adjustment to the supply events.

In our case, the forcing on the system is a combination of pulse frequency (F_{pulse}), pulse magnitude (M_{pulse}), pulse grain size distribution (GSD_{pulse}), and water discharge (Q_w). Due to the simple geometry and the lack of bedforms in a 1-D numerical model, the fluvial reaction to the forcing is restricted to the bedload transport rate (q_b), channel slope (S), and channel grain size distribution (GSD_{fluv}) (see Table 3). Pulse frequency and magnitude can be combined in the virtual pulse velocity (U_{pulse}), with the magnitude normalized by reach width (w_r) and length (l_r):

$$U_{pulse} = F_{pulse} \cdot M_{pulse}/(w_r \cdot l_r). \quad (12)$$

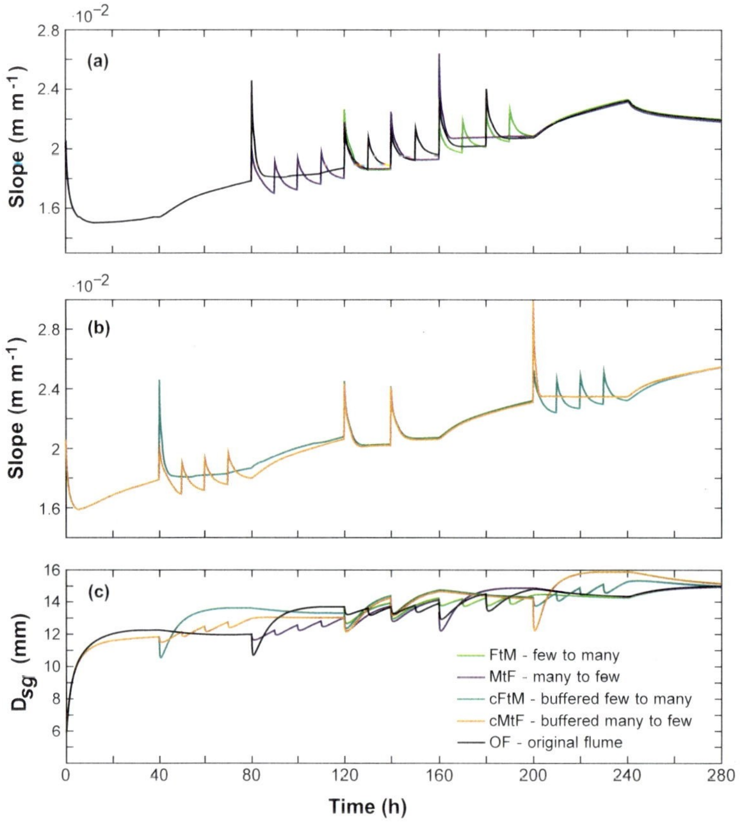

Figure 4. Comparison of slope and mean surface grain size D_{sg} from runs with different event sequencing (see Table 2) ($\sigma = 1.6$, factor of 2 increase of τ^*_{rm}). Event sequences are **(a)** rearranged pulse phases to the flume experiments and **(b)** a setup where the pulsed phases are buffered with constant-feed phases. Panel **(c)** shows D_{sg} for the runs in both **(a)** and **(b)**.

A characteristic pulse period time (T_{pp}) for a sediment supply regime is the inverse of the pulse frequency:

$$T_{pp} = 1/F_{pulse}. \tag{13}$$

We define a reach averaged fluvial export velocity $U_{fluv} = q_b/l_r$ in order to compare the sediment export of material to the pulsed sediment input. Considering a multi-event time frame $T_{sim}(\gg T_{pp})$ for each simulation, we can assume that the fluvial system will adjust to the external forcing over time, leading to an adjustment of the reach averaged fluvial sediment transport to match the external forcing of the virtual pulse velocity:

$$T_{sim} \gg T_{pp}: \quad U_{fluv} \approx U_{pulse}. \tag{14}$$

The time it takes to achieve an equilibrium state is highly dependent on the initial conditions of the simulations. Instead of using a process rate threshold to find an equilibrium time, we can infer a simulation-time independent relation between the supply regime (i.e., T_{pp}) and the state of the system in equilibrium (i.e., U_{fluv}, S, and GSD_{fluv}). This way we only have to run simulations for long enough to verify equilibrium with the respective supply regime (in our case 20 000 h), and then observe properties of the channel at the very end of the simulation, even though equilibrium might have been achieved earlier.

When comparing the system state in equilibrium between many different supply regimes, while keeping the initial fluvial reworking capacity constant (mainly geometry and Q_w), we can identify how the equilibrium conditions change under different supply regimes. Due to the fixed channel geometry in the 1-D model, only the bed slope (S) and the grain size distribution (GSD_{fluv}) can adjust to the change in the supply regime.

To better compare the temporal adjustment to sediment pulses of different magnitudes and frequencies, we non-dimensionalized the pulse period T_{pp} with a fluvial evacuation time T_{fe}:

$$T_{fe} = \frac{l_r w_r D_{fg}}{U_{fluv}} \left(\frac{D_{fg}}{D_{ag}}\right)^2. \tag{15}$$

This timescale is a representation of how long it would take to remove a layer of sediment as long and as wide as the flume (l_r and w_r) with the thickness of the median feed grain size D_{fg}, under the average transport rate U_{fluv}, multiplied by an estimate of the ratio of feed grain size to armoured grain

Table 3. List of forcing and reacting parameters, and timescales in our simulations with abbreviations and their dimension (T represents time and L represents length).

Forcing parameters		Reacting parameters		Timescales	
Pulse frequency	F_{pulse} (1 T^{-1})	Fluvial export velocity	U_{fluv} (L T^{-1})	Simulation time	T_{sim} (T)
Pulse magnitude	M_{pulse} (L^3)	Channel slope	S (L L^{-1})	Pulse period	T_{pp} (T)
Pulse grain size distr.	$\text{GSD}_{\text{pulse}}$ (L)	Surface grain size distr.	GSD_{fluv} (L)		
Water discharge	Q_{w} ($\text{L}^3\,\text{T}^{-1}$)			Derived from simulations:	
Derived parameters:				Fluvial evacuation time	T_{fe} (T)
Virtual pulse velocity	U_{pulse} (L T^{-1})			Armouring time	T_{ar} (T)

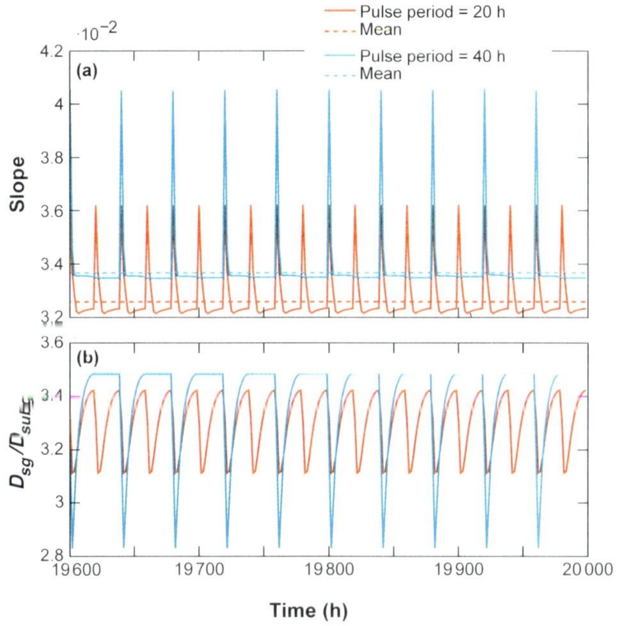

Figure 5. **(a)** Slope and **(b)** armouring ratio for the last 400 h of 2 out of 40 experiments using $\sigma = 1.6$. A run with a low event frequency is shown in blue, and a run with a high event frequency is shown in red. In these example runs the mean slope in equilibrium is higher for the run with longer pulse periods.

size $(D_{\text{fg}}/D_{\text{ag}})^2$, which can be interpreted as an inverse of the degree of potential bed armouring similar to $D_{\text{s90}}/D_{\text{s50}}$ (Recking, 2012). We developed Eq. (15) visually by matching the inflection of lines in Fig. 7a to $T_{\text{pp}}/T_{\text{fe}} = 1$.

Figure 7a shows the same data as Fig. 6b, but in non-dimensionalized time $T_{\text{pp}}/T_{\text{fe}}$. A ratio of $T_{\text{pp}}/T_{\text{fe}} < 1$ can be interpreted as a condition in which the pulsed input of material occurs faster than the fluvial removal of a D_{fg} thick theoretical layer of material under average transport conditions modified by armouring. At a ratio of $T_{\text{pp}}/T_{\text{fe}} > 1$, the sediment is fed in time steps longer than the time that the fluvial system needs to remove said theoretical layer.

Figure 7b shows the normalized armouring ratio $(\text{AR}_{\text{elp}}/\text{AR}_{\text{const}})$, which was obtained by dividing the armouring ratio $\text{AR}_{\text{elp}} = D_{\text{sg}}/D_{\text{subg}}$ at the end of the last pulse for each simulation with the armouring ratio at the end of the corresponding constant-feed run. Similarly to the fluvial evacuation time that we used to non-dimensionalize the slope adjustment, we used an armouring time T_{ar} to non-dimensionalize the grain size adjustment:

$$T_{\text{ar}} = \frac{l_{\text{r}} w_{\text{r}} D_{f90}}{U_{\text{fluv}}} 0.5 \left(\frac{D_{f90}}{D_{\text{a90}}}\right)^2. \tag{16}$$

This timescale represents how long it would take to remove a layer of sediment as long and as wide as the flume (l_{r} and w_{r}) with the thickness of the supplied D_{f90}, under the reach averaged transport rate $U_{\text{fluv}} = q_{\text{b}}/l_{\text{r}}$, multiplied by an estimate of the ratio of the sediment supply D_{f90} to the D_{a90} of an armoured bed. We developed Eq. (16) visually by matching the inflection of lines in Fig. (7b) to $T_{\text{pp}}/T_{\text{ar}} = 1$.

4.3 Interpretation of the equilibrium simulations

The condition of $T_{\text{pp}}/T_{\text{fe}} = 1$ constitutes a threshold in the slope adjustment to pulsed sediment supply. Pulse periods shorter than the fluvial evacuation time ($T_{\text{pp}}/T_{\text{fe}} < 1$) lead to equilibrium slopes similar to the constant feed equilibrium slopes ($S_{\text{mlp}} \approx S_{\text{const}}$). In the case of narrow GSDs ($\sigma < 0.4$), simulations with pulse periods longer than the fluvial evacuation time ($T_{\text{pp}}/T_{\text{fe}} > 1$) show an up to 20 % lower slope than in the constant-feed equivalent run ($S_{\text{mlp}} < S_{\text{const}}$). In contrast, runs with either very low frequency of supply events or wide GSD ($\sigma \geq 0.4$) show equilibrium slopes that are up to 30 % higher than the respective constant-feed runs ($S_{\text{mlp}} > S_{\text{const}}$). Simulations with a wider range of material ($\sigma > 1$) show a drop in $S_{\text{mlp}}/S_{\text{const}}$ right at $T_{\text{pp}}/T_{\text{fe}} = 1$, but then an increase in $S_{\text{mlp}}/S_{\text{const}}$ at longer pulse distances. We interpret conditions of lower S_{mlp} to be less armoured, as they coincide with lower values of $\text{AR}_{\text{elp}}/\text{AR}_{\text{const}}$ as shown in Fig. 7b. The simulations showing an intermittent decrease of S_{mlp} seem to be in a state in which material can be efficiently exported from the system without having intense armouring limiting the slope adjustment. These large pulses increase the slope rapidly, leading to high shear velocities which causes high transport rates.

We interpret the cause for increasing slope and armouring ratios for long pulse periods in the following way. A longer

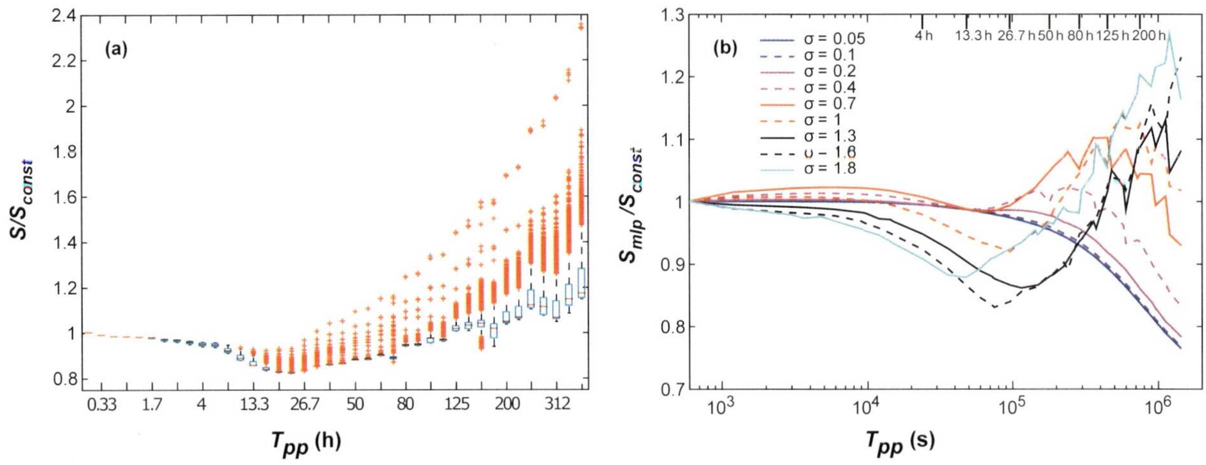

Figure 6. (a) Distribution of the ratios of slope during the last pulse to the constant-feed slope for all 40 runs with $\sigma = 1.6$. We normalized the values of the slope in the last pulse with the slope of the constant-feed run of the same grain size distribution width (σ), allowing for the comparison of the equilibrium slopes between runs with different σ. The red lines represent the normalized mean slopes ($S_{\text{mlp}}/S_{\text{const}}$), which we chose as the main indicator for the equilibrium state of the channel slope. Note that data from longer pulse periods T_{pp} will contain more data points for the box plots, as there are more slope values recorded during the longer time between pulses (sampling every 10 min). The red crosses are outliers in the distribution of slope values, illustrating the extreme slope values during the time right after the pulse was introduced into the channel. **(b)** Mean slope ratios for all runs grouped by width of GSD (σ).

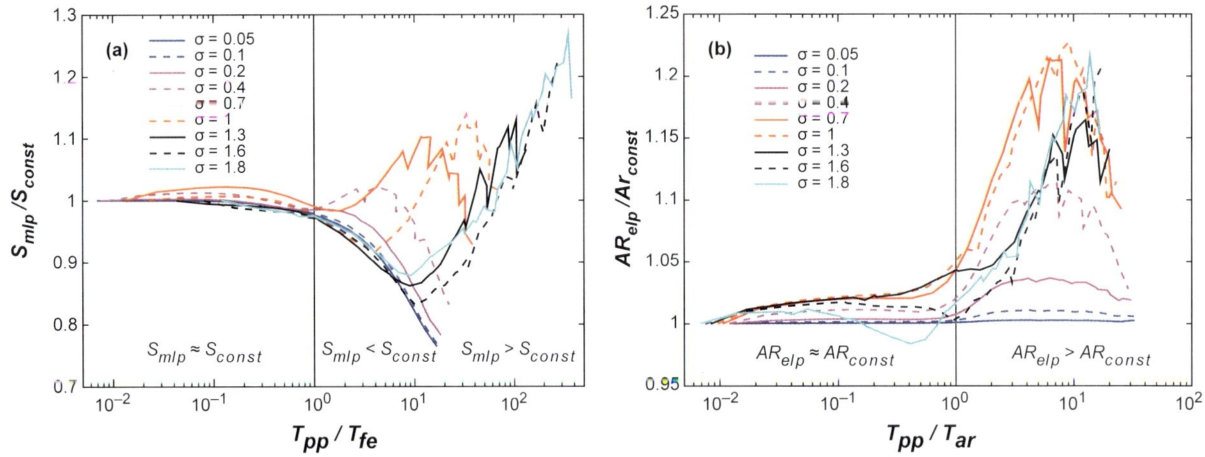

Figure 7. (a) Mean slope ratios in non-dimensional timescale. $T_{\text{pp}}/T_{\text{fe}} < 1$: model runs with a high-frequency sediment supply show similar equilibrium slopes and armouring ratios as conditions of constant sediment feed ($S_{\text{mlp}} \approx S_{\text{const}}$). $T_{\text{pp}}/T_{\text{fe}} > 1$: if the GSD is narrow ($\sigma < 0.4$) or the pulse period is not much longer than the fluvial evacuation time, we observe lower equilibrium slopes than in constant-feed runs ($S_{\text{mlp}} < S_{\text{const}}$). Runs with either very low frequency of supply events or wide GSD ($\sigma \geq 0.4$) show equilibrium slopes that are higher than the respective constant-feed runs ($S_{\text{mlp}} > S_{\text{const}}$). **(b)** Relative armouring ratio in non-dimensional timescale. $T_{\text{pp}}/T_{\text{ar}} > 1$: low frequency of supply events leads to an increase in armouring ratio compared to constant-feed runs ($\text{AR}_{\text{elp}} > \text{AR}_{\text{const}}$), especially for wide GSDs ($\sigma \geq 0.4$).

time between pulses ($T_{\text{pp}} > T_{\text{ar}}$) causes more intense armouring, which shows that the channel bed is starved of sediment between pulses, as finer grain fractions are removed from the surface. This leads to the development of an armouring layer, which restricts the incision into lower deposits, initially limiting the sediment output of the system ($U_{\text{fluv}} < U_{\text{pulse}}$). This imbalance between input and output of material leads to an increased sediment storage over time, which due to the fixed geometry of the channel leads to an increase in slope. This can increase the shear velocity and in return leads to higher sediment output rates. This response loop between armouring and slope adjustment will continue until the sediment output matches the long-term sediment supply ($U_{\text{fluv}} \approx U_{\text{pulse}}$). Note that due to the restricted geometry of our model, slope and grain size are the main parameters in the channel that can change in response to the sediment supply regime. It is possible that other morphological adjustments (e.g., channel width) could compensate for the transport rate disequilibrium in a similar fashion.

The non-dimensionalized presentation of the simulation results shows two distinct modes of adjustment of the fluvial system to episodic sediment supply regimes: (1) "constant-feed-like" behaviour in runs where supply events are of high frequency and low magnitude. Under this kind of forcing the equilibrium slopes and armouring ratios are similar to equilibrium conditions of runs with constant sediment feed. (2) "Pulse-dominated" behaviour occurred in runs where sediment was fed in low frequency and high magnitude events.

An interesting finding is that when $T_{pp} > T_{fe}$, an increase in slope only occurs in simulations that have grain size distributions wide enough to allow armouring to occur. In these cases, we could use the timescale of T_{ar} to determine if the armouring would be significant enough to prevent a decrease of the equilibrium slope. Even though armouring develops very quickly in both our simulations and the flume experiments, its long-term persistence in the time between sediment pulses is what governs the channel response. Hence, the grain size distribution in a series of supply events can be more important for the channel response in the long term than the frequencies and magnitudes of the individual events themselves.

It is notable that the channel response at $T_{pp} = T_{fe}$ does not change abruptly, but the system response slowly tilts to either pulse-dominated on the one end, or constant-feed-like on the other end of the spectrum. While all constant-feed-like channels (for each σ) have very similar equilibrium properties, all pulse-dominated channels are different in both slope and armouring ratios.

4.4 Implications of the equilibrium simulations

Applying the threshold of $T_{pp}/T_{fe} = 1$ between constant-feed-like and pulse-dominated supply regimes to the flume experiments is inconclusive. The unity of $T_{pp} = T_{fe} \approx 6$ h lies between the constant-feed runs and the highest frequency runs (four pulses with $T_{pp} = 10$ h), which means that we have no experimental constant-feed-like pulsed regime where $T_{pp} < T_{fe}$. Elgueta-Astaburuaga and Hassan (2017) found that the four-pulse phase caused a sediment transport response that was similar to the constant-feed runs, implying that T_{fe} would lie between 10 h and 20 h. The flume experiments were not executed long enough to reach equilibrium, which complicates the attribution of a specific channel response to a specific forcing. Besides recreating the 280 h flume experiment in the model, the 360 equilibrium simulations include four configurations that repeat the constant-feed and three pulse periods for 20 000 h. These four configurations only reached the equilibrium after about 12 000 h of simulated time, which is very long in comparison to the conditions of the flume experiments where each supply regime only lasted 40 h.

The numerical model was calibrated with only one set of flume experiments. As our study mainly focusses on comparing different simulation results from the same model, the applicability of our results to other flume studies or field cases is uncertain. However, we are confident that the numerical model is an adequate tool to gain insight on the effect of episodic sediment supply on fluvial channels in a general sense. For example if there was an inaccuracy in the calculation of the shear velocity, it would affect all model runs and thus be counterbalanced by relating the changed resulting slope to the constant-feed slope (S_{mlp}/S_{const}), and the changed armouring ratio to the constant-feed armour ratio (AR_{elp}/AR_{const}).

We tested this by executing two additional batches of simulations with a 25 % decrease and a 25 % increase in the total mass fed respectively. If the experimental design would strongly affect the threshold between constant-feed-like and pulse-dominated conditions, we would expect these simulations to plot differently than the data in Fig. 7b. But as shown in Supplement Fig. S5 for the case of $\sigma = 1.6$, the change in the slope ratio S_{mlp}/S_{const} is relatively insensitive to the total feed volume. We also expect this to be the case when changing the grain size distributions to include large particles that are initially immobile with the applied discharge. In such simulations, the slope would increase to a point where these initially immobile grain sizes become mobile at a very high equilibrium slope. As we compare all results to the corresponding constant-feed equilibrium slope of the same grain size distribution, these conditions might collapse on the existing data as well. However, it is possible that the channel parameters would become extreme in a way that the empirically derived transport function by Wilcock and Crowe would no longer be realistically applicable.

As we developed the fluvial evacuation time T_{fe} and the armouring timescale T_{ar} purely from observations in numerical simulations, their usefulness remains to be proven in the field. If such a threshold behaviour between episodic sediment supply event frequency and the fluvial adjustment of a channel exists, it should be possible to find signatures in channel morphology, sediment storage volume, or channel slope when comparing streams subject to different pulse periods and with different fluvial transport capacities. It is possible to use Eq. (15) with information about the long-term sediment supply volume, grain size supply, and average channel dimensions to calculate T_{fe} and thus infer the matching threshold pulse period where $T_{pp} = T_{fe}$. If the long-term sediment supply occurs in more frequent events than this threshold, the system can be assumed to experience constant-feed-like sediment supply. If the supply frequency is lower, we would expect to find a morphological signature of a pulse-dominated supply regime.

To provide an example application of the developed timescales, we applied Eqs. (15) and (16) to data from East

Table 4. Application of the fluvial evacuation time to the rapids reach in East Creek. The system is assumed to be in equilibrium with a matching fluvial transport rate and long-term sediment supply rate. The time of active fluvial transport is estimated to be $100\,\mathrm{h\,yr^{-1}}$. We approximated the long-term fluvial transport rate U_{fluv} with three years of data from a sediment trap below the reach. We assumed that the subsurface grain size measurements reflect the average supply GSD and the surface grain size measurements represent the long-term average state of armouring in the reach. Besides the original "East Creek" data and the two "threshold" pulse frequency fits, we assumed four more scenarios with doubled armoured grain sizes or doubled fluvial transport rates to give a rough estimate of error bounds.

	Fluvial parameters			Supply regime		Timescales			
	$U_{\mathrm{fluv}}(=U_{\mathrm{pulse}})$ ($\mathrm{m^3\,yr^{-1}}$)	D_{a50} (mm)	D_{a90} (mm)	M_{pulse} ($\mathrm{m^3}$)	F_{pulse} ($\mathrm{1\,yr^{-1}}$)	T_{fe} (yrs)	T_{ar} (yrs)	$T_{\mathrm{pp}}/T_{\mathrm{fe}}$	$T_{\mathrm{pp}}/T_{\mathrm{ar}}$
East Creek	0.75	57	150	0.75	1	2.23	3.58	0.45	0.28
Threshold T_{fe}	0.75	57	150	1.67	0.45^{a}	2.23	3.58	1	0.62
– with $2\times U_{\mathrm{fluv}}$	1.5	57	150	1.67	0.90^{a}	1.11	1.79	1	0.62
– with $2\times D_{\mathrm{a50}}$ and D_{a90}	0.75	114	300	0.42	1.80^{a}	0.56	0.89	1	0.62
Threshold T_{ar}	0.75	57	150	2.68	0.28^{b}	2.23	3.58	1.61	1
– with $2\times U_{\mathrm{fluv}}$	1.5	57	150	2.68	0.56^{b}	1.11	1.79	1.61	1
– with $2\times D_{\mathrm{a50}}$ and D_{a90}	0.75	114	300	0.67	1.12^{b}	0.56	0.89	1.61	1

For all calculations: supply $D_{\mathrm{f90}}=90$ mm, supply $D_{\mathrm{f50}}=32$ mm, channel width = 2.5 m, channel length = 72 m. Some frequencies calculated to match.
[a] $T_{\mathrm{pp}}=T_{\mathrm{fe}}$ and [b] $T_{\mathrm{pp}}=T_{\mathrm{ar}}$

Creek, which is a small creek in the Fraser watershed close to Vancouver in BC, Canada. The "rapids" channel section of this creek was used to design the flume experiments as a 1 to 6 Froude scaled model. Table 4 shows the resulting timescales in seven different scenarios that use the reported values for sediment supply, grain size distribution, and channel dimensions from Cienciala and Hassan (2013) and Papangelakis and Hassan (2016). In East Creek, we assume that of all the supplied sediment is contributed by annual events with a magnitude that matches the annual fluvial transport. As the calculated fluvial evacuation time T_{fe} is 2.23 years, this scenario implies that the system would behave in a constant-feed-like manner, which would only change if the supply events were more than 2.23 years apart on average (i.e., $T_{\mathrm{pp}}=2.23$ years), as shown in the "threshold T_{fe}" calculation. Due to the high uncertainty in our assumption that the measured values represent equilibrium conditions, we calculated two more scenarios with double transport rates ($2\times U_{\mathrm{fluv}}$) and double armoured grain size ($2\times D_{\mathrm{a50}}$ and D_{a90}). The last three calculations in Table 4 show which pulse frequency is needed to match the armouring timescale (T_{ar}). While we do not know if East Creek is in equilibrium with the sediment supply regime and the measurements used do not reflect long-term conditions, these calculations can still give a rough idea of whether a system is constant-feed-like in our classification of channel response to episodic supply regimes.

5 Summary and conclusions

We characterized an episodic sediment supply regime in terms of event frequency, magnitude, and supplied grain size distribution. To test the effect that different episodic sediment supply regimes can have on the morphology of a mountain stream, we developed a numerical model to recreate and extend simulations from flume experiments. The model performs well in recreating the flume experiments in both slope and grain size distributions (GSD), which are the two variables that represent morphological adjustment in our model. Channel width is fixed and bedforms are assumed to be absent, even though bedforms did occur in the flume experiments.

To understand the extent to which event succession plays a role in the flume experiments, we simulated alternative pulse successions of large-to-small events (i.e., infrequent-to-frequent) and small-to-large events (i.e., frequent-to-infrequent), while keeping the total sediment volume feed the same. These simulations show that different pulse frequency sequences have no strong effect on the long-term slope and GSD of the bed surface. In the short term large pulse events can dominate the channel response causing an abrupt increase in slope, while the effect of subsequent smaller events is subdued as the channel is still adjusted to the large pulse. If smaller events dominate at first, the channel adjusts more gradually.

In our second set of simulations, we imposed different episodic sediment supply regimes with the same total sediment supply volume on the same initial channel geometries with constant discharge. While being kept constant within a run, the episodic supply regimes differed in event frequencies, magnitudes, and GSD. We simulated 40 different event frequencies for which the sum of event magnitudes matched an overall equal total sediment supply. All 40 pulse configurations were calculated for 9 GSD that differed in the width of the distribution σ around the same geometric mean grain size. The channels adjusted to the episodic sediment supply until they reached an equilibrium state in which each successive pulse led to the same slope and grain size adjustment. We compared this state between runs and found a distinctive regime change when the time between pulses (T_{pp}) became lower than a fluvial evacuation time (T_{fe}), which we developed as a measure of the time it takes to remove a D_{s50} thick layer from the channel surface under average transport conditions, modified by a measure of potential armouring (see Eq. 15).

The condition of $T_{pp} < T_{fe}$ causes a constant-feed-like sediment supply regime, as the model runs show similar slopes and surface grain size distributions as constant-feed runs of the same GSD. When $T_{pp} > T_{fe}$ the sediment supply regime becomes pulse-dominated. Under these conditions, we observed a lower relative slope in cases where the GSD is narrow ($\sigma < 0.4$), as the long time between pulses in combination with a low armouring potential allows more erosion in the reach, ultimately lowering the equilibrium slope. If the GSD is wide enough to allow armouring ($\sigma \geq 0.4$), a stronger armouring layer can develop during the periods of selective transport of smaller grain sizes and bedload starvation. This limits the minimum slope and increases sediment storage (and thus slope) in the long term.

The application of the episodic supply regime classification to data from East Creek shows that the threshold to a pulse-dominated regime lies at the fluvial evacuation time of roughly 2.2 years. This creek probably receives sediment at a lower interval, which indicates a pulse-dominated regime. The armouring timescale lies around 3.5 years, indicating that if the long-term sediment supply was introduced over event frequencies between 2.2 and 3.5 years, it would be removed most efficiently and result in a lower slope.

Steeper channels than East Creek could show both a lower fluvial evacuation time (due to higher slope, smaller channel area) and a lower pulse frequency (more landslide dominated), which could make these channels more likely to be pulse-dominated. Further study of field cases is needed to strengthen the case for our classification of channel response types to episodic supply regimes. In natural rivers, there are further modes of adjustment that the system can undergo after receiving sediment pulses, for example changes in bed forms or the storage of excess material in sediment bodies along the channel. Still, the condition when a channel is receiving more material per pulse than what can be exported in the same time frame (i.e., the ratio of T_{pp}/T_{fe} is above 1), should be observable in natural rivers as irregularities in channel long profiles due to increased sediment storage. In our model, we only supplied grain sizes that were transportable by the imposed flow conditions. In field streams it can be that the biggest clasts (e.g., boulders) are only transportable by extreme flow events, which would further increase the slope of reaches with high sediment supply.

Author contributions. TM and MAH developed the research questions and designed the setup of the flume experiments, while the data was collected by TM and others. The numerical model was developed by TM under supervision of MAH. TM conducted the simulations and the related data analysis. The paper was written by TM and MAH.

Competing interests. The authors declare that they have no conflict of interest.

Acknowledgements. Tobias Müller was supported by a Four Year Doctoral Fellowship from the University of British Columbia and a travel grant from Stiftung für Kanadastudien. The flume experiments were supported by NSERC and CFI grants (to Marwan Hassan) and were conducted collaboratively by Maria Elgueta, Claudia von Flotow, and the first author under the supervision of Marwan Hassan. Computational resources were provided by WestGrid (https://www.westgrid.ca, last acces: 6 November 2018) and Compute Canada (https://www.computecanada.ca/, last acces: 6 November 2018). Carles Ferrer-Boix provided support in the development of an early version of the numerical model. We thank Eric Leinberger for improving the figures and Conor McDowell for comments on an early draft of this work. We are grateful for the comments and suggestions provided by three anonymous reviewers that helped to clarify and improve this paper. Special thanks go to Gary Parker, who provided critical comments and many suggestions that improved this paper, and together with Chenge An provided suggestions regarding the revision of the paper.

References

Ahnert, F.: Equilibrium, scale and inheritance in geomorphology, Geomorphology, 11, 125–140, https://doi.org/10.1016/0169-555x(94)90077-9, 1994.

An, C., Cui, Y., Fu, X., and Parker, G.: Gravel-bed river evolution in earthquake-prone regions subject to cycled hydrographs and repeated sediment pulses, Earth Surf. Proc. Land., 42, 2426–2438, https://doi.org/10.1002/esp.4195, 2017a.

An, C., Fu, X., Wang, G., and Parker, G.: Effect of grain sorting on gravel bed river evolution subject to cycled hydrographs: Bed load sheets and breakdown of the hydrograph boundary layer, J. Geophys. Res.-Earth, 122, 1513–1533, https://doi.org/10.1002/2016jf003994, 2017b.

Benda, L.: The influence of debris flows on channels and valley floors in the Oregon Coast Range, USA, Earth Surf. Proc. Land., 15, 457–466, https://doi.org/10.1002/esp.3290150508, 1990.

Benda, L. and Dunne, T.: Stochastic forcing of sediment supply to channel networks from landsliding and debris flow, Water Resour. Res., 33, 2849–2863, https://doi.org/10.1029/97WR02388, 1997a.

Benda, L., Hassan, M. A., Church, M., and May, C. L.: Geomorphology of Steepland Headwaters: The Transition from Hillslopes to Channels, J. Am. Water Resour. As., 41, 835–851, https://doi.org/10.1111/j.1752-1688.2005.tb03773.x, 2005.

Blom, A., Arkesteijn, L., Chavarrías, V., and Viparelli, E.: The equilibrium alluvial river under variable flow and its channel-forming discharge, J. Geophys. Res.-Earth, 122, 1924–1948, https://doi.org/10.1002/2017jf004213, 2017.

Brummer, C. J. and Montgomery, D. R.: Influence of coarse lag formation on the mechanics of sediment pulse dispersion in a mountain stream, Squire Creek, North Cascades, Washington, United States, Water Resour. Res., 42, W07412, https://doi.org/10.1029/2005wr004776, 2006.

Brunsden, D.: Applicable model of long term landform evolution, Z. Geomorphol. Supp., 36, 16–26, 1980.

Brunsden, D.: A critical assessment of the sensitivity concept in geomorphology, Catena, 42, 99–123, https://doi.org/10.1016/s0341-8162(00)00134-x, 2001.

Brunsden, D. and Thornes, J. B.: Landscape Sensitivity and Change, T. I. Brit. Geogr., 4, 463–484, https://doi.org/10.2307/622210, 1979.

Bull, W. B.: Geomorphic responses to climatic change, Oxford University Press, https://doi.org/10.2307/215397, 1991.

Chartrand, S. M., Hassan, M. A., and Radić, V.: Pool-riffle sedimentation and surface texture trends in a gravel bed stream, Water Resour. Res., 51, 8704–8728, https://doi.org/10.1002/2015wr017840, 2015.

Church, M., Hassan, M. A., and Wolcott, J. F.: Stabilizing self-organized structures in gravel-bed stream channels: Field and experimental observations, Water Resour. Res., 34, 3169–3179, https://doi.org/10.1029/98wr00484, 1998.

Cienciala, P. and Hassan, M. A.: Linking spatial patterns of bed surface texture, bed mobility, and channel hydraulics in a mountain stream to potential spawning substrate for small resident trout, Geomorphology, 197, 96–107, https://doi.org/10.1016/j.geomorph.2013.04.041, 2013.

Cui, Y. and Parker, G.: Numerical Model of Sediment Pulses and Sediment-Supply Disturbances in Mountain Rivers, J. Hydraul. Eng., 131, 646–656, https://doi.org/10.1061/(asce)0733-9429(2005)131:8(646), 2005.

Cui, Y., Parker, G., Lisle, T. E., Gott, J., Hansler-Ball, M. E., Pizzuto, J. E., Allmendinger, N. E., and Reed, J. M.: Sediment pulses in mountain rivers: 1. Experiments, Water Resour. Res., 39, 1239, https://doi.org/10.1029/2002wr001803, 2003.

Cui, Y., Parker, G., Braudrick, C., Dietrich, W. E., and Cluer, B.: Dam removal express assessment models (DREAM). Part 1: model development and validation, J. Hydraul. Res., 44, 291–307, https://doi.org/10.1080/00221686.2006.9521683, 2006.

Dietrich, W. E., Kirchner, J. W., Ikeda, H., and Iseya, F.: Sediment supply and the development of the coarse surface layer in gravel-bedded rivers, Nature, 340, 215–217, https://doi.org/10.1038/340215a0, 1989.

Elgueta, M. A.: Channel adjustment of a gravel-bed stream under episodic sediment supply regimes, Master's thesis, University of British Columbia, https://doi.org/10.14288/1.0165847, 2014.

Elgueta-Astaburuaga, M. A.: Effects of episodic sediment supply on channel adjustment of an experimental gravel bed, PhD thesis, University of British Columbia, https://doi.org/10.14288/1.0364558, 2018.

Elgueta-Astaburuaga, M. A. and Hassan, M. A.: Experiment on temporal variation of bed load transport in response to changes in sediment supply in streams, Water Resour. Res., 53, 763–778, https://doi.org/10.1002/2016wr019460, 2017.

Ferrer-Boix, C. and Hassan, M. A.: Influence of the sediment supply texture on morphological adjustments in gravel-bed rivers, Water Resour. Res., 50, 8868–8890, https://doi.org/10.1002/2013wr015117, 2014.

Ferrer-Boix, C. and Hassan, M. A.: Channel adjustments to a succession of water pulses in gravel bed rivers, Water Resour. Res., 51, 8773–8790, https://doi.org/10.1002/2015wr017664, 2015.

Ferrer-Boix, C., Chartrand, S. M., Hassan, M. A., Martín-Vide, J. P., and Parker, G.: On how spatial variations of channel width influence river profile curvature, Geophys. Res. Lett., 43, 6313–6323, https://doi.org/10.1002/2016gl069824, 2016.

Hassan, M. A. and Church, M.: Experiments on surface structure and partial sediment transport on a gravel bed, Water Resour. Res., 36, 1885–1895, https://doi.org/10.1029/2000wr900055, 2000.

Hassan, M. A. and Zimmermann, A. E.: Channel Response and Recovery to Changes in Sediment Supply, in: Gravel-Bed Rivers, John Wiley & Sons, Ltd, 464–473, https://doi.org/10.1002/9781119952497.ch33, 2012.

Hassan, M. A., Church, M., Lisle, T. E., Brardinoni, F., Benda, L., and Grant, G. E.: Sediment Transport and Channel Morphology of Small, Forested Streams, J. Am. Water Resour. As., 41, 853–876, https://doi.org/10.1111/j.1752-1688.2005.tb03774.x, 2005.

Hassan, M. A., Gottesfeld, A. S., Montgomery, D. R., Tunnicliffe, J. F., Clarke, G. K. C., Wynn, G., Jones-Cox, H., Poirier, R., MacIsaac, E., Herunter, H., and Macdonald, S. J.: Salmon-driven bed load transport and bed morphology in mountain streams, Geophys. Res. Lett., 35, L04405, https://doi.org/10.1029/2007gl032997, 2008a.

Hassan, M. A., Smith, B. J., Hogan, D. L., Luzi, D. S., Zimmermann, A. E., and Eaton, B. C.: 18 Sediment storage and transport in coarse bed streams: scale considerations, in: Gravel-Bed Rivers VI: From Process Understanding to River Restoration, edited by Habersack, H. P. H. and Rinaldi, M., 473–496, Elsevier, https://doi.org/10.1016/s0928-2025(07)11137-8, 2008b.

Hoey, T. B. and Ferguson, R.: Numerical simulation of downstream fining by selective transport in gravel bed rivers: Model development and illustration, Water Resour. Res., 30, 2251–2260, https://doi.org/10.1029/94wr00556, 1994.

Hoffman, D. F. and Gabet, E. J.: Effects of sediment pulses on channel morphology in a gravel-bed river, Geol. Soc. Am. Bull., 119, 116–125, https://doi.org/10.1130/b25982.1, 2007.

Howard, A. D.: Equilibrium and time scales in geomorphology: Application to sand-bed alluvial streams, Earth Surf. Proc. Land., 7, 303–325, https://doi.org/10.1002/esp.3290070403, 1982.

Johnson, J. P. L., Aronovitz, A. C., and Kim, W.: Coarser and rougher: Effects of fine gravel pulses on experimental step-pool channel morphodynamics, Geophys. Res. Lett., 42, 8432–8440, https://doi.org/10.1002/2015gl066097, 2015.

Lisle, T. E. and Church, M.: Sediment transport-storage relations for degrading, gravel bed channels, Water Resour. Res., 38, 1-1–1-14, https://doi.org/10.1029/2001wr001086, 2002.

Lisle, T. E., Pizzuto, J. E., Ikeda, H., Iseya, F., and Kodama, Y.: Evolution of a sediment wave in an experimental channel, Water Resour. Res., 33, 1971–1981, https://doi.org/10.1029/97wr01180, 1997.

Lisle, T. E., Cui, Y., Parker, G., Pizzuto, J. E., and Dodd, A. M.: The dominance of dispersion in the evolution of bed material waves in gravel-bed rivers, Earth Surf. Proc. Land., 26, 1409–1420, https://doi.org/10.1002/esp.300, 2001.

Luzi, D. S.: Sediment transport and morphological response of a semi-alluvial channel: insights from a Froude scaled laboratory model, PhD thesis, University of British Columbia, https://doi.org/10.14288/1.0165910, 2014.

Madej, M. A.: Temporal and spatial variability in thalweg profiles of a gravel-bed river, Earth Surf. Proc. Land., 24, 1153–1169, https://doi.org/10.1002/(SICI)1096-9837(199911)24:12<1153::AID-ESP41>3.0.CO;2-8, 1999.

Madej, M. A.: Development of channel organization and roughness following sediment pulses in single-thread, gravel bed rivers, Water Resour. Res., 37, 2259–2272, https://doi.org/10.1029/2001wr000229, 2001.

Madej, M. A. and Ozaki, V.: Channel response to sediment wave propagation and movement, Redwood Creek, California, USA, Earth Surf. Proc. Land., 21, 911–927, https://doi.org/10.1002/(sici)1096-9837(199610)21:10<911::aid-esp621>3.0.co;2-1, 1996.

Miller, D. J. and Benda, L. E.: Effects of punctuated sediment supply on valley-floor landforms and sediment transport, Geol. Soc. Am. Bull., 112, 1814–1824, https://doi.org/10.1130/0016-7606(2000)112<1814:eopsso>2.0.co;2, 2000.

Nelson, P. A., Venditti, J. G., Dietrich, W. E., Kirchner, J. W., Ikeda, H., Iseya, F., and Sklar, L. S.: Response of bed surface patchiness to reductions in sediment supply, J. Geophys. Res.-Earth, 114, F02005, https://doi.org/10.1029/2008jf001144, 2009.

Papangelakis, E. and Hassan, M. A.: The role of channel morphology on the mobility and dispersion of bed sediment in a small gravel-bed stream, Earth Surf. Proc. Land., 41, 2191–2206, https://doi.org/10.1002/esp.3980, 2016.

Parker, G.: Selective Sorting and Abrasion of River Gravel. I: Theory, J. Hydraul. Eng., 117, 131–147, https://doi.org/10.1061/(asce)0733-9429(1991)117:2(131), 1991.

Parker, G.: Transport of Gravel and Sediment Mixtures, in: Sedimentation Engineering, American Society of Civil Engineers, vol. 110, 165–251, https://doi.org/10.1061/9780784408148.ch03, 2008.

Parker, G., Hassan, M., and Wilcock, P.: 10 Adjustment of the bed surface size distribution of gravel-bed rivers in response to cycled hydrographs, in: Gravel Bed Rivers VI: From Process Understanding to River Restoration, Elsevier, 241–285, https://doi.org/10.1016/s0928-2025(07)11127-5, 2007.

Podolak, C. J. P. and Wilcock, P. R.: Experimental study of the response of a gravel streambed to increased sediment supply, Earth Surf. Proc. Land., 38, 1748–1764, https://doi.org/10.1002/esp.3468, 2013.

Pryor, B. S., Lisle, T., Montoya, D. S., and Hilton, S.: Transport and storage of bed material in a gravel-bed channel during episodes of aggradation and degradation: a field and flume study, Earth Surf. Proc. Land., 36, 2028–2041, https://doi.org/10.1002/esp.2224, 2011.

Recking, A.: Influence of sediment supply on mountain streams bedload transport, Geomorphology, 175-176, 139–150, https://doi.org/10.1016/j.geomorph.2012.07.005, 2012.

Roberts, R. G. and Church, M.: The sediment budget in severely disturbed watersheds, Queen Charlotte Ranges, British Columbia, Can. J. Forest Res., 16, 1092–1106, https://doi.org/10.1139/x86-189, 1986.

Sklar, L. S., Fadde, J., Venditti, J. G., Nelson, P., Wydzga, M. A., Cui, Y., and Dietrich, W. E.: Translation and dispersion of sediment pulses in flume experiments simulating gravel augmentation below dams, Water Resour. Res., 45, W08439. https://doi.org/10.1029/2008wr007346, 2009.

Thomas, M. F.: Landscape sensitivity in time and space – an introduction, Catena, 42, 83–98, https://doi.org/10.1016/s0341-8162(00)00133-8, 2001.

Venditti, J. G., Dietrich, W. E., Nelson, P. A., Wydzga, M. A., Fadde, J., and Sklar, L.: Effect of sediment pulse grain size on sediment transport rates and bed mobility in gravel bed rivers, J. Geophys. Res.-Earth, 115, F03039, https://doi.org/10.1029/2009jf001418, 2010.

Viparelli, E., Sequeiros, O. E., Cantelli, A., Wilcock, P. R., and Parker, G.: River morphodynamics with creation/consumption of grain size stratigraphy 2: numerical model, J. Hydraul. Res., 48, 727–741, https://doi.org/10.1080/00221686.2010.526759, 2010.

von Flotow, C.: Temporal adjustments of a streambed following an episodic sediment supply regime, Master's thesis, University of British Columbia, https://doi.org/10.14288/1.007421, 2013.

Wilcock, P. R.: Toward a practical method for estimating sediment-transport rates in gravel-bed rivers, Earth Surf. Proc. Land., 26, 1395–1408, https://doi.org/10.1002/esp.301, 2001.

Wilcock, P. R. and Crowe, J. C.: Surface-based Transport Model for Mixed-Size Sediment, J. Hydraul. Eng., 129, 120–128, https://doi.org/10.1061/(asce)0733-9429(2003)129:2(120), 2003.

Wolman, M. G. and Miller, J. P.: Magnitude and Frequency of Forces in Geomorphic Processes, J. Geol., 68, 54–74, https://doi.org/10.1086/626637, 1960.

Wong, M. and Parker, G.: One-dimensional modeling of bed evolution in a gravel bed river subject to a cycled flood hydrograph, J. Geophys. Res.-Earth, 111, F03018, https://doi.org/10.1029/2006jf000478, 2006.

Zimmermann, A. E., Church, M., and Hassan, M. A.: Video-based gravel transport measurements with a flume mounted light table, Earth Surf. Proc. Land., 33, 2285–2296, 2008.

9

Topographical change caused by moderate and small floods in a gravel bed ephemeral river – a depth-averaged morphodynamic simulation approach

Eliisa S. Lotsari[1,5], Mikel Calle[2], Gerardo Benito[2], Antero Kukko[3,4], Harri Kaartinen[3,5], Juha Hyyppä[3], Hannu Hyyppä[4], and Petteri Alho[3,5]

[1]Department of Geographical and Historical Studies, University of Eastern Finland, Joensuu, Yliopistokatu 2, P.O. Box 111, 80101 Joensuu, Finland

[2]National Museum of Natural Sciences, Spanish Research Council (CSIC), Madrid, Calle de Serrano 117, 28006 Madrid, Spain

[3]Department of Remote Sensing and Photogrammetry, Finnish Geospatial Research Institute, National Land Survey of Finland, Kirkkonummi, Geodeetinrinne 2, 02430 Masala, Finland

[4]Department of Built Environment, Aalto University, Espoo, Vaisalantie 8, P.O. Box 15800, 00076 Aalto, Finland

[5]Department of Geography and Geology, University of Turku, 20014 Turku, Finland

Correspondence: Eliisa S. Lotsari (eliisa.lotsari@uef.fi)

Abstract. In ephemeral rivers, channel morphology represents a snapshot at the end of a succession of geomorphic changes caused by floods. In most cases, the channel shape and bedform migration during different phases of a flood hydrograph cannot be identified from field evidence. This paper analyses the timing of riverbed erosion and deposition of a gravel bed ephemeral river channel (Rambla de la Viuda, Spain) during consecutive and moderate- (March 2013) and low-magnitude (May 2013) discharge events, by applying a morphodynamic model (Delft3D) calibrated with pre- and post-event surveys by RTK-GPS points and mobile laser scanning. The study reach is mainly depositional and all bedload sediment supplied from adjacent upstream areas is trapped in the study segment forming gravel lobes. Therefore, estimates of total bedload sediment mass balance can be obtained from pre- and post-field survey for each flood event. The spatially varying grain size data and transport equations were the most important factors for model calibration, in addition to flow discharge. The channel acted as a braided channel during the lower flows of the two discharge events, but when bars were submerged in the high discharges of May 2013, the high fluid forces followed a meandering river planform. The model results showed that erosion and deposition were in total greater during the long-lasting receding phase than during the rising phase of the flood hydrographs. In the case of the moderate-magnitude discharge event, deposition and erosion peaks were predicted to occur at the beginning of the hydrograph, whereas deposition dominated throughout the event. Conversely, the low-magnitude discharge event only experienced the peak of channel changes after the discharge peak. Thus, both type of discharge events highlight the importance of receding phase for this type of gravel bed ephemeral river channel.

1 Introduction

The hydrology of ephemeral rivers is dominated by occasional large flash floods that cause morphological change (Tooth, 2000; Benito et al., 2011). Due to their catastrophic nature the costs of these floods include major economic, social and environmental aspects (Petersen, 2001). These impacts are caused by both hydro- and morphodynamics during discharge events. To reduce the emergency costs and enhance the preventive flood mitigation measures, understanding the forces of flow and related channel change throughout discharge events is important. Thus, so that it is possible to temporally allocate the flood mitigation measures most efficiently, more understanding of river morphodynamics during discharge events is needed.

Most geomorphological studies have been concentrated on large discharge events due to their impacts on river channel change, related river environments and human infrastructure (Greenbaum and Bergman, 2006; Grodek et al., 2012; Nardi and Rinaldi, 2015; Hooke, 2016b). However, moderate and low flows have also been shown to cause great morphological changes in gravel bed river channels (Calle et al., 2015; Hooke, 2016a), as a small discharge over long time spans can substantially rework the sediment and fluvial bedforms produced by greater floods (Greenbaum and Bergman, 2006).

According to Hooke and Mant (2000), the pattern and magnitudes of fluvial morphological changes show the best relationship with the magnitude of peak discharge. However, during flash floods, it is difficult to perform sediment transport or bedform migration measurements to detect the timing of topographical changes, i.e. whether the greatest changes occur for example due to the peak discharge, the slope of the rising limb, or the length of the receding limb. Case studies have reported most erosion during the rising and peak flow phases in perennial rivers (e.g. Gendaszek et al., 2013), but similar knowledge for ephemeral river channels is still limited. It is important to understand the capacities of ephemeral rivers for sediment deposition and flooding due to the combined effects of water flow and sediment transport during flood situations. If the total erosion and deposition is greater during the receding phase than the rising phase, the receding phase should not be ignored when planning flood mitigation measures.

Conventionally, ephemeral river channel changes associated with specific flood events are interpreted on the basis of the post-flood bedform and grain size distribution (e.g. Euler et al., 2017). New insight on morphodynamic changes and their driving parameters during flash floods can be gained by applying simulation methods. Simulations may provide information about the channel dynamics from the times when it has not been possible to perform measurements (Lotsari et al., 2014a) and thus increase our understanding of sediment dynamics during flood events (Hooke et al., 2005). However, morphodynamic modelling of gravel migration has so far been more common in perennial braided gravel bed rivers (e.g. Williams et al., 2016b) or laboratory flumes (Kaitna et al., 2011) than in ephemeral rivers. There are examples of simulating sediment transfer in ephemeral channels (Graf, 1996), although most of the few recent simulations have been performed in sandy reaches (Billi, 2011; Lucía et al., 2013) and on alluvial fans (Pelletier et al., 2005). Ephemeral rivers with gravel-sized bedload particles have rarely been simulated. One of the few existing example studies were carried out by Hooke et al. (2005), simulating morphological changes during flash floods with a cellular automata model. Their model worked well with simulations using moderately large discharges during clear water conditions, but discharge events with sediment loads had some tendency for excess deposition (Hooke et al., 2005). Further examination of different discharge events with modelling approaches is thus required.

It is possible to calibrate morphodynamic models based on the pre- and post-flood topographies of different consecutive flood situations. This causes uncertainties to the modelling, as calibration and validation data are measurable only during dry channel conditions. However, it is the best available way to gain understanding on the temporal evolution of the ephemeral river channels during flooding. In particular, the differences and similarities in model performance between different flood magnitudes can be detected (Lotsari et al., 2014a; Williams et al., 2016a). Recently, the measurement techniques for deriving this calibration data for morphodynamic modelling have increased. One of them is accurate laser scanning (mobile and terrestrial), which enables the channel topography before and after flooding to be captured in detail (Milan et al., 2007; Vaaja et al., 2011; Calle et al., 2015; Kasvi et al., 2015; Kukko et al., 2015). Laser scanning enables rapid measurements of gravel bed rivers at the sub-grain-level resolution (Milan et al., 2007). In ephemeral rivers the quality of topographical data can be very good and the uncertainties are less than in perennial rivers, either with laser scanning or traditional RTK-GPS measurements, because these rivers can be surveyed when the riverbed is dry. The high uncertainties in topographical measurements of sub-water areas in gravel bed perennial rivers have been related particularly to the high bedload velocities and temporal variability in bedload (Williams et al., 2015). Mobile laser scanning (MLS), for example from a vehicle or backpack, provides a practical scanning angle to survey channel banks, bar lobes, and vertical surfaces in an ephemeral river (Vaaja et al., 2011; Kasvi et al., 2015; Kukko et al., 2015). Also, data collection over large channel bars has been enhanced with these methods. Digital elevation models produced from airborne laser scanning data have also been applied for detecting the geomorphic effects of different discharge events (Hauer and Habersack, 2009; Croke et al., 2013; Thompson and Croke, 2013; Nardi and Rinaldi, 2015) and for recording and calibrating sediment transport models (Rodriguez-Lloveras et al., 2015). These reveal the reorganization of the channel morphology due to flood events even over large ar-

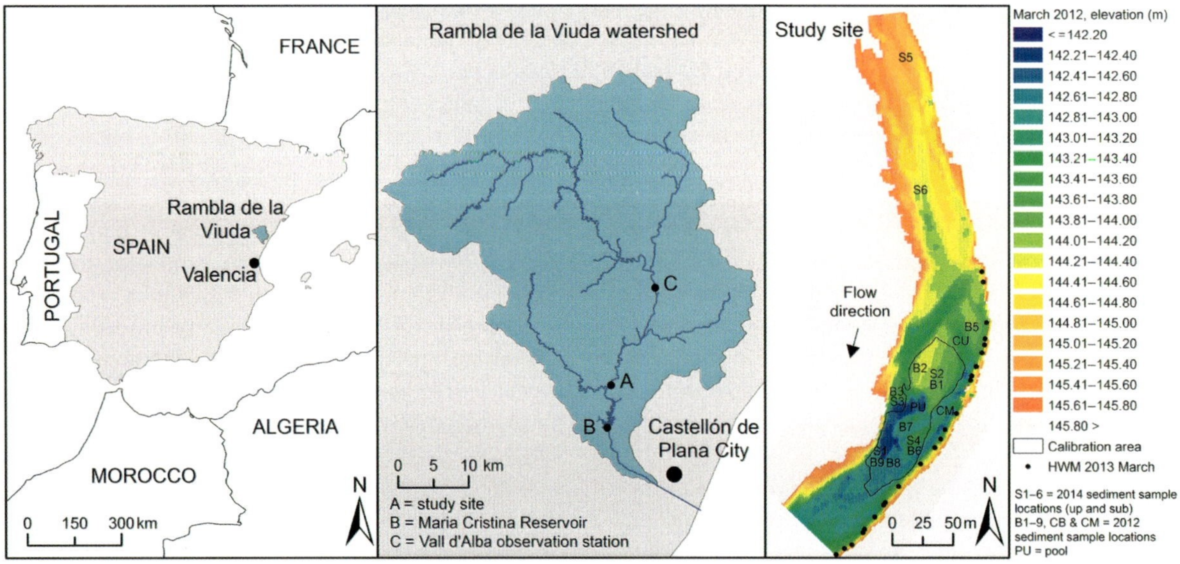

Figure 1. The study site of Rambla de la Viuda, eastern Spain. The elevation on March 2012 is derived mainly from mobile laser scanning data, but the edges are from the 2009 national digital elevation model. The calibration area was defined based on the data coverage of March 2013. The high-water mark (HWM) measurement locations of 2013 March are shown (their values are plotted in Fig. 3). Also, the sediment sample locations are shown (see also Table 2 and Fig. 2).

eas with great detail (Thompson and Croke, 2013). These accurate topography data, together with RTK-GPS measurements, enable more detailed calibration of morphodynamic models than before.

This paper analyses the evolution of a gravel bed ephemeral river channel (Rambla de la Viuda, Spain) during consecutive and moderate- (March 2013) and low-magnitude (May 2013) discharge events by applying a morphodynamic modelling (Delft3D) approach. Based on the simulations, we analyse (1) the timing of riverbed erosion and deposition in relation to the flow hydrograph phases during moderate- and low-magnitude discharge events, (2) the hydraulic characteristics (e.g. shear stress) explaining these channel and bedform morphodynamics, and (3) the prevailing fluvial processes, and related sediment transport routing, during these different magnitude discharge events in a gravel bed ephemeral stream. This study is based on both accurate topographical measurements with RTK-GPS survey and MLS, before and after each discharge event, and morphodynamic simulation (2-D implementation of Delft3D). Thus, the model is calibrated with data from a moderate-magnitude event and then validated based on the consecutive low-magnitude event. With this simulation method it is possible to improve and deepen the analyses made earlier based on only the survey work (e.g. Calle et al., 2015).

2 The study area

Rambla de la Viuda is an ephemeral stream with a catchment area of 1523 km^2, located in eastern Spain (Fig. 1). The river has a braided pattern associated with a high sediment supply (Calle et al., 2015). The river is prone to flash floods, and the stream flow occurs on an average of 20 days per year. Due to the high infiltration, only intensive rains, i.e. a run-off threshold of 65 mm, cause flow in the river corridor (Segura-Beltrán and Sanchis-Ibor, 2013). This run-off threshold may vary between seasons due to soil moisture and the frequency of rains. The rains producing floods in this catchment are caused by mesoscale convective complexes (autumn and spring heavy rains), and those may be up to 100 km in diameter. These rains producing the floods are related to Atlantic fronts, which do not give much spatial variability.

The study site is located 8 km upstream of the dam of the María Cristina reservoir. The simulation area is circa 440 m long (length along the channel) and includes an advancing gravel lobe bar front, which is the lowest 200 m of the simulation area. This lowest 200 m long reach is applied as the calibration area (Fig. 1). This area was selected essentially because it was possible to estimate the bedload sediment input with certainty (Calle et al., 2015). This is due to the gravel extraction as part of river mining activities in recent decades, which has produced a flat valley bottom on which gravel bars were deposited. Recently, new bars have been deposited by floods, and gravel lobes with a sharp front have been formed in the downstream edge of the area. The gravel bars and lobes of the simulation area are non-armoured or weakly armoured (see also Table 2) and move freely during low-magnitude flows.

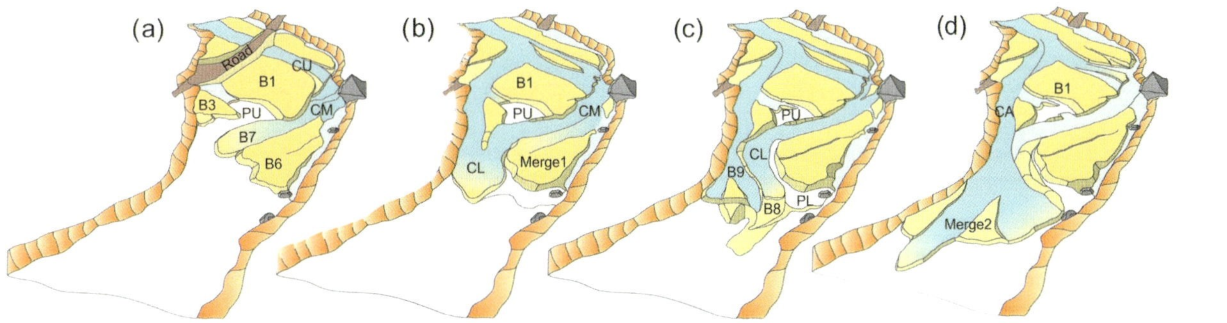

Figure 2. The conceptual graphic presented by Calle et al. (2015: Fig. 10, www.schweizerbart.de/journals/zfg): the pre-stage of 2012 **(a)**, high flood stage during the March 2013 event **(b)**, low flood stage during the May 2013 event **(c)**, and final stage after May 2013 **(d)**. CU: upper channel; CM: middle channel; CL: lower channel; PU: upper pool; PL: lower pool; B1–9: gravel bars (see Calle et al., 2015).

The two events under study occurred in spring 2013. These were caused by two rain events, which were recorded at 5 min intervals at the precipitation and gauging station provided by the Automatic System of Hydrologic Information (SAIH-Jucar) at Vall d'Alba (Fig. 1 and Calle et al., 2015). The first rain event (March 2013) started on 4 March lasting for 3 days with a total of 70 mm. The second rain event (April–May 2013) started on 27 April and lasted 4 days with a total of 72 mm. These rain events caused flows that lasted 13 days (started at 12:00 GMT+1 on 5 March) and 9 days (started 14:10 on 30 April), respectively. The peak discharges of 23 (at 11:05 on 6 March) and 12.5 $m^3 s^{-1}$ (at 9:05 on 1 May) were registered at the gauging station of Vall D'Alba, respectively. These two discharge events transported 12–41 mm sized gravel (D_{50} values) according to the measurements (see Sect. 3.3 below), which had been performed in the areas of topographical changes. These findings have also been published in Calle et al. (2015). The movement of these gravels caused the development of the bar fronts (Calle et al., 2015).

Calle et al. (2015) described the main morphological changes in the same river reach (i.e. around the calibration area of Fig. 1) caused by these moderate (March 2013) and small (May 2013) discharge events based on multi-temporal MLS and RTK-GPS surveys before and after the floods (Fig. 2). They also related the observed morphological and sediment textural changes with hydraulic parameters (flow velocity, depth, and discharge) estimated using a two-dimensional implementation of hydrodynamic model (Delft3D) and investigated whether the combination of the applied techniques is a reliable method to study the morphodynamics of a flood event. It was shown that MLS surveys and additional RTK-GPS surveys are suitable for a dryland river environment. In addition, a two-dimensional hydrodynamic simulation was able to estimate the hydraulic characteristics associated with the discharges. Change detection and spatial grain size distribution analyses after flood showed a high availability of material (up to D_{84} of 32–45 mm) and absence of a well-developed armoured layer (Calle et al., 2015). Thresholds for sediment transport were proven to fit with the Hjulström graph (1935) in this ephemeral environment. However, simulations did not include topographical change and its influence on the hydraulic parameters during the floods and could not answer how topographic changes evolved during rising limb, peak stage, and receding limb of the hydrographs. In this paper, we investigate the ephemeral river further with morphodynamic modelling. We also answer the above-mentioned questions and deepen the previous analyses that were based on only topographical measurements.

3 Morphodynamic simulation approach

3.1 Boundary conditions: discharges and water levels

Two flow hydrographs recorded in 2013 were defined as the upstream boundary condition and the water levels as the downstream boundary condition for the hydro- and morphodynamic model (2-D module of Delft3D-FLOW). As no direct measurements were possible in the study reach during the discharge events, such as those demonstrated for the case of perennial rivers (Williams et al., 2013), the input data and calibration procedures differ from the traditional ones carried out for simulating perennial rivers. At the study reach, the hydrograph peak discharge was estimated from continuous lines of flotsam (i.e. high-water marks, HWMs) emplaced by the floodwater. The hydrograph shape of the study site can be assumed to be similar to Vall d'Alba gauge station, due to the widespread continuous character of the rain events. The hydrograph shape was also verified to be similar to Vall d'Alba observation station by later water level data, which were captured with the water level sensors installed at the study site in late 2014 (see Supplement). The HWM left by the March 2013 discharge event provided evidence that the peak discharge was greater at the study site than that measured at Vall d'Alba (Figs. 1 and 3). The hydrographs were rescaled by using different multipliers to match the peak

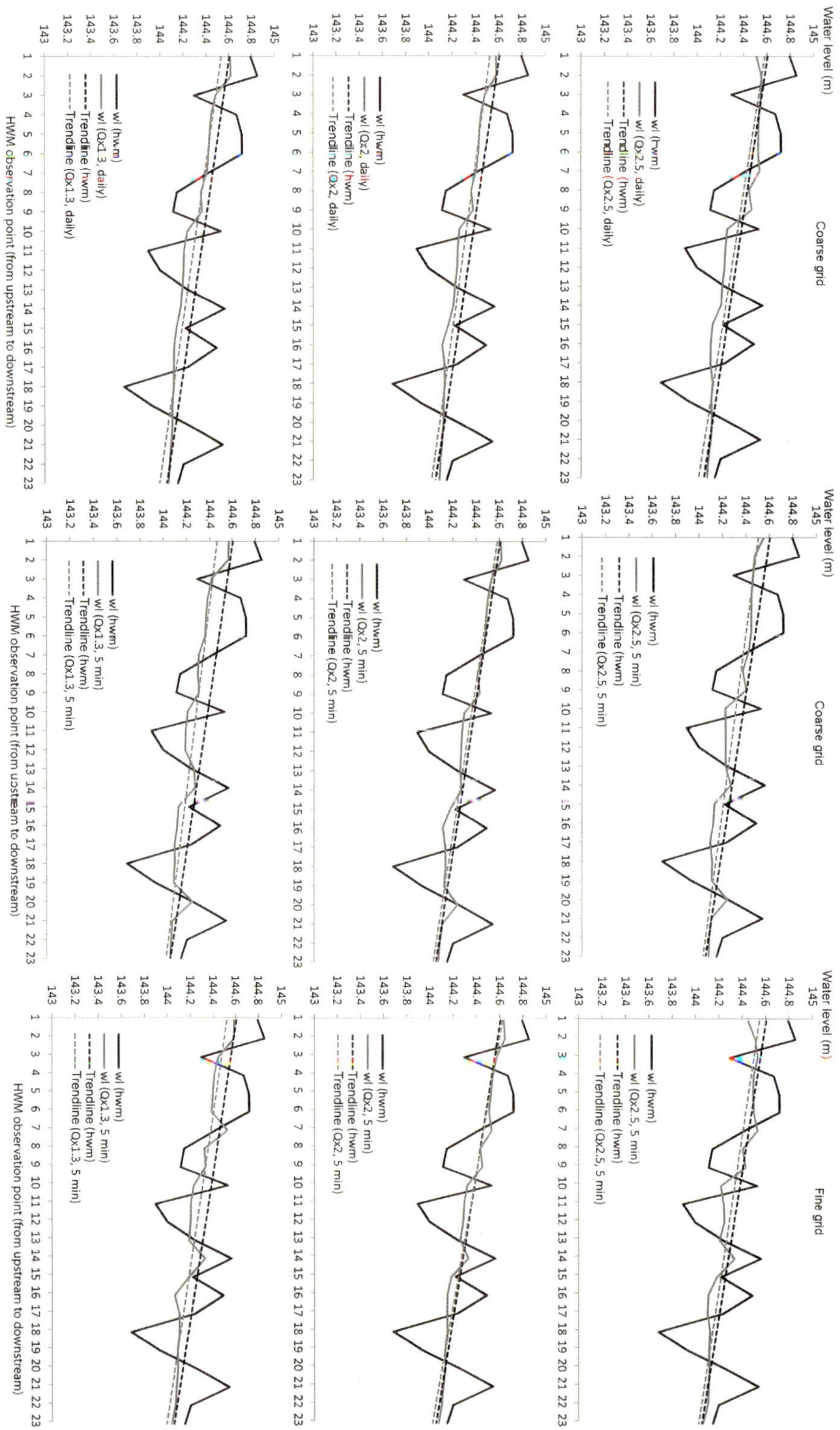

Figure 3. The hydrodynamic calibration results based on the coarse grid (1.51–5.31 m cells) and fine grid (0.76–3.03 m cells). In this figure, the high-water marks (HWMs) and simulated water levels are presented. In addition, the trend lines fitted along these measurements and simulation results are shown. These calibration results of the fine grid were previously presented in Calle et al. (2015) by the same authors. Qx1.3, Qx2, and Qx2.5 refer to the discharges scaled to match the peak discharge at the study site. Daily and 5 min refer to the observation interval of discharges at Vall d'Alba station. The abbreviation "wl" is water level based on HWMs and simulations and "trendline" is the trend line fitted in the measured HWMs and the simulation results. The location of these HWM measurement points have been shown in Fig. 1.

discharge calculated during the calibration procedure of the model (see Sect. 3.4 below and Calle et al., 2015).

Water level was defined, based on the actual HWM measurements along the longitudinal profile (Calle et al., 2015). The HWM measurements varied slightly along the longitudinal profile, and the trend line, which was fitted between these field measurements, showed an accurate water level slope. Note that these HWMs were measured at the valley side and it might not have reflected the water level across the whole channel as bed changes might have locally influenced water level. Therefore, the trend line constituted a good indicative of water slope and level. The downstream-most value of this water level trend line was applied as the peak water level value of the March 2013 discharge event (Fig. 3, and see also Sect. 3.4). The changes in water level over time were calculated for both March and May 2013 events based on the relative change (%) of the discharge between each discharge measurement time interval.

3.2 Channel geometry

The channel topography was measured, using MLS and RTK-GPS (Calle et al., 2015). Three topography data sets were applied. These were (1) initial topography (MLS, March 2012), (2) calibration topography between the floods (RTK-GPS, March 2013), and (3) final validation topography after the floods (MLS, June 2013). The MLS was performed in March 2012 (pre-flood situation) and in June 2013 (after the two discharge events of spring 2013) (Calle et al., 2015 and Table 1). The MLS was performed both with a backpack and on a platform mounted on a 4 × 4 vehicle. A detailed description of the MLS data set can be found in Calle et al. (2015) and Table 1. According to Calle et al. (2015) and Kukko et al. (2015) the RMSE in elevation was 18.2 mm, and it was 36 mm for the 3-D position of the targets. In addition, the edges of the bar lobes, which had been moved by the March 2013 discharge event, were measured with RTK-GPS just after the discharge event in March 2013 (Trimble 4700, 1 Hz, accuracies specified by the manufacturer: XY 0.02 m, Z 0.05 m, absolute systematic measurement error ≤ 0.054 m). Thus, the error estimates of the topographic data can be found from Calle et al. (2015). RTK-GPS data represented the geometry after the March 2013 event and before the May 2013 event and were therefore applied as a calibration data set. This enabled better estimation of the model performance compared to only using one data set after the two consecutive flood events.

The 2-D implementation of the Delft3D morphodynamic model required channel geometry in grid format. Orthogonalized curvilinear grids of two different resolutions were created from both measured topographies, one with "coarse" 1.51–5.31 m cells and one with "fine" 0.76–3.03 m cells (i.e. circa half of the coarser grid cell sizes). These cell sizes were selected for testing the impacts of cell sizes on simulation results, but also due to their computational effectiveness. Cell sizes smaller than the fine resolution (i.e. 0.76–3.03 m, which is the smallest applied resolution range of the grid) did not enhance the results and only increased the computational time. Thus, curvilinear grids of two resolutions, coarse 1.51–5.31 m cells and fine 0.76–3.03 m cells, were created from the topography.

The initial input channel topography for all of the simulations was defined by adding the MLS 2012 measurement points to the curvilinear grids, both to the coarse and fine grids, and averaging the point values for the grid cells. There was also a DEM available from 2009 (from airborne laser scanning), which had a 1 m resolution. After adding the MLS data, this 2009 DEM was added to the grid cells, located outside the laser scanning perimeter, in order to cover the higher banks. The coarser resolution of the 2009 DEM did not affect the simulation results of the channel because the high water levels of the 2013 spring events barely reached these higher bank elevations.

For calibrating and validating the modelling results, topographies were also created from the March 2013 and June 2013 measurements so that those geometries had a similar resolution to the initial input channel geometry. The model outputs after each of the discharge events were then compared to these topography grids created from the March and June 2013 measurements. The grid-form geometry of March 2013, i.e. calibration topography between the floods, was created by adding only the GPS measurements to the grid cells, and the rest of the channel was excluded from the area to be analysed during calibration (Fig. 1). The grid-form geometry of June 2013, i.e. the final validation topography after the May 2013 flood, was defined by first adding the laser scanning data set, which had been processed to include both backpack and car MLS data, to the grid cells. Then the 2009 DEM was added to the grid to cover the higher bank areas. However, only the same area, as applied in the case of March 2013 (i.e. the gravel bar lobe area), was used in the validation of the model performance. These observed elevation and volumetric changes between the events were compared to the simulated changes. For calculating the volumetric changes the curvilinear grid topographies were converted into regular grids. To minimize the errors, a 0.5 m regular grid cell size was selected, as the original cells were mostly divisible by that value.

3.3 Grain sizes

Spatially varying grain sizes were measured from the gravel lobe area prior to the first flood and between the floods, using the Wolman (1954) sampling method (Table 2). However, this method did not recover the differences between upper layer and sub-layer sediment distribution and what kind of particles the 2013 discharge events had moved. Therefore, the gravel moved by the 2013 spring events were measured using a US standard gravelometer (US SAH-97 handheld particle analyser) in summer 2014 (Table 2). This was

Table 1. The properties of the MLS measurements of 2012 and 2013 (adapted from Calle et al. 2015).

	2012 March	2013 June
Laser sensor/GPS sensor	FARO Photon 120	FARO Focus 3-D 120 S
Receiver	NovAtel DL4+	NovAtel Flexpak6
Inertial measurement unit	Honeywell HG1700 AG11	Northrop Grumman UIMU-LCI
IMU frequency	100 Hz	200 Hz
Referencing system	GPS (1 Hz)	GNSS (1 Hz)
Wavelength	785 nm	905 nm
Scan frequency	61 Hz	95 Hz
Point acquisition	488 Hz	488 Hz
Angular resolution	0.045° (0.8 mrad)	0.07° (1.2 mrad)
3-D RMSE	0.034 m	0.019 m

Table 2. The grain sizes measured in 2012–2014 (see their locations from Fig. 1). WCM, WCU, and WB1-7 were measured in 2012, WB8 and 9 were measured in March 2013, and SUB1–6 and UP 1–6 were measured in June 2014. W: Wolman; C: channel; B: bar; SUB: sub-layer; UP: upper layer. Most of the samples were within the calibration area, and their locations can be seen in Fig. 1 and also in Fig. 11 of Calle et al. (2015). Due to the spatial scarcity of the 2013 measurements (only two samples) they were not applied as the surface grain sizes for the model. The armour ratio was calculated following Lisle and Madej (1992), i.e. ratio of the surface to subsurface D_{50}.

	Upper layer		Sub-layer	
Sample	D_{50} (mm)	D_{84} (mm)	D_{50} (mm)	Armour ratio
WCM	18	29	–	–
WCU	30	101	–	–
WB1	22	41	–	–
WB2	17	33	–	–
WB3	17	31		
WB5	12	21	–	–
WB6	17	37	–	–
WB7	14	22	–	–
WB8	18	35	–	–
WB9	20	31	–	–
UP1 and SUB1	22.5	39	18.5	1.216
UP2 and SUB2	35.5	79	21.3	1.667
UP3 and SUB3	18.8	31	14.3	1.314
UP4 and SUB4	18.0	30.5	14.4	1.250
UP5 and SUB5	31.2	55.4	18.6	1.677
UP6 and SUB6	41.2	80	16.7	2.467
Average UP1–6	26.3	–	–	–
Average SUB1–6	17.1	–	–	–
Average UP and SUB1–6	21.1	–	–	–

possible as no discharge had occurred between May 2013 and summer 2014. These measurements represented different active forms, from bars to the channel bed, which had evolved during the 2013 spring floods. The bulk sieving measurements (ca. 80 kg from 1.1 × 1.1 m area, 0–10 cm from the surface) were performed at six different locations, and from both the upper layer (UP1–6) and sub-layer (SUB1–6, ca. more than 10 cm below the surface) to evaluate armouring (Fig. 1 and Table 2). The upper layer–sub-layer contact was established following the criteria described by Bunte and Abt (2001) considering the size of the largest particles' embedded depth. The difference in the average D_{50} values of upper-layer samples between 2012 (18.5 mm) and 2014 (21.1 mm) measurements was only 1.6 mm (Table 2).

The D_{50} and D_{84} grain sizes of 2012 and March 2013 were first used for the calibration and initial testing of the hydrodynamic model (see Sect. 3.4). For the calibration of the morphodynamic model, the spatial distribution of the D_{50} grain sizes of both 2012 and 2014 was then applied. The grain size values were assigned to equivalent morphological elements defined in the geomorphological map. These grain sizes were transferred to each cell of the curvilinear morphodynamic model's grid. Different input grain size distributions were applied in the model tests: (1) spatially varying upper layer grain sizes, (2) spatially varying sub-layer grain sizes, and (3) constant average grain sizes (average of upper layer, sub-layer, or both upper layer and sub-layer). The values were applied to the whole active layer of the riverbed. The D_{50} grain size values were used in the model.

3.4 Hydrodynamic simulations

The same calibrated hydrodynamic model (2-D implementation of the Delft3D-FLOW) was applied as in Calle et al. (2015). The flow was simulated as depth-averaged flow by using the well-known Navier–Stokes and shallow-water equations. Thus, the modelled fluid was considered vertically homogeneous. These well-known equations were found in the user manual of the model (Deltares, 2011).

The hydrodynamic model was calibrated by adjusting the Manning's n values so that the simulated water levels matched with the measured high-water-level marks and their water level trend line of the 6 March 2013 discharge peak situation (Fig. 3 and Calle et al., 2015). The hydrodynamic model was calibrated by first applying daily discharge values as the upstream boundary condition. The n values were defined based on Limerinos (1970) equations for D_{50} and D_{84} grain sizes (Table 3). It is noteworthy that vegetation was almost non-existent in the channel. There was only scarce vegetation in the upstream part and higher up in the banks, i.e. out of reach of the 2013 discharge events. Therefore, the roughness values were possible to define by the particle size on the active river channel. The Limerinos equation was applied for the whole range of water levels (i.e. hydraulic radiuses) for the March 2012 and March 2013 grain sizes of D_{50} and D_{84}. These represented the best of the preceding conditions of the March 2013 and May 2013 flows. The Limerinos calculations were performed for a cross section located at the downstream side of the simulation area. This cross section was defined from 2012 geometry, i.e. pre-flood geometry, which remained unchanged during the 2013 discharge events. At the 144.151 m water level, which was the measured HWM elevation at this location, the wetted perimeter of the cross section was 52.09 m and the hydraulic radius was 0.95 m (maximum flow depth 1.28 m). Each of the geomorphological elements, and thus also grid cells, received their own Manning's n value during the calibration procedure. Note that the assignment of n values was static throughout the simulations, i.e. the spatial distribution of the roughness values was not changed between simulation time steps.

The calibration was first made for the coarser grid of 1.51–5.31 m cell sizes. Because the discharge estimated at the study site was greater than the one recorded at the Vall d'Alba gauge station, three different daily discharges ("Qx1.3", "Qx2" and "Qx2.5") were tested during the calibration of the hydrodynamic model. The "Qx1.3" was defined based on the increase in watershed area (30 %) between the study area and the Vall d'Alba observation station. This Qx1.3 discharge simulation matched the HWMs using the highest n values calculated with the Limerinos 1970 equation, i.e. for shallow flow (Table 3). In this Qx1.3 discharge, each of the observations was multiplied by 1.3. The "Qx2" (observations multiplied by 2) water surface elevation matched the HWMs using an average n values calculated from Limerinos (1970) equations. Also, the effects of bedform (from +0.01 to +0.015) and bank roughness (+0.02) had been added to these average n values (Chow, 1959; Acrement and Schneider, 1989). The simulated water levels also matched with HWMs when the "Qx2.5" discharge and n values representing the high-flow stage values (i.e. lower n values on the value range) calculated from Limerinos (1970) equations were selected. Similar n value ranges were gained for different water level situations, when either the equation for D_{84} or D_{50} grain sizes was applied. Based on the Limerinos equation calculations, the Qx2 was expected to work the best due to its average nature and inclusion of bedform and bank roughness effects.

When the calibration against the HWMs and their trend line was successful with these daily discharges and coarse grid, the same Manning's n values were applied for simulations with a hydrograph with a 5 min measurement interval. The "Qx1.3", "Qx2" and "Qx2.5" versions of these 5 min hydrographs were tested. These simulations also reproduced the water levels, and the best fit was obtained with "Qx2" (Table 3). Thus, no more adjustments were done for the n values. Finally, the 5 min hydrograph simulations were performed with the fine grid (0.76–3.03 m cells, Fig. 3). These simulation results corresponded to the observations better than if the coarse grid cell sizes were used (as also shown in Calle et al., 2015). Finally, both March and May 2013 discharge events were simulated. Similar multipliers (Qx1.3, Qx2, Qx2.5) and Manning values were applied for both events. Both these discharge events were simulated as unsteady flow and by applying the bathymetry derived from 2012 topographical measurements. The Qx2 discharge was proven to best produce the observed hydrodynamics (Fig. 3; see also Calle et al., 2015). All in all, 21 simulations were run during the calibration of the hydrodynamic model.

3.5 Morphodynamic simulations

Calibration of the morphodynamic model involved adjusting the model parameterization so that the model output geometries matched the measured bedform geometries as well as possible. The model applied an "upwind" bed update scheme, in which the elevation of the bed was dynamically updated at each computational time step. In all simulations, the initial boundary conditions of the input and output sediment transport load amount were defined as 0. This is because the flood events started from 0, and no significant transport, i.e. evolution of bedforms, was observed upstream and downstream of the simulation area. The model then calculated the transport based on the selected transport equations and using equilibrium concentration for carrying input

Table 3. The Manning's *n* values and discharges that gave the best simulation results when compared to the measured HWMs and their trend line during the calibration. Qx1.3, Qx2, and Qx2.5 stand for Vall D'Alba's discharges multiplied by 1.3, 2, and 2.5, respectively. This same table is presented in Calle et al. (2015).

Element type	*n* (Qx1.3)	*n* (Qx2)	*n* (Qx2.5)
Channel/active bar	0.063–0.065	0.05–0.053	0.03–0.04
Extract area	0.07	0.06	0.03
Block	0.068	0.058	0.05
Exposed area	0.07	0.06	0.033–0.04
Scarce vegetation bar	0.065	0.055	0.035–0.04
Vegetated bar	0.07	0.06	0.04
Pleistocene terrace	0.07	0.06	0.05
Rocky bank	0.07	0.06	0.035
Vegetated bank	0.07	0.06	0.05
Opencast mine	0.07	0.06	0.04

Table 4. The morphodynamic simulations and applied parameters. The fine grid size is 0.76–3.03 and the coarser size is 1.51–5.31 m. EH: Engelund–Hansen; MPM: Meyer-Peter and Müller. Events: 1 = only the March 2013 event was simulated; 2 = both the March and May 2013 events were simulated. The simulations that were selected for the channel change analyses are bolded.

Simulation	Transport equation	Discharge events	Transverse slope	Grid size	Grain size
1	EH	1	1.5	coarse	varying, 2012
2	EH	1	1.5	fine	varying, 2012
3	EH	1	1.5	coarse	varying, upper, 2014
4	EH	1	1.5	coarse	varying, sub, 2014
5	EH	1	1.5	coarse	varying, average upper+sub, 2014
6	EH	1	1.5	coarse	constant, average upper, 2014
7	EH	1	1.5	coarse	constant, average sub, 2014
8	EH	1	1.5	coarse	constant, average upper+sub, 2014
9	**EH**	**2**	**3**	**fine**	**varying, upper, 2014**
10	**EH**	**2**	**1.5**	**fine**	**varying, upper, 2014**
11	MPM	2	1.5	fine	varying, upper, 2014

sand sediment fractions. This was similar to earlier studies carried out in a gravel bed river (Williams et al., 2016b) and was selected based on earlier experiences of the river (Calle et al., 2015) and because no suspended load measurements were available and the sand sediment fraction was almost non-existent in the river. Based on the experience in the study site, the load should be in equilibrium particularly in erosional areas of the study site. Because the bed level gradient affects the bedload transport, the slope in the initial direction of the transport (referred to as the longitudinal bed slope) and the slope in the direction perpendicular to that (referred to as the transverse bed slope) were utilized. The transverse slope affects transport towards the downslope direction (Deltares, 2011). The Bagnold (1966) equation was applied for the longitudinal slope and Ikeda (1982), as presented by Van Rijn (1993), was applied for the transverse slope.

Altogether, 61 morphodynamic simulations were needed for calibrating the morphodynamic model. These 61 simulations included simulations with Qx1.3, Qx2, and Qx2.5 discharge hydrographs. During these simulations the match of the simulated water levels to HWMs was also checked. The roughness values were found to be still valid. Out of these 61 morphodynamic simulations, 11 simulations performed during calibration were selected to be presented in the Supplement of this paper (see also Table 4). The reason for their selection was that these 11 simulations showed the effects of grain size (before (2012) and after (2014) floods, and grain sizes from different layers (2014)), grid size (coarse: 1.51–5.31 m; fine: 0.76–3.03 m), transverse slope (user-defined coefficients in the bedload transport equations: default 1.5 and increased to 3) and transportation equations (Engelund–Hansen, EH; Meyer-Peter and Müller, MPM) on model performance (Table 4, and Supplement). These parameter tests were selected for the calibration procedure, as these had earlier been found to be important for Delft3D (2-D implementation) model simulations, albeit in perennial rivers (e.g. Kasvi et al., 2014). The morphodynamic simulation results were compared to the measured topographies within the calibration area, in particular to the volumetric change of riverbed and displacement of the lobe front. Due to the

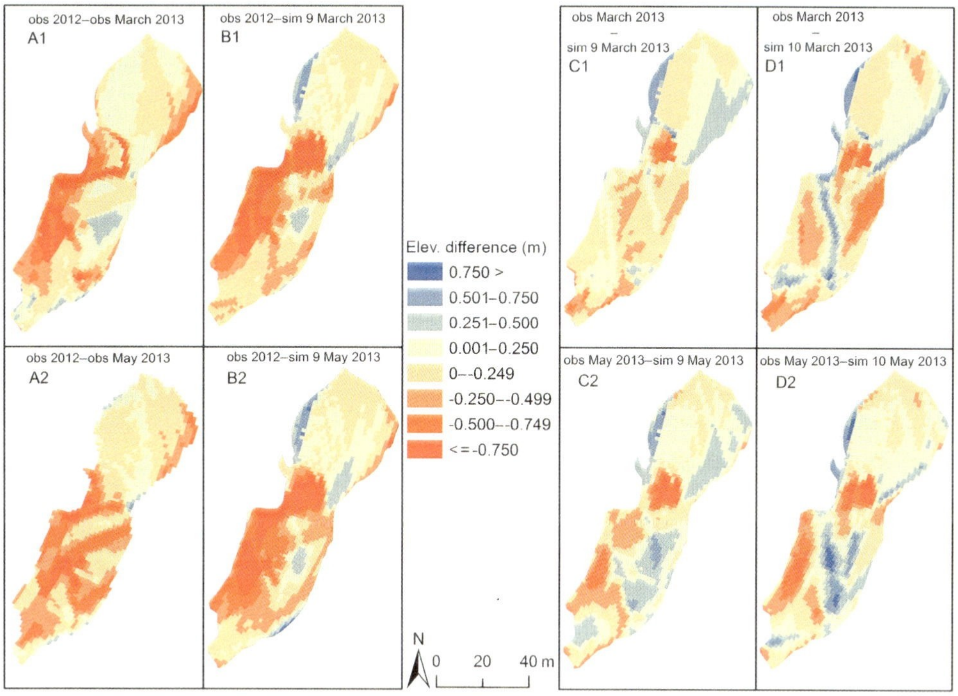

Figure 4. The comparison between the observed and simulated (simulations 9 and 10) elevations. The negative values (red) mean that the observed (A) or simulated (B: simulation 9) March (1) or May (2) 2013 topographies were higher (i.e. deposition had occurred) than was observed in 2012. The negative values also mean that simulated elevations of March or May 2013 were higher than was observed in March or May 2013 with simulations 9 (C) and 10 (D). Simulation 9 corresponded slightly better the observations than simulation 10, and thus its results are shown in more detail. The pool area, in the middle of the area, was the only area where simulation results clearly showed more deposition than observed (see also PU in Fig. 2).

better correspondence of simulations with Qx2 discharges to the observed channel evolution, all these selected morphodynamic simulations had these input data. The longitudinal slope did not affect the results, and the default value of 1.0 was selected to be used. The model was first calibrated with the March 2013 discharge event (simulations 1–11) and then validated with the May 2013 discharge event (simulations 9–11).

Here we present the summary of the calibration results, which can be found in more detail in the Supplement of this paper. The best simulation results in relation to the surveyed volumetric changes were achieved with the fine-grid simulations. The coarser grid size overestimated incision and was therefore discarded. The grain size and its spatial variation affected the model results greatly and needed the largest number of tests (Supplement). When the spatially varying upper-layer sediments of 2014 (representing the sediments moved by the two discharge events) were used (simulation 10), the volumetric changes were less and fitted the observations better than if spatially varying 2012 "before floods" grain sizes were applied (simulation 2). The bedforms were also best represented when upper layer sediments were applied (simulations 9 and 10).

The deposition and erosion amounts resembled the observed elevations, and the lobe movement distances were the best when the transverse slope was increased (simulation 9). The best correspondence of the simulated and measured spatial bedform pattern was obtained with simulations 9 and 10. However, the default (1.5) transverse slope of simulation 10 resulted in a more excavated thalweg, when compared to simulation 9 with an increased transverse slope (Fig. 4).

Earlier simulations performed for ephemeral rivers, e.g. Hooke et al. (2005) and Graf (1996), had applied the Bagnold (1966) total load equation. However, the Meyer-Peter and Müller (MPM, 1948) equation had previously been proven to perform well, i.e. better than, for example, Bagnold (1980) and Parker (1990) equations in flash-flood simulations and in ephemeral gravel river channels (Reid et al., 1996; Cao et al., 2010). The widely applied equations of "Engelund–Hansen 1967" (EH) and "Meyer-Peter and Müller 1948" (MPM) were selected from the standard Delft3D model's assortment of equations (see Deltares, 2011) because they had been developed by using initially non-armoured bed conditions and were the most appropriate for the study area, according to the D_{50} grain sizes. The MPM was expected to perform the best, as it had originally been developed for particles of 0.4–29 mm overall diameter. The EH was selected as the total load equation due to its proven performance in a variety of environments (e.g. Kasvi et al., 2014), even though it had originally been tested on

Table 5. The volumetric changes and lobe movement of observations and the two selected simulations (also see the Supplement). Gravel storage was calculated by subtracting the erosion from deposition.

		First event (March 2013)						
		Erosion		Deposition		Gravel storage		
Sim	Grid	Total m^3	Difference compared to observed %	Total m^3	Difference compared to observed %	Total m^3	Difference compared to observed %	Lobe movement (m)
obs	fine	154.56		984.72		830.16		58.5
9	fine	164.69	+7	1119.28	+14	954.59	+15	59.9
10	fine	178.69	+16	1190.85	+21	1012.16	+22	52.1
		Second event (May 2013)						
Obs	fine	63.67		1087.59		1023.93		87.1
9	fine	182.53	+187	1222.32	+12	1039.79	+2	61.3
10	fine	160.09	+151	1289.38	+19	1129.29	+10	79.5

up to 0.93 mm median particle sizes (Engelund and Hansen, 1967). The transport equation had a crucial role in the simulation results. The EH (simulations 9 and 10) was superior in reproducing the channel morphology. MPM resulted in much smaller and incorrect transport values (simulation 11) during both March and May 2013 discharge events.

Thus, simulations 9 and 10 produced the best correspondence with the observed riverbed changes and bed formations (Fig. 4). These were used for analysing the temporal riverbed evolution during the moderate- and low-magnitude discharge events.

4 The temporal evolution of Rambla de la Viuda

4.1 The evolution during the moderate-magnitude discharge event

All in all, lobe movement of ca. 60 m was observed during the March 2013 discharge event and an additional ca. 30 m was observed during the May 2013 discharge event. The bedload sediment mass balance was possible to estimate. Deposition was greater than erosion based on measurements (i.e. in the calibration area) (Table 5). Hydrographs and erosion and deposition rates were also plotted against time for simulations 9 and 10 (Fig. 6), which had reproduced the best observed morphodynamics within the calibration area (see Sect. 3.6). Sediment erosion and deposition were calculated from the total channel bed change amount for each hour from this calibration area. In addition, the bed elevation changes between the key time steps, i.e. times of clear changes in discharge or bed evolution, were also defined from the whole simulation area (Fig. 6: simulation 9 as an example).

Both simulations experienced more deposition than erosion during the March 2013 event (Table 5 and Fig. 5). The greatest total erosion and deposition (in cubic metres) occurred at the beginning of this discharge event (Fig. 5). However, the initial changes were slightly smaller when the default transverse slope was applied (simulation 10). The peak of channel bed changes occurred at the beginning of the moderate-magnitude discharge event, i.e. before the discharge peak. This suggests a possible positive hysteresis phenomenon. Significant changes were observed in small peaks at 9 and 71 h, which give an idea of the high system response to steep changes in the discharge, rather than a progressive increase (as seen in the absolute peak of 23 h). The deposition in simulations 9 and 10 started very slowly, decreasing after 13 h, but the decrease in erosion was slightly faster. However, all in all, the deposition and erosion remained high for a long time and only started clearly declining after 71 h, i.e. the secondary discharge peak. The erosion and deposition volumes followed the changes in receding discharge more faithfully than during the rising phase, occurring simultaneously to the changes in discharge. Thus, the morphodynamics of both simulations 9 and 10 followed the changes in discharge better during the falling stages of the discharge peak, i.e. when the discharge was 30 $m^3 s^{-1}$ or less (Fig. 5). During the rising phase, there were temporal differences between changes in discharge and changes in deposition and erosion. When the default transverse slope was applied (simulation 10), the erosion and deposition continued to be high longer in the receding phase than when the transverse slope was increased to 3 (simulation 9). The great deposition amount during the receding phase was mainly due to the propagation of the bar lobe front.

When detecting the channel changes spatially during this March 2013 moderate flow event from the whole simulation area (based on simulation 9, as an example), the channel evolved and bar lobes advanced on both sides of the large lateral bar due to the diverted flows (B1 location in Figs. 1 and 2, and B in Fig. 6). However, the main flow and bar lobe movement occurred on the right bank side of the channel (B in Fig. 6) and diagonal bar aggradation took place

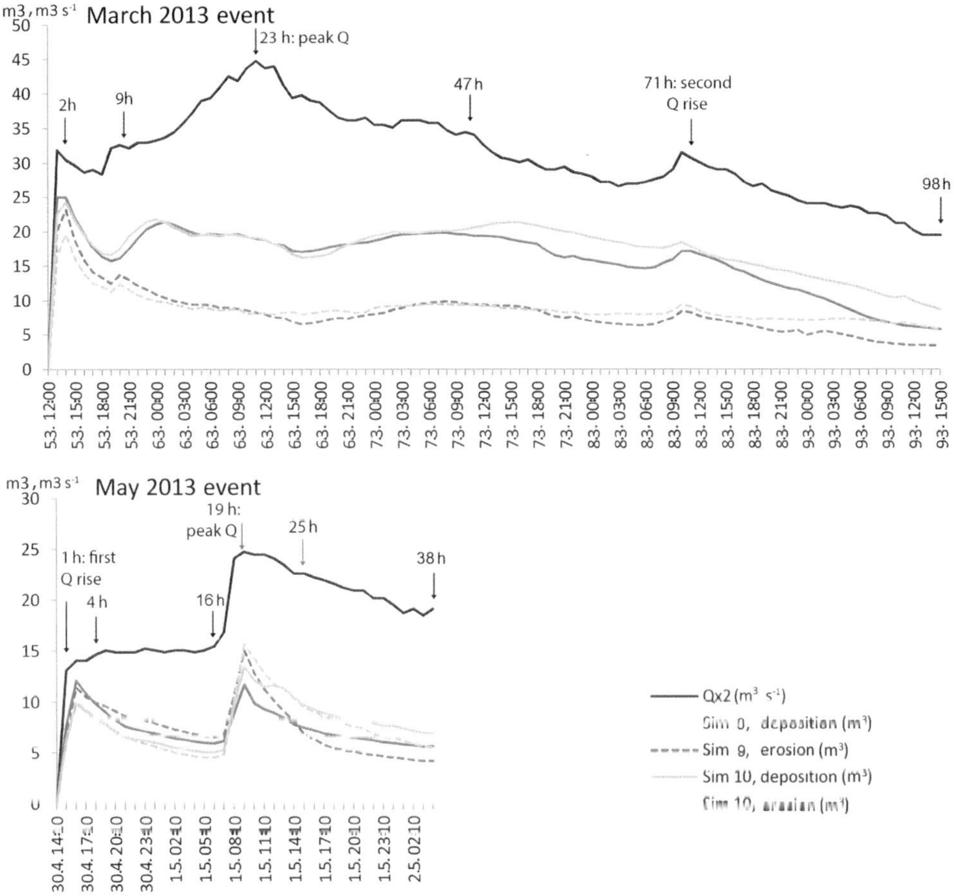

Figure 5. The hourly volumetric changes during the moderate- and low-magnitude flow events of March and May 2013. The results are based on simulations 9 and 10. The graphs show the erosion and deposition from the beginning of the flow events (12:00 on 5 March and 14:10 on 30 April) until the time when the erosion and deposition had declined and levelled out during the receding phases. The key time steps are pointed out as hours from the beginning of the flow events.

in the downstream part of the study site. Both topographical observations and model simulations showed the development of the diagonal bar alongside the cut-off of the bar (B). The model results showed that the initial cut-off and simultaneous initiation of the diagonal bar took place during the rising limb, more precisely the first couple of hours of the flood event (Fig. 6). However, during these first hours of the flood event, the changes in the riverbed were local, and the greatest elevation changes occurred especially in the bar lobe area and right upstream of the large lateral bar B1 (A location in Fig. 6). It is noteworthy that the diagonal bar would not have developed to its full extent without the long receding phase of the flood hydrograph. Spatially, the most changes occurred throughout the whole simulation area during the 24 h following the discharge peak, i.e. between 6 and 7 March. The diagonal bar formation was slightly greater in the model outcomes than based on observations. Despite this, the model showed potential in producing the channel development following the established theories of gravel bed evolution. Similar to the diagonal bar movement in the downstream part of the study site (B in Fig. 6), changes also occurred in another lateral bar further upstream (A in Fig. 6). This bar experienced excavation on the right bank side and the propagation of lobe movement downstream particularly after the discharge peak. Thus, throughout the study area the initial hours of the channel changes caused the selection of the flow and sediment transport routes where the most erosion would take place later during the moderate-magnitude discharge event.

In addition to the braiding display of the gravel bars, i.e. alternating bars and pools, the river channel also has a broader meandering planform. The simulated area can be considered as one bend. The downstream lobe area locates downstream of the bend's apex, and the greatest erosion occurred during the lowering flood phase on the right bank side, which is the inner bank of the bend at the inlet area to the bend (A). This also followed the results of other meandering river studies about erosion locations (Lotsari et al., 2014b). Thus, during the higher flow stage the flow routing and channel started acting more like in a meandering river. However, during lower

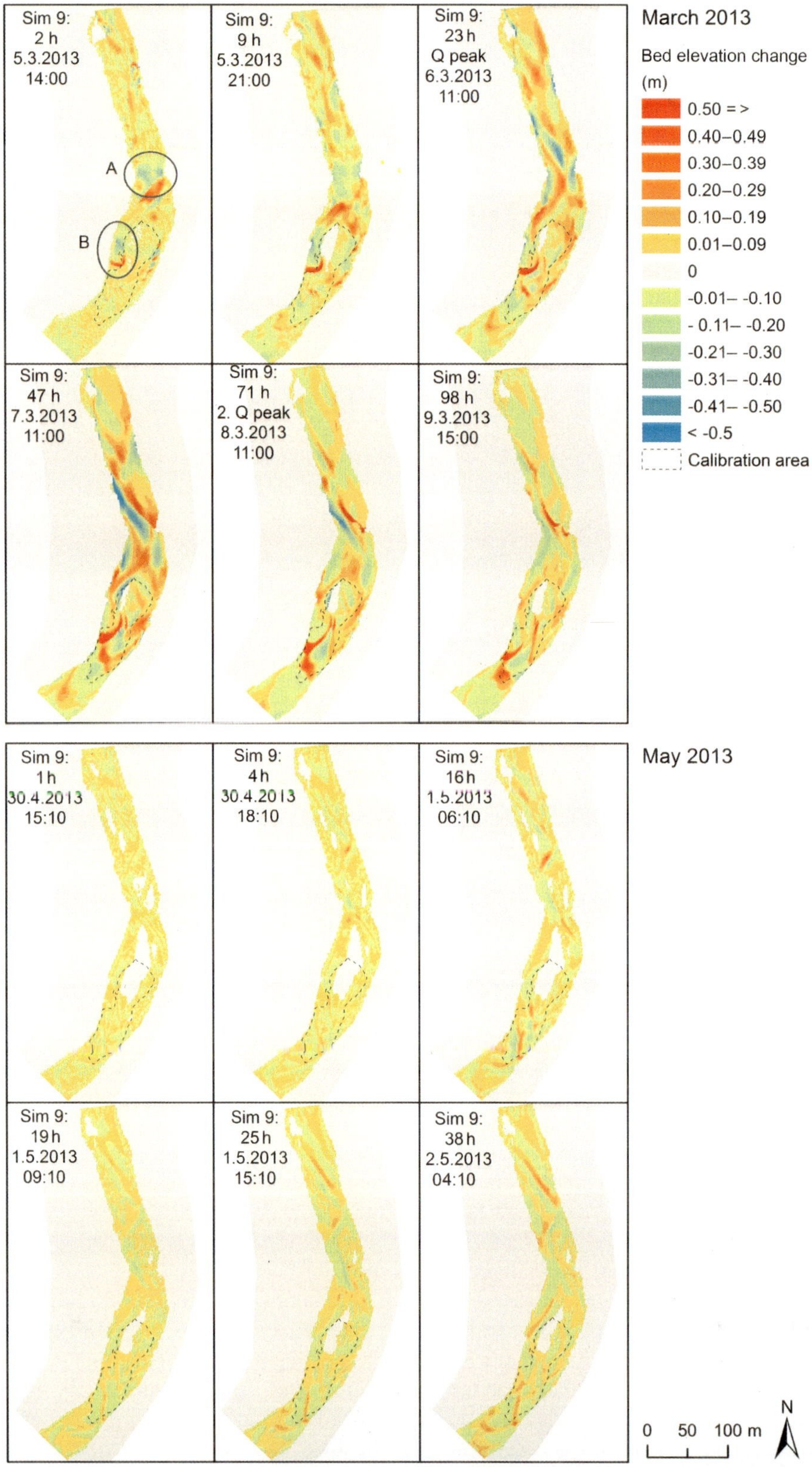

Figure 6. The modelled bed elevation changes between the key time steps of March and May 2013 events (see Fig. 5). The changes of the 5 March 2017 at 14:00 are from the beginning of the discharge event.

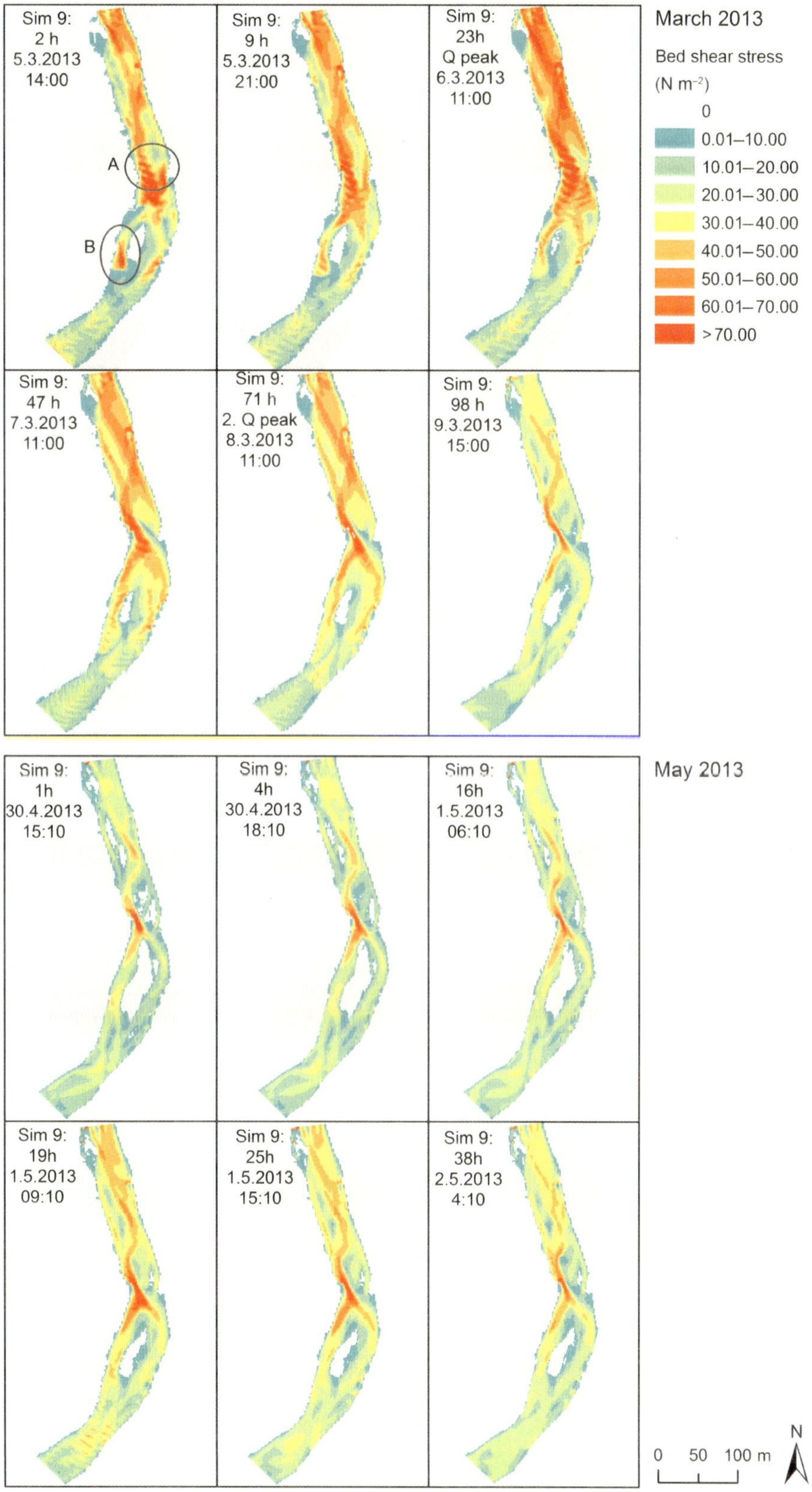

Figure 7. The bed shear stress of simulation 9 during moderate and small discharge events. The key time steps are presented (see Fig. 5).

flow stage, such as during the initial rising flood stages, when not all of the bars were covered with water, the fluvial processes and related channel changes resembled braided channel development.

4.2 The evolution during the low-magnitude discharge event

During the May 2013 discharge event, the hourly changes in erosion and deposition within the calibration area followed the discharge evolution more than during the March discharge event (Fig. 5). The discharge event of May 2013 had two peaks, of which the latter discharge peak was greater. The erosion and deposition peak occurred approximately an hour after this greatest peak discharge had been reached. This suggests a possible negative hysteresis phenomenon. Thus, the deposition and erosion peaks did not occur immediately at the beginning of the discharge event, as in the case of the March 2013 event. The erosion became greater than the deposition 4 h after the beginning of the discharge event (simulation 9). The deposition dominated again 6 h after the sediment transport peak (at 25 h). During the March discharge event, the erosion was never greater than the deposition. Note also that the discharge remained constant after both peaks, whereas a decrease in the deposition and erosion rates was observed (Fig. 5).

When considering the whole simulation area (based on simulation 9, as an example), the main flow path was again on the right bank side of the channel, where the greatest changes also occurred during this low-magnitude discharge event (B in Fig. 6). The quick rise in discharge during the first hour and between the 16th and 19th hours of this discharge event did not cause spatially great changes to the riverbed, as most of the changes were within ± 10 cm throughout the simulation area. Thus, the spatial morphodynamics of this low-magnitude event differed from the moderate-magnitude discharge event. The channel acted more like a braided river during this May 2013 low-magnitude event, as there were continuous small bed elevation changes throughout the event. Thus, there were not such clear high erosion or deposition periods, but continuous steady changes. The bar lobes progressed downstream, as the sediment was transported from the proximal side of the bars to their distal side.

4.3 The flow characteristics during the moderate and small discharge events

The bed shear stress (based on simulation 9, as an example) during the first hours of the March 2013 moderate-magnitude discharge event revealed that the bend apex (A in Fig. 7) and right bank side of the downstream lateral bar (B in Fig. 7) experienced the most fluid forces. These explained the locations of the greatest changes and the initial cut-off of the lateral bar (B). It is noteworthy that these locations were the initial high erosion and bedload transport locations, but the high bed shear stresses occurred spatially more widely during the rising stage (ninth hour) and the peak flow situation (23rd hour). Even though the changes in the river channel had been great between the 24 h following the discharge peak (Fig. 6), the fluid forces (bed shear stress) had already started to concentrate on the thalwegs of the channel (Fig. 7). It is noteworthy that throughout the receding phase of the moderate-magnitude March 2013 discharge event the greatest shear stresses occurred mainly in thalwegs. Thus, the initial forces of the flood were greater throughout the channel area, but later concentrated on the channel routings formed by the initial stages of the flood discharges, particularly at the right bank side at the apex (A) during the receding phase of the flood.

The spatial distribution of the bed shear stresses of the low-magnitude discharge event (May 2013) differed from that of the moderate-magnitude event (Fig. 7). Throughout the discharge event, the greatest fluid forces concentrated in the thalwegs, which had been formed by the preceding moderate-magnitude discharge event of March 2013. Particularly, it could be seen that the bed shear stresses were the greatest at the inner bank side at the apex (A), and then the main flow route was at the right bank side (B) in the lobe area. The greatest bed shear stresses occurred during the second rise of the low-magnitude discharge event, i.e. between 16 and 19 h from the beginning of the event. Thus, similarly as revealed in Fig. 5, the transport followed the changes in discharge more during this low-magnitude May 2013 event than during the March 2013 event (Figs. 6 and 7).

5 Discussion

5.1 Uncertainties related to the simulations

The morphodynamic simulation of the ephemeral river provides a good quantification of the sequence of channel changes described by Calle et al. (2015). It also extends the analyses performed based on only the topographical data and is the only way to understand topographical changes during the flash floods. The reliability of the presently applied and also future models, which are calibrated against the events under interest, can be improved with the quality and temporal density of the available calibration topography, i.e. pre- and post-flood bedform geometries. We had two highly accurate MLS data sets and one RTK-GPS topographical data set, which is more than in many other studies. In recent studies performed in perennial rivers, where topographical measurements have been sparse, the greatest 2-D morphodynamic model uncertainties have related to the channel topographies (Sanyal, 2017). The high uncertainties in topographical measurements of sub-water areas in gravel bed perennial rivers have been related particularly to the high bedload velocities and temporal variability in bedload (Williams et al., 2015). However, despite the high-quality data from two events at Rambla de la Viuda, we think that further research with multiple yet-to-come events needs to be run to assess the repeatability and validation of the model even better. For example, at Rambla de la Viuda, large floods have not yet occurred since the beginning of the MLS measurement approaches. As has also been stated earlier (Verhaar et al., 2008; Lotsari et al., 2015), the roughness conditions defined for small discharge events might not be suitable for simulating extreme events. Therefore, the work and refinement of the model will continue, and the applicability of the model for larger floods will be tested, when validation data are available.

Transport rates (suspended and bedload) and flow measurements, which are measured during the events for calibrating models of perennial rivers, are always difficult and dangerous to perform in this particular ephemeral river. The uncertainties of the present model approach thus relate to the lack of sediment transport, flow, and topographical data during the events. However, the selected initial boundary conditions seem to be congruent with the flooding mechanisms of Rambla de la Viuda and also with other published works, i.e. Williams et al. (2016b), for loosely consolidated sand and gravel. Thus, the modelled flow carried each sand sediment fraction (suspended), which had been adapted to the local flow conditions at inflow boundary, and the model assumed that very little accretion or erosion was experienced near the model boundaries. Based on measurements, the presence of sand size particles and their concentration in the study site were also almost non-existent, and no channel changes occurred at the downstream boundary of the simulation area. Thus, this equilibrium load condition was considered valid, and it was also the only option for the present modelling approach, as no input suspended load measurements were available. Similar to the Williams et al. (2016b) state, the model results, i.e. the modelled deposition and erosion, when compared to observations, could have possibly been enhanced if the input suspended sediment load observations would having been available. However, according to Sanyal (2017) the sediment transport is always inherently approximate in nature, and sediment load added to the model causes uncertainties to the results, despite detailed sediment load measurements have been used as model input.

Despite this lack of data during the flood events, we had a good control on sediment volume and gravel particle size moving downstream as the forefront lobe prograded over a flat valley bottom (as gravel bed had been mined). Total volume input and total transport rates observed in an earlier study of Rambla de la Viuda by Calle et al. (2015), had already proved the high availability of sorted gravel particles and were the basis of the decisions made while building up the model. As already mentioned, the channel changes at the output downstream boundary of the studied reach were zero. In addition, the simulation result supports the hypothesis of Calle et al. (2015) that moderate- and low-magnitude events reworked sediment locally within the reach. This means that these flows were not able to establish a sediment connection upstream, i.e. between larger reaches, and the transported sediment originated from the erosion of adjacent areas.

In addition, we were able to use discharge and precipitation stations further upstream of the study area, and thus also a hydrograph with a known shape (see Supplement). Sometimes only the HWMs are known, and for example, Cao et al. (2010) analysed the bedload transport by using a symmetrical hydrograph. This could be why their simulations showed less transport during the receding phase than ours. An et al. (2017a, b) and Viparelli et al. (2011) had raised the issue related to the cyclic hydrographs and hydrograph boundary layer (HBL), which defines the spatial region within which riverbed topography and grain size respond to the hydrograph. They state that the HBL causes problems, if the material is poorly sorted. The calibration area of the present study was in the downstream part of the simulation area, i.e. outside of the influence of the possible HBL. The gravel of Rambla de la Viuda was well sorted and we can also therefore state that the HBL was not an issue for the calibration of our model.

As the curvilinear grids of fine 0.76–3.03 m and coarse 1.51–5.01 m cells were tested for morphodynamic simulations, the finer grid was proven to better predict the channel evolution, including both flow and bar lobe movement paths. Grids of circa 3 m cell sizes have been found by Williams et al. (2016b) to be suitable for Delft3D model applications. Furthermore, the grain sizes and their spatial distribution affected the morphodynamic simulation results greatly. Hooke et al. (2005) have shown that grain sizes have a major effect on the morphological impacts of floods. Simulation results may be enhanced when using techniques that are able to spatially obtain more complete and detailed grain size data (e.g. Wang et al., 2013) from the surface layers, instead of lower-resolution site sampling such as the Wolman method. Earlier studies have stated that a calculation of the bedload transport rate of mixtures should be based on the availability of each size range in the surface layer (Parker, 1990), and that during the calibration of the Delft3D model uniform sediment cannot be an acceptable assumption to study the long-term response of natural gravel bed braided rivers (Singh et al., 2017). The best simulation results of Rambla de la Viuda were indeed gained when spatially varying upper surface layer sediment sizes were applied. However, further analyses of spatial and vertical grain size distribution effects on ephemeral gravel bed river simulations are suggested. The next step should be to apply the surface grain size variation derived from the laser scanning data to analyse these effects of grain size distribution in further detail (Casas et al., 2010a, b; Kukko et al., 2015).

The patterns of channels, bars, and braiding properties are highly sensitive to the bed slope parameterization, and particularly transverse slope affects the sediment transport in long-term simulations (Schuurman et al., 2013; Kasvi et al., 2014; Singh et al., 2017). Also, during these rather short-term flood event simulations at gravel bed Rambla de la Viuda, the effects of the increasing transverse slope on channel evolution were evident, and the simulation results resembled the observed topographies more. Kasvi et al. (2014) had noticed that the higher the transverse bed slope effect (i.e. value 3), the more filling of pools in a meandering river. Similar to Kasvi et al. (2014), the pool area of Rambla de la Viuda experienced more deposition with the increased transverse slope value (Fig. 4). This pool area was the only clearly wrongly simulated location, when this high transverse slope value was applied, but otherwise the lobe movement and elevation best resembled the observations. Thus, in a gravel bed river with short flow events, adjustments are clearly needed in this parameter in order to replicate the observed changes while simulating with the currently available sediment transport equations. Thus, similar to the perennial river simulations (e.g. Kasvi et al., 2014), the transverse slope was important for the simulated changes.

Despite the MPM equation having been earlier successfully applied in ephemeral gravel bed rivers (Reid et al., 1996), the movement was minimal in Rambla de la Viuda and occurred only in the beginning of the moderate event. Therefore, after these tests with the MPM equation, the application of the MPM equation was discarded from the final analyses. The transport rate is imposed as bedload transport due to currents in both EH and MPM equations, which are used in Delft3D-FLOW. The MPM is more advanced than EH as the MPM includes a critical shear stress for transport (Deltares, 2011). However, this could be one reason for less movement, if critical stresses for transport required in the MPM equation were not exceeded during the lower discharges of the hydrographs. According to Barry et al. (2004) the formulae containing a transport threshold typically exhibit poor performance when compared to the observed gravel bed river's bedload data. These equations include MPM (1948) and Bagnold (1960) equations, which have been applied in ephemeral channels. In addition, Singh et al. (2017) state that during variable discharge conditions, the MPM formula cannot produce the braided river pattern because MPM is not suitable if the value of the Shields parameter falls outside its applicability range. These above-mentioned facts could explain why the MPM equation did not move particles during the lower discharges and during the receding phase of the hydrographs of Rambla de la Viuda, where flow is temporally highly variable. The application of the MPM equation could also partly explain why the receding phase's transport was less in the study of Cao et al. (2010), than in the present study. Despite being unable to produce the correct morphodynamics caused by the moderate- or low-discharge events of Rambla de la Viuda, the MPM equation could have potential in the simulation of greater floods of ephemeral gravel bed rivers, as Reid et al. (1996) and Cao et al. (2010) have shown. However, further research on this matter should also be carried out with higher discharges at the river reach.

Engelund and Hansen (1967) developed their equation, so that the effects of dunes, e.g. transport from stream side to lee side and form drag, were included. This could be the reason why the EH worked the best at the Rambla de la Viuda, where clear sharp lobe edges occur, and the local slope varies a lot. In addition, the EH is a total load equation (Engelund and Hansen, 1967), whereas, for example, the MPM equation calculates bedload. There are no universal sediment transport formulas that can cover all coarse sediment transport conditions from low discharges to peak discharges (Li et al., 2016; Singh et al., 2017). The grain sizes, which had been moved by the 2013 discharge events, were on the upper limit of the available sediment transport equations. A matter of further research is also to detect whether the transport equations would able to move the sediments at higher discharges, if in reality the higher discharges would possibly transport even larger sediment particles.

However, despite the challenges of applying morphodynamic simulation approaches to ephemeral river channel change analyses, the results are promising and show the importance of both rising and receding phases. The observations of sediment transport during flow hydrographs would still be important, in addition to the continuous time series of pre- and post-flood topography, but at the moment, their measurement is still a complex and risky issue. By continuing the long-term observations of the ephemeral channel changes with the high-precision equipment, more reliable calibration and validation of the models may be gained.

5.2 Moderate- and low-magnitude flow events as channel modifiers

Our study has shown that geomorphic responses to the two analysed discharge events differed. Previously this uniqueness of geomorphic response has been shown for perennial rivers (Pitlick, 1993), where topographical and sedimentary data have also been applied as initial conditions. Hooke (2016a) stated that the flow events of similar magnitude can have differing effects, depending on the state of the system, as the long-term evolution of the ephemeral river channel and its material greatly influence the response to the stream flow. Some events are more erosional and some are more depositional (Hooke, 2016a). Moreover, Hooke et al. (2005) noted the importance of simulating and analysing the feedback effects of consecutive events. At Rambla de la Viuda, the riverbed morphology formed by the March 2013 event influenced the later channel changes during the May 2013 flow. For example, the flow was diverted to the right bank side during the first flood (of March 2013), which also therefore acted as the main channel for the May 2013 flow, which was lower in magnitude.

The simulation results of the Rambla de la Viuda showed that the differences between rising, peak, and receding phases of a moderate-magnitude discharge event are very important in an ephemeral river environment. A higher total number of channel changes occurred during the receding phase than in the early stages of the discharge events.

Deposition dominated due to the progradation of the frontal bar lobe, particularly on the right bank side of the channel. Thus, the continuous channel changes were similar to those for braided perennial rivers (Lotsari et al., 2014a). However, the channel changes differed from a recent study of Gendaszek et al. (2013), who studied the gravel perennial riverbed changes during moderate- (65 $m^3 s^{-1}$) and high-flow (159 $m^3 s^{-1}$) events. They found that most erosion occurred during the rising and the peak flow phases but did not mention great changes during the receding phase. They found only some scour during sustained high flows following the flood peak (Gendaszek et al., 2013). Noteworthy is that they applied one sensor per reach, and thus the site selection could have greatly affected their results.

Ferguson (1993) stated the potential in numerical modelling of the coupling between geometry, flow, and bedload transport, if it can be applied successfully to braided channels. The results were promising at Rambla de la Viuda, which also has a braided pattern. According to Wheaton et al. (2013) the chute cut-off mechanism, already described by Ferguson (1993), is the most common braiding mechanism, but the cut-off is not only an erosional process, but more the result of deposition during the construction of diagonal bars. In our study area, there was a situation resembling a chute cut-off, as the channel was cut more on the right bank side of the bar than before. The modelling was capable of producing this observed chute cut-off from the right bank side during the moderate discharge event (March 2013). The high values of bed shear stresses are related to this initial chute cut off (B location in Figs. 6 and 7). Both erosion and deposition are related to these changes, as both topographical observations and model simulations showed the development of the diagonal bar alongside the cut-off of the bar. However, the diagonal bar formation was slightly greater in the model outcomes than based on observations. Despite this, the model showed potential in producing the channel development following the established theories of gravel bed evolution. In addition, the model showed that the initial cut-off and simultaneous initiation of the diagonal bar took place during the rising limb, but the diagonal bar would not have developed to its full extent without the long receding phase of the flood hydrograph. Williams et al. (2015) had found that the choking is the main process for braiding development in their studied perennial river. However, we were not able to observe choking processes within our study reach. Further analyses of this braiding process are necessary. In particular, studies from longer river reaches of ephemeral rivers would be important.

In addition to these braiding processes, the fluvial processes at higher discharges resembled the ones in a meandering river bend more. The greatest erosion and bed shear stresses occurred on the right inner bank side at the inlet area of the bend (A in Figs. 6 and 7) during the peak and receding flood phases of the moderate-magnitude (March 2013) event. This also followed the results of the erosion and high-velocity core locations of perennial meandering rivers (Dietrich and Smith, 1983; Lotsari et al., 2014b). However, during a lower flow stage, such as during the initial rise of the floods and the whole May 2013 discharge event, the spatial distribution of channel changes resembled braided channel development more. Therefore, the morphodynamics of the low-magnitude May 2013 event differed from the preceding moderate-magnitude discharge event. During the May 2013 low-magnitude event, there were continuous steady small bed elevation changes throughout the event. It is noteworthy that throughout the receding phase of the moderate-magnitude March 2013 discharge event and the whole May 2013 event, the greatest shear stresses occurred mainly at the channel routings formed by the initial stages of the moderate-magnitude flood discharges.

It is noteworthy that during the moderate-discharge event the erosion and deposition peak occurred much earlier than the discharge peak. However, the low-magnitude discharge event experienced the greatest channel changes an hour after the discharge peak. Our model results (e.g. Figs. 7 and 8) suggest the possible existence of hysteresis in the rate of bedform changes, being positive in the case of moderate-magnitude flow (bedform change peak occurs before flood peak) and negative in the case of low-magnitude flow (bedform change peak occurs during or after flood peak). The hysteresis phenomenon has been described well in sediment transport studies, and their effect is due, among other factors, to sediment depletion or surface gravel consolidation in the channel (Reid et al., 1985) or a long-lasting portion of the baseflow during the recession limb (Walling, 1974). Cao et al. (2010) have shown, based on their 1-D simulations, that bedload transport in an ephemeral river can have similarities to a perennial river. However, even though perennial rivers may have sharp rising phases in their discharge hydrographs (e.g. Long, 2009), they more likely have a greater initial threshold for particle movement by bed armouring than ephemeral rivers (Reid et al., 1996; Hassan et al., 2009). However, the armouring of perennial rivers can decrease as bedload concentrations increase (Müller and Pitlick, 2013). Hysteresis of both kinds have also been shown in perennial rivers regarding turbidity, but their flashy storm hydrographs have more often caused an anticlockwise (i.e. negative) hysteresis phenomenon (Lloyd et al., 2016). This would indicate that ephemeral rivers act more similarly to perennial rivers during their low-magnitude flow events. Even though further research is needed, the results indicate that the greater the discharge event's magnitude in an ephemeral river, the more different the channel evolution and its timing are compared to perennial braided gravel bed rivers.

In our study reach, 30 $m^3 s^{-1}$ was the threshold discharge for the channel changes to become gradual and to follow the discharge changes during the receding phase. According to Hooke (2016a), the threshold values of hydraulic conditions for erosion and deposition in ephemeral channels vary between sites, mostly due to the size and nature of the bed material. Threshold discharge for the deposition and erosion of 15 mm particles has been observed to be 35 and 1 $m^3 s^{-1}$, respectively, and of 50 mm particles 14 and 40 $m^3 s^{-1}$, respectively (Hooke, 2016a). The average upper and sub-layer D_{50} grain sizes at the Rambla de la Viuda were 26.3 and 17.1 mm, respectively, and the maximum D_{50} grain size was 40.2 mm. The discharges of the Rambla de la Viuda were overall within the ranges of Hooke's thresholds for the movement of these sized particles, and the simulations clearly showed their transport. In addition, the threshold discharge of 30 $m^3 s^{-1}$ for starting the channel changes of the Rambla de la Viuda to follow the discharge changes during the receding phase coincide well with Hooke's (2016a) threshold analysis of deposition. However, the threshold could also be site dependent, such as for the Spanish Mediterranean region, where both Hooke et al. (2016a) and the present study were performed.

6 Conclusions

A 2-D morphodynamic model was successfully implemented for inferring the morphological changes caused by moderate- and low-magnitude discharge events. The morphological starting and final conditions were derived from high-resolution laser scanning topography, bedform mapping, and sediment texture analysis of both surface and subsurface layers on the main morphological bars and channel units. The flow hydrograph was based on a continuous (5 min) record from an upstream gauge station, which was rescaled to the peak flow estimated at the study reach based on flotsam evidences. The model parameters were calibrated to best fit the pre- and post-topographic field evidence. The following conclusions can be made about the morpho- and hydrodynamics in an ephemeral river during moderate- and low-magnitude discharge events:

1. The 2-D implementation of morphodynamic model was proven to work during moderate- and small-magnitude flash-flood events in an ephemeral river. The spatially varying grain size data, the applied transverse slope parameter value, and the streamflow hydrograph were the most important factors affecting the simulation results of the bedform and channel evolution. The selection of sediment transport equations was also critical to match the sediment mass balance, concluding that the total load equation (EH) worked the best. When modelling events of moderate and low magnitude in ephemeral rivers 50 to 100 m wide, a resolution of at least 1 m is recommended (i.e. one cell is 1–2 % of the total width) because it is able to show the steep lobe forefront scarp.

2. Morphodynamic models can shed light on the channel evolution during flash-flood events. Both rising and receding phases of discharge events were predicted to be important for bar movement and channel evolution and thus should not be ignored while planning flood mitigation measures. The erosion and deposition can be greater during the long-lasting receding phase than in the rising phase of moderate- and low-magnitude discharge events, despite the typical hydrograph shape of a flash flood. The receding phase also contributed greatly to the shaping of the bedforms and channel pattern.

3. The deposition and erosion peak rates were predicted to occur at the beginning of the moderate-magnitude discharge events (e.g. March 2013), whereas deposition dominated throughout the event, i.e. even during the rising phase. Conversely, the low-magnitude discharge event (e.g. May 2013) only experienced the greatest channel changes after the discharge peak.

4. These different predicted erosion and deposition patterns suggest that the timing of the channel change peak, compared to the timing of the discharge peak, differs between different magnitude discharge events. These stress the importance of previous flood history (timing, succession, and magnitude) in understanding the geomorphic response of gravel bed ephemeral rivers. The peak of channel changes during the moderate flood occurred before the discharge peak. Conversely, the low-magnitude discharge event experienced the peak of channel changes after the discharge peak, suggesting a rupture of the upper layer before sediment was moved. These differences between the events were at least evident with these two events and the applied parameterization. However, further studies of this possible hysteresis effect are needed from multiple discharge events.

5. The results showed that the fluid forces of the initial stages of the moderate-magnitude discharge event caused the initiation of chute cut-off and diagonal bar formation and defined the flow routing of the rest of the moderate-magnitude discharge event. The flow during the following low-magnitude discharge event followed these same thalwegs caused by the preceding moderate-magnitude event.

6. The clearest difference in the predicted riverbed changes between the rising and receding phases were that erosion and deposition characteristics followed the temporal discharge changes more during the receding phase than during the rising phase. The threshold discharge, below which the channel bed changes started following the discharge changes temporally, was around 30 $m^3 s^{-1}$. Noteworthy is that this ephemeral channel acted like the braided river channel during these lower-flow conditions, but when the bars were submerged in higher discharges the high fluid forces followed the meandering river planform of Rambla de la Viuda. Thus, there were different fluvial processes working at different flow stages, particularly during the moderate-magnitude discharge event.

7. Preventive measures against flash-flood-induced channel changes should take into account moderate- and low-magnitude flows. This has received support from earlier studies (e.g. Hooke, 2016a, b) as the receding phase can cause substantial channel changes, it plays a major role in inducing damages and modifications to the river environments, and it needs to be acknowledged while planning flood mitigation measures.

Author contributions. EL performed most of the writing and all of the model simulations. MC and GB also greatly contributed in the writing of the paper and enabled the study with their projects, as they initiated the studies of Rambla de la Viuda. All the other authors contributed to the writing process by commenting on the paper and its content. All authors contributed in collecting the field data. Laser scanning was carried out by AK, HK, and PA. The laser scanning data were processed by AK and HK. JH and HH contributed by providing funding for the study and they also worked on the development of the laser scanning approaches. GB, MC, and EL measured the sedimentological data. GPS measurements of topography and high-water marks were performed by GB and MC.

Competing interests. The authors declare that they have no conflict of interest.

Acknowledgements. We would like to thank the Institut Cartografic Valencia, Servicio Automatico de Informacion Hidrologica of the Confederacion Hidrográfica del Júcar for supporting data. Financial support is provided by the Spanish "Ministerio de Economia y Competitividad" (projects CGL2011-29176 and CGL2014-58127-C3-1-R), the Academy of Finland (Extreme and annual fluvial processes in river dynamics – ExRIVER (grant number 267345), the Centre of Excellence in Laser Scanning Research – CoE-LaSR (grant number 272195), and the Strategic Research Council project "COMBAT" (grant number 293389)), Maj and Tor Nessling Foundation (Äärimmäisten ja vuotuisten fluviaaliprosessien vaikutukset jokidynamiikkaan; grant number 2013067), and the Spanish Ministry of Economy and Competitiveness through the projects PALEOMED (CGL2014-58127-C3-1-R) and EPHIMED (CGL2017-86839-C3-1-R). Finally, we would like to thank the editors and the three reviewers for comprehensive reviews, which greatly helped us to improve the paper.

References

Acrement, G. J. and Schneider, V. R.: Guide for Selecting Manning's Roughness Coefficients for Natural Channels and Flood Plains, United States Geological Survey Water-Supply Paper, 2339, 1–38, 1984.

An, C., Fu, X., Wang, G., and Parker, G.: Effect of grain sorting on gravel bed river evolution subject to cycled hydrographs: Bed load sheets and breakdown of the hydrograph boundary layer, J. Geophys. Res.-Earth, 122, 1513–1533 https://doi.org/10.1002/2016JF003994, 2017a.

An, C., Cui, Y., Fu, X., and Parker, G.: Gravel-bed river evolution in earthquake-prone regions subject to cycled hydrographs and repeated sediment pulses, Earth Surf. Proc. Land., 42, 2426–2438, https://doi.org/10.1002/esp.4195, 2017b.

Bagnold, R. A.: An Approach to the Sediment Transport Problem from General Physics, Geological Survey Professional Paper 422-I, United States Government Printing Office, Washington, 1966.

Bagnold, R. A.: An empirical correlation of bedload transport rates in flumes and natural rivers, P. Roy. Soc Lond. A Mat., 372, 453–473, https://doi.org/10.1098/rspa.1980.0122, 1980.

Barry, J. I., Buffington, J. M., and King J. G.: A general power equation for predicting bed load transport rates in gravel bed rivers, Water Resour. Res., 40, W10401, https://doi.org/10.1029/2004WR003190, 2004.

Benito, G., Thorndycraft, V. R., Rico, M. T., Sánchez-Moya, Y., Sopeña, A., Botero, B. A., Machado, M. J., Davis, M., and Pérez-González, A.: Hydrological response of a dryland ephemeral river to southern African climatic variability during the last millennium, Quaternary Res., 75, 471–482, https://doi.org/10.1016/j.yqres.2011.01.004, 2011.

Billi, P.: Flash flood sediment transport in a steep sand-bed ephemeral stream, Int. J. Sediment Res., 26, 193–209, https://doi.org/10.1016/S1001-6279(11)60086-3, 2011.

Bunte, K. and Abt, S. R.: Sampling surface and subsurface particle-size distributions in wadable gravel and cobble bed streams for analyses in sediment transport, hydraulics, and streambed monitoring, Gen. Tech. Rep. RMRS-GTR-74, Fort Collins, CO, US Department of Agriculture, Forest Service, Rocky Mountain Research Station, 1–428, 2001.

Calle, M., Lotsari, E., Kukko, A., Alho, P., Kaartinen, H., Rodriguez-Lloveras, X., and Benito, G.: Morphodynamics of an ephemeral gravel-bed stream combining mobile laser scanner, hydraulic simulations and geomorphological indicators, Z. Geomorphol. Supp., 59, 33–57, https://doi.org/10.1127/zfg_suppl/2015/S-59196, 2015.

Cao, Z., Hu, P., and Pender, G.: Reconciled bedload sediment transport rates in ephemeral and perennial rivers, Earth Surf. Proc. Land., 35, 1655–1665, https://doi.org/10.1002/esp.2005, 2010.

Casas, A., Lane, S. N., Yu, D., and Benito, G.: A method for parameterising roughness and topographic sub-grid scale effects in hydraulic modelling from LiDAR data, Hydrol. Earth Syst. Sci., 14, 1567–1579, https://doi.org/10.5194/hess-14-1567-2010, 2010a.

Casas, M. A, Lane, S. N., Hardy, R. J., Benito, G., and Whiting, P. J.: Reconstruction of subgrid-scale topographic variability and its effect upon the spatial structure of three-dimensional river flow, Water Resour. Res., 46, W03519, https://doi.org/10.1029/2009WR007756, 2010b.

Chow, V. T.: Open Channel Hydraulics, McGraw-Hill Book Company, New York, USA, 1959.

Croke, J., Fryirs, K., and Thompson, C.: Channel–floodplain connectivity during an extreme flood event: implications for sediment erosion, deposition, and delivery, Earth Surf. Proc. Land., 38, 1444–1456, https://doi.org/10.1002/esp.3430, 2013

Deltares: Delft3D-FLOW, Simulation of Multi-dimensional Hydrodynamic Flows and Transport Phenomena, Including Sediments, User Manual: Hydro-Morphodynamics, Delft, the Netherlands, 2011.

Dietrich, W. E. and Smith, J. D.: Influence of the point bar on flow through curved channels, Water Resour. Res., 19, 1173–1192, 1983.

Engelund, F. and Hansen, E.: A Monograph on Sediment Transport in Alluvial Streams, Technical University of Denmark, Teknisk Forlag, Copenhagen, 1967.

Euler, T., Herget, J., Schlömer, O., and Benito, G.: Hydromorphological processes at submerged solitary boulder obstacles in streams, Catena, 157, 250–267, https://doi.org/10.1016/j.catena.2017.05.028, 2017.

Ferguson, R. I.: Understanding braiding processes in gravel-bed rivers: progress and unsolved problems, in: Braided Rivers, Geological Society Special Publication No. 75, edited by: Best, J. L. and Bristow, CS, 73–87, 1993.

Gendaszek, A. S., Magirl, C. S., Czuba, C. R., and Konrad, C. P.: The timing of scour and fill in a gravel-bedded river measured with buried accelerometers, J. Hydrol., 495, 186–196, https://doi.org/10.1016/j.jhydrol.2013.05.012, 2013.

Graf, W. L.: Transport and deposition of plutonium-contaminated sediments by fluvial processes, Los Alamos Canyon, New Mexico, Geol. Soc. Am. Bull., 108, 1342–1355, https://doi.org/10.1130/0016-7606(1996)108<1342:TADOPC>2.3.CO;2, 1996.

Greenbaum, N. and Bergman, N.: Formation and evacuation of a large gravel-bar deposited during a major flood in a Mediterranean ephemeral stream, Nahal Me'arot, NW Israel, Geomorphology, 77, 169–186, https://doi.org/10.1016/j.geomorph.2006.01.016, 2006.

Grodek, T., Jacoby, Y., Morin, E., and Katz, O.: Effectiveness of exceptional rainstorms on a small Mediterranean basin, Geomorphology, 159/160, 156–168, https://doi.org/10.1016/j.geomorph.2012.03.016, 2012

Hassan, M. A., Marren, P. M., and Schwartz, U.: Bar structure in an arid ephemeral stream, Sediment. Geol., 221, 57–70, https://doi.org/10.1016/j.sedgeo.2009.07.012, 2009.

Hauer, C. and Habersack, H.: Morphodynamics of a 1000-year flood in the Kamp River, Austria, and impacts on floodplain morphology, Earth Surf. Proc. Land., 34, 654–682, https://doi.org/10.1002/esp.1763, 2009.

Hjulström, F.: Studies of the Morphological Activity of Rivers as Illustrated by the River Fyris – Inaugural Dissertation, Almqvist & Wiksells, 1935.

Hooke, J.: Morphological impacts of flow events of varying magnitude on ephemeral channels in a semiarid region, Geomorphology, 252, 128–143, https://doi.org/10.1016/j.geomorph.2015.07.014, 2016a.

Hooke, J.: Geomorphological impacts of an extreme flood in SE Spain, Geomorphology, 263, 19–38, https://doi.org/10.1016/j.geomorph.2016.03.021, 2016b.

Hooke, J. M. and Mant, J. M.: Geomorphological impacts of a flood event on ephemeral channels in SE Spain, Geomorphology, 34, 163–180, https://doi.org/10.1016/S0169-555X(00)00005-2, 2000.

Hooke, J. M., Brookes, C. J., Duane, W., and Mant, J. M.: A simulation model of morphological, vegetation and sediment changes in ephemeral streams, Earth Surf. Proc. Land., 30, 845–866, https://doi.org/10.1002/esp.1195, 2005.

Ikeda, S.: Incipient Motion of Sand Particles on Side Slopes, J. Hydr. Eng. Div.-ASCE, 108, 95–114, 1982.

Kaitna, R., Chiari, M., Kerschbaumer, M., Kapeller, H., Zlatic-Jugovic, J., Hengl, M., and Huebl, J.: Physical and numerical modelling of a bedload deposition area for an Alpine torrent, Nat. Hazard. Earth Sys., 11, 1589–1597, https://doi.org/10.5194/nhess-11-1589-2011, 2011.

Kasvi, E., Alho, P., Lotsari, E., Wang, Y., Kukko, A., Hyyppä, H., and Hyyppä, J.: Two-dimensional and three-dimensional computational models in hydrodynamic and morphodynamic reconstructions of a river bend: sensitivity and functionality, Hydrol. Process., 29, 1604–1629, https://doi.org/10.1002/hyp.10277, 2014.

Kasvi, E., Vaaja, M., Kaartinen, H., Kukko, A., Jaakkola, A., Flener, C., Hyyppä, H., Hyyppä, J., and Alho, P.: Sub-bend scale flow–sediment interaction of meander bends – A combined approach of field observations, close-range remote sensing and computational modelling, Geomorphology, 238, 119–134, https://doi.org/10.1016/j.geomorph.2015.01.039, 2015.

Kukko, A., Kaartinen, H., and Zanetti, M.: Backpack Personal Laser Scanning System for Grain-Scale Topographic Mapping, 46th Lunar and Planetary Science Conference, The Woodlands, TX, USA, 2015.

Li, Z., Cao, Z., Liu, H., and Pender, G.: Graded and uniform bed load sediment transport in a degrading channel, Int. J. Sediment Res., 31, 376–385, https://doi.org/10.1016/j.ijsrc.2016.01.004, 2016.

Limerinos, J. T.: Determination of the Manning coefficient from measured bed roughness in natural channels, Geological survey water-supply paper, 1898-B, US Department of the Interior, prepared in cooperation with the California Department of Water Resources, United States Government Printing Office, Washington, 1970.

Lisle, T. E. and Madej, M. A.: Spatial variation in armouring in a channel with high sediment supply, in: Dynamics of Gravel-bed Rivers, edited by: Billi, P., Hey, R. D., Thorne, C. R., and Tacconi, P., John Wiley & Sons Ltd., Chichester, 277–293, 1992.

Long, A. J.: Hydrograph separation for karst watersheds using a two-domain rainfall–discharge model, J. Hydrol., 364, 249–256, https://doi.org/10.1016/j.jhydrol.2008.11.001, 2009

Lotsari, E., Wainwright, D., Corner, G. D., Alho, P., and Käyhkö, J.: Surveyed and modelled one-year morphodynamics in the braided lower Tana River, Hydrol. Process., 28, 2685–2716, https://doi.org/10.1002/hyp.9750, 2014a.

Lotsari, E., Vaaja, M., Flener, C., Kaartinen, H., Kukko, A., Kasvi, E., Hyyppä, H., Hyyppä, J., and Alho, P.: Annual bank and point bar morphodynamics of a meandering river determined by high-accuracy multitemporal laser scanning and flow data, Water Resour. Res., 50, 5532–5559, https://doi.org/10.1002/2013WR014106, 2014b.

Lotsari, E., Thorndycraft, V., and Alho, P.: Prospects and challenges of simulating river channel response to future climate change, Prog. Phys. Geogr., 39, 483–513, https://doi.org/10.1177/0309133315578944, 2015.

Lloyd, C. E. M., Freer, J. E., Johnes, P. J., and Collins, A. L.: Using hysteresis analysis of high-resolution water quality monitoring data, including uncertainty, to infer controls on nutrient and sediment transfer in catchments, Sci. Total Environ., 543, 388–404, https://doi.org/10.1016/j.scitotenv.2015.11.028, 2016

Lucía, A., Recking, A., Martín-Duque, J. F., Storz-Peretz, Y., and Laronne, J. B.: Continuous monitoring of bedload discharge in a small, steep sandy channel, J. Hydrol., 497, 37–50, https://doi.org/10.1016/j.jhydrol.2013.05.034, 2013.

Meyer-Peter, E. and Müller, R.: Formula for bed load transport, Proceedings of the 2nd Meeting, Int. Assoc. Hydraulic Structures Res. (IAHR), Stockholm, Sweden, 39–64, 1948.

Milan, D. J., Heritage, G. L., and Hetherington, D.: Application of a 3D laser scanner in the assessment of erosion and deposition volumes and channel change in a proglacial river, Earth Surf. Proc. Land., 32, 1657–1674, https://doi.org/10.1002/esp.1592, 2007.

Müller, E. R. and Pitlick, J.: Sediment supply and channel morphology in mountain river systems: 1. Relative importance of lithology, topography, and climate, J. Geophys. Res.-Earth, 118, 2325–2342, https://doi.org/10.1002/2013JF002843, 2013.

Nardi, L. and Rinaldi, M.: Spatio-temporal patterns of channel changes in response to a major flood event: the case of the Magra River (central–northern Italy), Earth Surf. Proc. Land., 40, 326–339, https://doi.org/10.1002/esp.3636, 2015.

Parker, G.: Surface-based bedload transport relation for gravel rivers, J. Hydraul. Res., 28, 417–436, https://doi.org/10.1080/00221689009499058, 1990.

Pelletier, J. D., Mayer, L., Pearthree, P. A., House, P. K., Demsey, K. A., Klawon, J. E., and Vincent, K. R.: An integrated approach to flood hazard assessment on alluvial fans using numerical modeling, field mapping, and remote sensing, Geol. Soc. Am. Bull., 117, 1167–1180, https://doi.org/10.1130/B25544.1, 2005.

Peterson, M. S.: Impacts of Flash Floods, in: Coping With Flash Floods, edited by: Gruntfest, E. and Handmer, J., NATO Science Series (Series 2. Environmental Security), vol. 77, Springer, Dordrecht, 2001.

Pitlick, J.: Response and recovery of a subalpine stream following a catastrophic flood, Geol. Soc. Am. Bull., 105, 657–670, 1993.

Reid, I., Frostick, L. E., and Layman, J. T.: The incidence and nature of bedload transport during flood flows in coarse-grained alluvial channels, Earth Surf. Proc. Land., 10, 33–44, https://doi.org/10.1002/esp.3290100107, 1985.

Reid, I., Powell, D. M., and Laronne, J. B.: Prediction of bed load transport by desert flash floods, J. Hydraul. Eng., 122, 170–173, https://doi.org/10.1061/(ASCE)0733-9429(1996)122:3(170), 1996.

Rodriguez-Lloveras, X., Bussi, G., Francés, F., Rodriguez-Caballero, E., Solé-Benet, A., Calle, M., and Benito, G.: Patterns of runoff and sediment production in response to land-use changes in an ungauged Mediterranean catchment, J. Hydrol., 531, 1054–1066, https://doi.org/10.1016/j.jhydrol.2015.11.014, 2015.

Sanyal, J.: Predicting possible effects of dams on downstream river bed changes of a Himalayan river with morphodynamic modelling, Quatern. Int., 453, 48–62, https://doi.org/10.1016/j.quaint.2017.03.063, 2017.

Schuurman, F., Marra, W. A., and Kleinhans, M. G.: Physics-based modeling of large braided sand-bed rivers: bar pattern formation, dynamics and sensitivity, J. Geophys. Res.-Earth, 118, 2509–2527, https://doi.org/10.1002/2013jf002896, 2013.

Segura-Beltrán, F. and Sanchis-Ibor, C.: Assessment of channel changes in a Mediterranean ephemeral stream since the early twentieth century, The Rambla de Cervera, eastern Spain, Geomorphology, 201, 199–214, https://doi.org/10.1016/j.geomorph.2013.06.021, 2013.

Singh, U., Crosato, A., Giri, S., and Hicks, M.: Sediment heterogeneity and mobility in the morphodynamic modelling of gravel-bed braided rivers, Adv. Water Resour., 104, 127–144, https://doi.org/10.1016/j.advwatres.2017.02.005, 2017.

Thompson, C. and Croke, J.: Geomorphic effects, flood power, and channel competence of a catastrophic flood in confined and unconfined reaches of the upper Lockyer valley, southeast Queensland, Australia, Geomorphology, 197, 156–169, https://doi.org/10.1016/j.geomorph.2013.05.006, 2013.

Tooth, S.: Process, form and change in dryland rivers: a review of recent research, Earth-Sci. Rev., 51, 67–107, https://doi.org/10.1016/S0012-8252(00)00014-3, 2000.

Vaaja, M., Hyyppä, J., Kukko, A., Kaartinen, H., Hyyppä, H., and Alho, P.: Mapping Topography Changes and Elevation Accuracies Using a Mobile Laser Scanner, Remote Sens., 3, 587–600, https://doi.org/10.3390/rs3030587, 2011.

van Rijn, L. C.: Principles of Sediment Transport in Rivers, Estuaries and Coastal Seas, 1993 Edn., Aqua Publications, the Netherlands, 1993.

Verhaar, P. M., Biron, P. M., Ferguson, R. I., and Hoey, T. B.: A modified morphodynamic model for investigating the response of rivers to short-term climate change, Geomorphology, 101, 674–682, https://doi.org/10.1016/j.geomorph.2008.03.010, 2008.

Viparelli, E., Gaeuman, D., Wilcock, P., and Parker, G.: A model to predict the evolution of a gravel bed river under an imposed cyclic hydrograph and its application to the Trinity River, Water Resour. Res., 47, W02533, https://doi.org/10.1029/2010WR009164, 2011.

Walling, D. E.: Suspended sediment and solute yields from a small catchment prior to urbanisation, in: Fluvial Processes in Instrumented Watersheds, edited by: Gregory, K. J. and Walling, D. E., Special Publication 6, Institute of British Geographers, 169–192, 1974.

Wang, Y., Liang, X., Flener, C., Kukko, A., Kaartinen, H., Kurkela, M., Vaaja, M., Hyyppä, H., and Alho, P.: 3D Modeling of Coarse Fluvial Sediments Based on Mobile Laser Scanning Data, Remote Sens., 5, 4571–4592, https://doi.org/10.3390/rs5094571, 2013.

Wheaton, J., Brasington, J., Darby, S. E., Kasprak, A., Sear, D., and Vericat, D.: Morphodynamic signatures of braiding mechanisms as expressed through change in sediment storage in a gravel-bed river, J. Geophys. Res.-Earth, 118, 759–779, https://doi.org/10.1002/jgrf.20060, 2013.

Williams, R. D., Brasington, J., Hicks, M., Measures, R., Rennie, C. D., and Vericat, D.: Hydraulic validation of two-dimensional simulations of braided river flow with spatially continuous ADCP data, Water Resour. Res., 49, 5183–5205, https://doi.org/10.1002/wrcr.20391, 2013.

Williams, R. D., Rennie, C. D., Brasington, J., Hicks, D. M., and Vericat, D.: Linking the spatial distribution of bed load transport to morphological change during high-flow events in a shallow braided river, J. Geophys. Res.-Earth, 120, 604–622, https://doi.org/10.1002/2014JF003346, 2015.

Williams, R. D., Brasington, J., and Hicks D. M.: Numerical Modelling of Braided River Morphodynamics: Review and Future Challenges, Geography Compass, 10, 102–127, https://doi.org/10.1111/gec3.12260, 2016a.

Williams, R. D., Measures, R., Hicks, D. M., and Brasington J.: Assessment of a numerical model to reproduce event-scale erosion and deposition distributions in a braided river, Water Resour. Res., 52, 6621–6642, https://doi.org/10.1002/2015WR018491, 2016b.

Wolman, M. G.: A method of sampling coarse river-bed material, Transactions of the American Geophysical Union, 35, 951–956, 1954.

10

Validation of digital elevation models (DEMs) and comparison of geomorphic metrics on the southern Central Andean Plateau

Benjamin Purinton and Bodo Bookhagen

Institute of Earth and Environmental Science, Universität Potsdam, Potsdam, Germany

Correspondence to: Benjamin Purinton (purinton@uni-potsdam.de)

Abstract. In this study, we validate and compare elevation accuracy and geomorphic metrics of satellite-derived digital elevation models (DEMs) on the southern Central Andean Plateau. The plateau has an average elevation of 3.7 km and is characterized by diverse topography and relief, lack of vegetation, and clear skies that create ideal conditions for remote sensing. At 30 m resolution, SRTM-C, ASTER GDEM2, stacked ASTER L1A stereopair DEM, ALOS World 3D, and TanDEM-X have been analyzed. The higher-resolution datasets include 12 m TanDEM-X, 10 m single-CoSSC TerraSAR-X/TanDEM-X DEMs, and 5 m ALOS World 3D. These DEMs are state of the art for optical (ASTER and ALOS) and radar (SRTM-C and TanDEM-X) spaceborne sensors.

We assessed vertical accuracy by comparing standard deviations of the DEM elevation versus 307 509 differential GPS measurements across 4000 m of elevation. For the 30 m DEMs, the ASTER datasets had the highest vertical standard deviation at > 6.5 m, whereas the SRTM-C, ALOS World 3D, and TanDEM-X were all < 3.5 m. Higher-resolution DEMs generally had lower uncertainty, with both the 12 m TanDEM-X and 5 m ALOS World 3D having < 2 m vertical standard deviation. Analysis of vertical uncertainty with respect to terrain elevation, slope, and aspect revealed the low uncertainty across these attributes for SRTM-C (30 m), TanDEM-X (12–30 m), and ALOS World 3D (5–30 m). Single-CoSSC TerraSAR-X/TanDEM-X 10 m DEMs and the 30 m ASTER GDEM2 displayed slight aspect biases, which were removed in their stacked counterparts (TanDEM-X and ASTER Stack).

Based on low vertical standard deviations and visual inspection alongside optical satellite data, we selected the 30 m SRTM-C, 12–30 m TanDEM-X, 10 m single-CoSSC TerraSAR-X/TanDEM-X, and 5 m ALOS World 3D for geomorphic metric comparison in a 66 km^2 catchment with a distinct river knickpoint. Consistent m/n values were found using chi plot channel profile analysis, regardless of DEM type and spatial resolution. Slope, curvature, and drainage area were calculated and plotting schemes were used to assess basin-wide differences in the hillslope-to-valley transition related to the knickpoint. While slope and hillslope length measurements vary little between datasets, curvature displays higher magnitude measurements with fining resolution. This is especially true for the optical 5 m ALOS World 3D DEM, which demonstrated high-frequency noise in 2–8 pixel steps through a Fourier frequency analysis. The improvements in accurate space-radar DEMs (e.g., TanDEM-X) for geomorphometry are promising, but airborne or terrestrial data are still necessary for meter-scale analysis.

1 Introduction

Digital elevation models (DEMs) provide hydrologists and geomorphologists with powerful tools to explore the linkages between fundamental geomorphic processes and landforms. Previously, DEMs have been used to test hypotheses of landscape evolution at local and regional scales using geomorphic metrics (e.g., Howard et al., 1994; Burbank et al., 1996; Whipple and Tucker, 1999; Montgomery et al., 2001; Dietrich et al., 2003; Roering et al., 2007). Modern geomorphologists use the quantitative subdiscipline of geomorphometry (Pike et al., 2009; Wilson, 2012; Sofia et al., 2016) to explore how tectonic, climatic, and lithologic signals can be inferred from DEMs (e.g., Snyder et al., 2000; Wobus et al., 2006; DiBiase et al., 2010; Bookhagen and Strecker, 2012; Kirby and Whipple, 2012; Scherler et al., 2015; Clubb et al., 2016; Olen et al., 2016), but questions remain to what extent transient responses can be recorded in landscape morphology (e.g., DiBiase et al., 2012) and how channel networks and hillslopes can independently act as records of basin transience (e.g., Ouimet et al., 2009; Hurst et al., 2012; Clubb et al., 2016; Forte et al., 2016). Such studies rely on accurate DEMs for the calculation of geomorphic metrics (e.g., slope and curvature) and extraction of geomorphic features (e.g., channels, hillslopes, hilltops). In spite of this, DEM elevation error reporting (Fisher and Tate, 2006; Reuter et al., 2009) – often carried out with limited control data – only accounts for absolute pixel elevation accuracy and does not include higher-order DEM derivatives (e.g., slope and curvature), geomorphic metrics, or landscape features of interest to geomorphologists. This problem is especially acute given that relatively small elevation errors will propagate and grow in the first (slope) and second (curvature) derivatives, potentially obscuring geomorphometric results (e.g., Wechsler, 2007).

Remotely sensed DEMs – referred to throughout this study as DEMs, as opposed to the often-used term digital terrain model (DTM) for bare-earth models with vegetation and structures removed – are generated from data that are originally distorted through sensor, terrain, and atmospheric conditions, leading to misrepresentations (error) in the final product (Smith and Sandwell, 2003; Fisher and Tate, 2006; Nuth and Kääb, 2011). These datasets are commonly received in gridded format – rather than point cloud, triangulated irregular network (TIN), or other recently developed adaptive formats (e.g., Liu et al., 2014) – resulting in a defined measurement interval (grid resolution) that may oversimplify fine landscape variability. Thus, the geomorphic scales of interest must be taken into consideration when selecting the appropriate DEM (e.g., Hengl, 2006). For instance, while channel profiles over long reaches are readily analyzed on 90 m resolution data, hillslopes with considerably smaller extents require higher 1–30 m resolution data capable of identifying individual hillslopes and ridges (Grieve et al., 2016a, b, c). Furthermore, DEM biases specific to a given sensor should be considered prior to analysis, especially when using satellite-derived DEMs in steep topography (e.g., Paul and Haeberli, 2008; Nuth and Kääb, 2011; Pipaud et al., 2015).

Since the release of the first global DEM by the United States Geological Survey (USGS) in 1996 (GTOPO30) at a resolution of 30 arcsec ($\sim$ 1 km), advances in remote sensing technology – particularly satellite observation – and processing capabilities have steadily improved the accuracy and increased the resolution of DEMs. The 2003 public release of the 90 m Shuttle Radar Topography Mission (SRTM) DEM with coverage from 56° S to 60° N (Farr et al., 2007) ushered in a new age of near-global digital topographic analysis (Wilson, 2012). With the 2009 release of the 30 m Advanced Spaceborne Thermal Emission and Reflection Radiometer Global DEM (ASTER GDEM; METI/NASA/USGS, 2009), and more recent releases of the improved ASTER GDEM version 2 (ASTER GDEM2; Tachikawa et al., 2011), SRTM C-band 30 m (SRTM-C), and up-sampled Advanced Land Observing Satellite (ALOS) World 3D 30 m (AW3D30), geomorphologists now have open access to many 30 m near-global DEMs. In addition to these public 30 m datasets, higher-resolution (< 15 m) DEMs from a variety of satellite sources are becoming increasingly available through commercial purchase or research agreements as edited products (e.g., ALOS World 3D and TanDEM-X WorldDEM), optical pairs for stereogrammetric processing (e.g., ALOS Panchromatic Remote-sensing Instrument for Stereo Mapping, or PRISM), and radar scenes for interferometric processing (e.g., TerraSAR-X/TanDEM-X).

The SRTM-C and ASTER GDEM2 have reported vertical accuracies of $\sim$ 5–20 m depending on terrain characteristics (e.g., Mukherjee et al., 2013; Rexer and Hirt, 2014), with some biases reported related to slope and aspect of the terrain (e.g., Berthier et al., 2006; Nuth and Kääb et al., 2011; Shortridge and Messina, 2011). While these accuracies allow long-term (decadal) tracking of glacial elevation changes (e.g., Racoviteanu et al., 2007; Paul and Haeberli, 2008), higher-resolution local DEMs from optical and radar sources have proven more accurate (<5 m vertical error) than these global products for glacial studies in steep terrain, particularly on shorter timescales (e.g., Berthier et al., 2007; Berthier and Toutin, 2008; Jaber et al., 2013; Neckel et al., 2013; Pandey and Venkataraman, 2013; Holzer et al., 2015; Rankl and Braun, 2016; Neelmeijer et al., 2017). However, to date no studies have assessed the accuracy of the current generation of sub-15 m, satellite-derived DEMs with regards to geomorphometry. These measurements, unlike glacial studies, rely on the derivatives of elevation (e.g., slope and curvature) and their spatial context, not absolute or relative height changes. Furthermore, glacial studies are typically conducted on lower-slope terrain and compare area-wide measurements, allowing some uncertainties to average out. Conversely, geomorphic studies examining channels and hillslopes in steeper terrain may be more impacted by remote-sensing errors and artifacts (e.g., from shadowing, sensor angle, foreshortening), and geomorphic metrics like slope and curvature rely on the accuracy of nearby pixels, for example within a 3 × 3 pixel moving window.

The application of light detection and ranging (lidar) by ground and aerial methods is often used to generate meter- to sub-meter-scale elevation point clouds and gridded DEM datasets at smaller areal extents than satellite-derived DEMs (e.g., Passalacqua et al., 2015). Lidar has revolutionized geomorphology with more accurate representations of the land surface and has led to new insights and discoveries in the realm of mass and energy transport laws (Dietrich et al., 2003; Roering et al., 2007), channel initiation (Passalacqua et al., 2010a, b), surface flow routing (Shelef and Hilley, 2013), erosion (Perroy et al., 2010), and landslide and fault scarp mapping (e.g., Roering et al., 2013; Tarolli, 2014). While coarser DEMs have proven useful in exploring mountain belt hypsometry and linkages between climate, erosion, and tectonics at basin or regional scales (e.g., Montgomery et al., 2001; DiBiase et al., 2010; Bookhagen and Strecker, 2012), their utility in analyzing process-level geomorphology and assessing critical hillslope parameters is limited and lidar is often deemed necessary (Roering et al., 2013; Tarolli, 2014; Passalacqua et al., 2015). Despite this, the limited spatial extent (~ 1 km^2) and high effort and cost of obtaining lidar are prohibitive factors to its application at basin or regional scales (10–1000 km^2).

Previous studies examining the effect of DEM resolution on geomorphic metrics and features have primarily used resampled or re-gridded lidar data (e.g., Tarolli and Tarboton, 2006; Tarolli and Dalla Fontana, 2009; Grieve et al., 2016c). Here we are interested not in high-quality resampled data, but rather data at their original resolution collected from different sensors, without any higher-resolution information from resampling. Advances in sub-15 m DEM availability and accuracy from a number of satellites necessitates investigation of their advantages over 30 m public DEMs in representing derivatives of elevation for channel and hillslope analysis in lieu of lidar.

This study presents a multi-DEM validation and comparison for the southern Central Andes in NW Argentina in an arid landscape with no vegetation cover, ideal for remote sensing. DEM validation is presented by (i) reporting the vertical accuracy of a number of satellite-derived DEMs at resolutions of 5–30 m from open-access portals, commercial sources, and research agreements and (ii) carrying out channel profile analysis and geomorphic metric comparisons for a 66 km^2 catchment with a defined channel knickpoint to assess the quality of these DEMs for tectonic geomorphology. Through this analysis we demonstrate state of the art wide areal coverage, satellite-derived DEM availability for geomorphometry.

2 Study area

The Puna de Atacama Plateau in NW Argentina (Fig. 1a) is the southern extension of the low-relief high-elevation internally drained Central Andean Plateau (also referred to as the Altiplano–Puna Plateau), extending for over 1500 km and reaching widths of over 350 km in the Central Andes (Allmendinger et al., 1997). Due to the plateau's hyper-arid climate caused by orographic blocking and regional atmospheric circulation patterns (Bookhagen and Strecker, 2008; Rohrmann et al., 2014), there is an absence of cloud and vegetation cover on the Puna de Atacama Plateau, creating ideal conditions for remote sensing of the bare-earth surface. As the Puna de Atacama Plateau is largely uninhabited and erosion rates are very low (Bookhagen and Strecker, 2012), the study site is a pristine environment experiencing little change from year to year, thus minimizing differences between DEMs collected years apart. Topographic expression is diverse on the plateau, with flat salars (salt flats) having near-zero relief at 5–10 km scales, surrounded by steep volcanoes and mountain ranges with > 2 km of relief at 2–5 km scales. This morphology is readily apparent around the Pocitos Basin, centered on the Salar de Pocitos (basin elevation at ~ 3600 m) and bordered by mountains such as the Nevado Queva, which reach elevations of over 6000 m (Fig. 1b). Within the Pocitos Basin, we focus geomorphometric analysis on the 66 km^2 Quebrada Honda catchment, with 1.2 km of relief (Fig. 1c). The Quebrada Honda was chosen for its size, coverage across available DEMs, uniform Paleozoic metasedimentary lithology, and the presence of a knickpoint 7 km upstream of the outlet, dividing the basin into transient and steady-state geomorphic regimes.

3 Data and methods

3.1 dGPS data

Vertical accuracy of optical and radar DEMs was assessed using a differential GPS (dGPS) dataset spanning ~ 4000 m of elevation and covering an area of ~ 50 000 km^2 centered on the Pocitos Basin (Fig. 1a). Of 333 555 total raw dGPS measurements collected during field campaigns from 2013 to 2016, 307 509 kinematically corrected points with vertical and horizontal accuracies < 0.5 m were selected for the final control on DEM vertical accuracy. Data were projected to the EGM96 geoid vertical and WGS84 horizontal datums in the UTM coordinate system zone 19S. This point measurement dataset was rasterized to the resolution and extent of each DEM. Multiple measurements within a DEM pixel were averaged and pixels without measurements were set to no data. This led to a reduction in the number of individual measurements used to assess DEM vertical accuracy, but it accounted for multiple measurements per pixel to provide a robust validation. Details of measurement collection and kinematic correction of the raw dGPS files using the Universidad Nacional

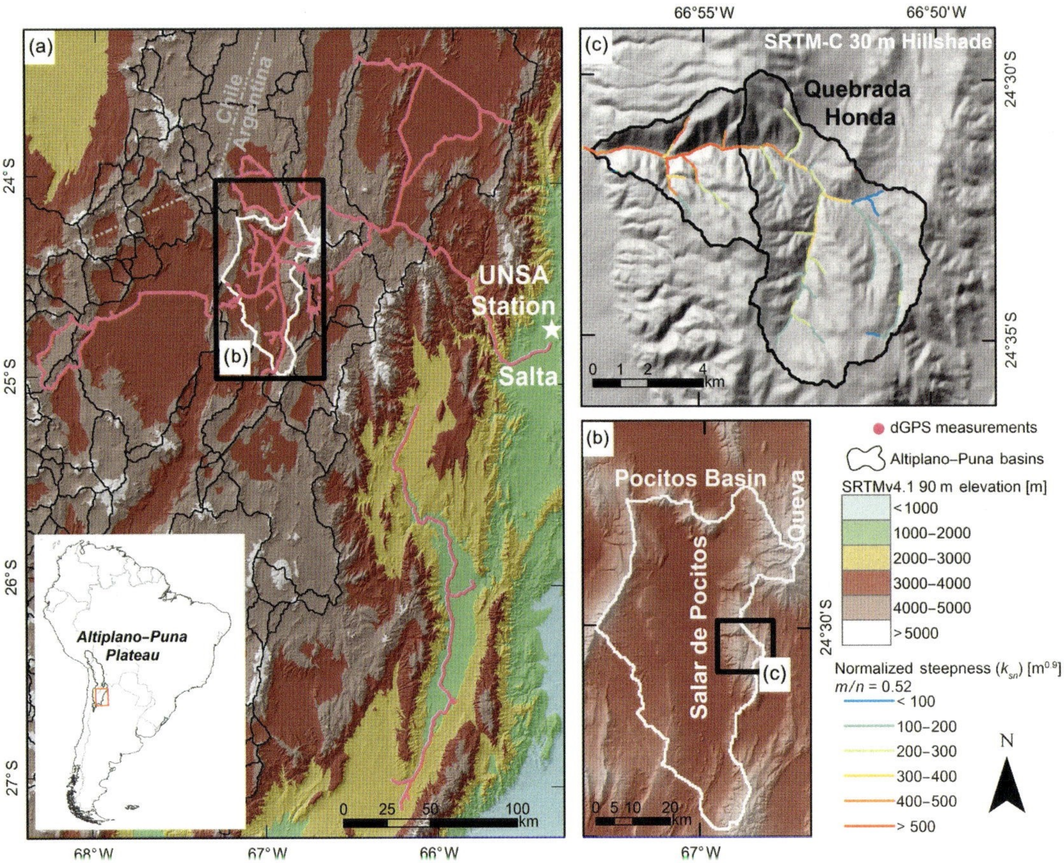

Figure 1. **(a)** Topographic overview of the study area in the southern Central Andes. Displayed in pink are 307 509 dGPS measurements. UNSA base station (white star) for dGPS kinematic correction located in Salta, Argentina. Inset shows South American continent with international borders and internally drained Central Andean Plateau. Study focus is the Pocitos Basin **(b)**, where elevation ranges from 3600 m on the flat salar to 6000 m on surrounding peaks. Geomorphometric analyses focus on the Quebrada Honda **(c)** catchment, draining an area of 66 km^2 from 5000 m of elevation down to 3800 m. A knickpoint 7 km upstream divides the basin into an upper and lower section with differing morphology (Fig. 2). The transition is observable along the trunk as normalized channel steepness (k_{sn}) averaged along 300 m reaches on the SRTM-C 30 m DEM increases to values > 500. The m/n reference value of 0.52 is calculated using chi plot analysis. Elevations in panels **(a)** and **(b)** are from the 90 m SRTMv4.1 DEM (Jarvis et al., 2008).

de Salta (UNSA) permanent station in Salta (Fig. 1a) can be found in the Supplement.

3.2 DEM datasets

DEMs collected from a number of public, commercial, and research agreement sources are listed in Table 1. All were referenced to the same datums (EGM96–WGS84) and projected onto UTM 19S using bilinear interpolation. The EGM96 undulation was subtracted from the TanDEM-X and single-CoSSC TDX DEMs to adjust their vertical datum (Baade and Schmullius, 2016). DEMs were co-registered to a common control – the 30 m SRTM-C, selected for its excellent geolocation accuracy (Rodriguez et al., 2006) – using affine parameters by up- or down-sampling the SRTM-C to the resolution of the DEM of interest and iteratively shifting to reduce the RMSE of the elevation difference using the MATLAB function *imregister*. By aligning datasets to one another, we minimize elevation uncertainty versus the dGPS measurements caused by slight offsets in geolocation of the DEMs (e.g., Nuth and Kääb, 2011). Co-registration was not carried out on the TanDEM-X data, as the SRTM-C was already used during initial processing steps to provide elevation corrections (Wessel, 2016). Additional information on each dataset listed in Table 1 is found in the Supplement, including datasets that were not included in the rest of the study due to lower resolution, lack of coverage, or quality issues (SRTMv4.1 90 m, SRTM-X 30 m, RapidEye 12 m, SPOT6 5 m, ALOS PRISM tri-stereopair 10 m, TerraSAR-X pairs 10 m, and single-CoSSC TerraSAR-X/TanDEM-X processed to 5 m).

ASTER stacking

The ASTER radiometer has collected along-track stereopairs with near-infrared cameras looking toward the nadir

Table 1. List of DEMs used for comparisons and geomorphic analyses.

Dataset (short name)	Data type	Resolution (m)	Source	Notes
SRTM C-band (SRTM-C)	Radar, edited global product	30	Public, https://lta.cr.usgs.gov/SRTM1Arc	Collected in February 2000, released 2014, previously only US coverage.
ASTER GDEM Version 2 (ASTER GDEM2)	Optical, edited global product	30	Public, https://asterweb.jpl.nasa.gov/gdem.asp	Released 2011, update of ASTER GDEM1 released 2009. Generated by automated processing and stacking of ASTER L1A stereopairs by NASA and METI[a].
ASTER L1A stereopair stack (ASTER stack)	Optical, raw stereopairs	30	Public, http://reverb.echo.nasa.gov/reverb/	Stacked DEM generated for this study by stereogrammetric processing of eight raw L1A stereopairs (Sect. 3.2, ASTER stacking).
ALOS World 3D (AW3D5 and AW3D30)	Optical, edited global product	5/30	Public (30 m), http://www.eorc.jaxa.jp/ALOS/en/aw3d30/ Commercial (5 m), http://aw3d.jp/en/	5 m DEM released 2015 as highest-resolution commercial global DEM, with down-sampled 30 m research version released 2016.
Single-CoSSC[b] TerraSAR-X/TanDEM-X (CoSSC TDX)[a]	Radar, raw interferograms	10	Research agreement, http://terrasar-x.dlr.de/	DEMs generated for this study by single CoSSC TerraSAR-X/TanDEM-X radar pairs. Same data used by DLR to generate the stacked TanDEM-X DEM in 2015
TanDEM-X DEM (TanDEM-X 12 m and TanDEM-X 30 m)	Radar, edited global product	12/30	Research agreement, http://tandemx-science.dlr.de/	Final 12 m DEM generated by stacking of overlapping CoSSC TerraSAR-X/TanDEM-X radar pair DEMs and down-sampled 30 m version, both from DLR[c].

[a] Ministry of Economy, Trade and Industry (METI) of Japan.
[b] Co-registered Single Look Slant Range Complex (CoSSC) raw interferometric product from DLR.
[c] Commercial 12 m product available as WorldDEM™ from AIRBUS.

(Band 3N) and backwards (Band 3B) between 83° S and 83° N since 1999 (Tachikawa et al., 2011). Using these stereopairs, a 30 m ASTER global DEM has been generated by automatic stereo correlation, stacking, and averaging of over 1.2 million scenes. The stacking of multiple lower-quality DEMs from the same source is a common technique, also undertaken to generate the 12 m TanDEM-X (from single-CoSSC TDX DEMs) and 5 m ALOS World 3D (from ALOS PRISM optical tri-stereopair DEMs). The 2011 release of the ASTER GDEM version 2 (ASTER GDEM2) used in the present study represented a vast improvement in quality (Tachikawa et al., 2011), but remaining noise is caused by issues with cloud cover, water masking, the smaller stereo correlation kernel, and misregistration of scenes prior to stacking (Nuth and Kääb, 2011). We seek to improve on the ASTER GDEM2 using eight raw ASTER L1A 3N/B stereopairs downloaded with variable overlap from the Pocitos Basin. Using stereogrammetric processing methods we generated eight 30 m DEMs from these stereopairs. Details of DEM generation along with RMSE of ground control and tie points are presented in the Supplement (Table S1). Each L1A DEM was co-registered to the SRTM-C, manually masked for outliers (locations where clouds or haze in the imagery caused abrupt > 1000 m steps in the final DEM), and differenced with the SRTM-C. Pixels were weighted with a bi-square scheme based on their correlation with the SRTM-C, and a weighted average of the overlapping DEMs was used to generate a higher-quality 30 m ASTER stack.

3.3 Elevation accuracy assessment

To assess DEM vertical accuracy, we first performed a pixel-by-pixel comparison of rasterized dGPS (vertical uncertainty < 0.5 m) and DEM elevation values after co-registration to the SRTM-C. Our preferred metric for DEM vertical accuracy is the mean ± 1-sigma (σ) standard deviation (SD) (Li, 1988; Fisher and Tate, 2006). Specifically, we are interested in the SD of DEM elevation versus dGPS height as our quality metric. Plotted histograms of uncertainty distribution were normalized by their respective mean offsets so the SD could be visually compared. Differences of ±30 m were filtered out as outliers caused by bad data and processing errors, and the percentage reduction in number of measurements from this filtering is reported along with the pre filtering mean and SD. While many other studies suggest additional statistical tests (e.g., Höhle and Höhle, 2009), our simplified method allows us to move into further analysis of the derivatives of elevation. We have compared the full error distribution, but note that the mean and SD capture the essence. The key is the spatial correlation or consistency of the DEM data because geomorphometric studies use the spatial content of DEMs and higher-order derivatives and not absolute elevation values.

In a second step, we examined error distributions with respect to terrain slope, aspect, and elevation for the DEMs, also normalized by mean offset, with ±30 m outliers excluded. Measurements were separated into 50–100 m elevation bins (depending on the full elevation range of the dataset), slopes were calculated from their eight neighboring pixels using the D8 algorithm and binned by 1°, and aspect (also D8 calculated) was binned by 10°, with north at 0° and east at 90°. Vertical uncertainty was plotted in each bin as a box plot showing the median, 25–75th percentile range, and first and 99th percentile outlier cutoffs.

3.4 Geomorphometric analysis

For a robust assessment of DEM quality, we go beyond pixel-by-pixel vertical accuracy comparisons by comparing longitudinal channel profiles and derived geomorphic metrics in the 66 km^2 Quebrada Honda catchment (Fig. 2). Here, a defined channel knickpoint separates downstream steep and upstream gentle-sloped terrain, and consistent climate and lithology allows us to test hypotheses of basin-wide adjustment to river gradients. We focused on a subset of the highest-quality DEMs with the aim of providing an assessment of the effects of different sensors and resolutions (e.g., SRTM-C 30 m, TanDEM-X 12 m and 30 m, and AW3D5 5 m) and DEM stacking (e.g., CoSSC TDX 10 m and TanDEM-X 12 m) on geomorphometry.

3.4.1 Channel profile analysis

Hydrological and geomorphic modeling is an important application of DEMs (Wilson, 2012), and channel network extraction is a necessary step prior to channel profile analysis. Although a number of recently developed methods for channel extraction via channel-head identification now exist (see Hooshyar et al., 2016, for a review), these methods have all been developed on high-resolution lidar data, with control datasets of field-mapped channel heads (Clubb et al., 2014) or channel networks (Passalacqua et al., 2010a, b). We are thus wary to apply these methods to our coarser satellite-derived data with no control from the field and no lidar data for a relative performance assessment. Instead, we use the simplistic threshold area approach (e.g., Tarboton et al., 1991), choosing the common reference area of 1 km^2 where breaks in area–slope scaling indicate the changeover to dominantly alluvial channel processes (Montgomery and Foufoula-Georgiou, 1993). Misrepresentation of channel location from this method is entirely restricted to the highest catchment reaches where the channel head lies and does not affect the majority of the downstream channel. The consistent use of area thresholding at the same reference area across DEMs allows the direct comparison of channel profile results in our study.

Advances in longitudinal channel profile analysis driven by accurate DEMs have elucidated changes in boundary conditions recorded in channel slope and upstream-propagating knickpoints (e.g., Wobus et al., 2006; Kirby and Whipple, 2012). The stream power incision model (SPIM) of landscape evolution provides the theoretical basis for relating channel slope and drainage area (see Kirby and Whipple, 2012, or Lague, 2014, for background and limitations of SPIM). Applied to a channel profile in steady state ($dz/dt = 0$), we find the relationship

$$S_C = \left(\frac{U}{K}\right)^{\frac{1}{n}} A^{\frac{-m}{n}}, \qquad (1)$$

where U is uplift, K is erodibility, A is local drainage area, S_C is local channel slope, and m and n are site-specific constants that scale the relative influences of climate and tectonics. Important to constrain here is the m/n ratio, used to normalize channel steepness across differently sized drainage areas for the mapping of regional patterns of deformation, climatic influence, and/or lithologic boundary conditions (e.g., Wobus et al., 2006; Kirby and Whipple, 2012; Forte et al., 2016). While Eq. (1) is derived for steady state, its integration (or the use of area–slope plots) to determine the m/n ratio and assess relative differences in channel steepness is a geometric consideration of local channel behavior and can thus be applied in nonequilibrium settings, like the Quebrada Honda. Here, we utilize the recently developed integration method of chi plot analysis (Perron and Royden, 2013) to estimate m/n from our DEMs (see Supplement for details).

Following channel extraction, we first applied the least-squares R^2 maximization chi plot technique of Perron and Royden (2013) to the Quebrada Honda trunk stream (Schwanghart and Scherler, 2014). This method attempts to linearize the entire channel to one best-fit line in chi space and does not provide robust uncertainty estimates for m/n, as linear regression is performed through serially correlated values of chi distance and elevation (Perron and Royden, 2013). Because of this, we also employed the piece-wise fitting m/n selection algorithm developed by Mudd et al. (2014) on the 30 m SRTM-C, 10 m CoSSC TDX, and 5 m AW3D5 DEMs (representing a cross section of DEM resolutions and sen-

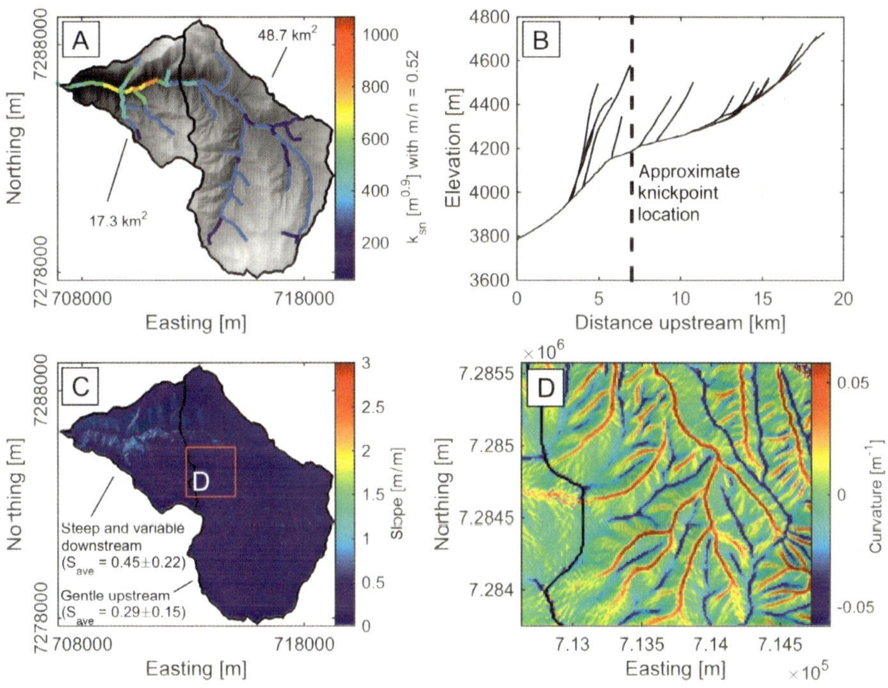

Figure 2. Topographic overview of Quebrada Honda from the 12 m TanDEM-X DEM. **(a)** Normalized channel steepness (k_{sn}) averaged over 300 m reaches using $m/n = 0.52$ with upstream and downstream drainage areas indicated by black outlines. All tributaries with a drainage area > 1 km^2 are plotted. Note oversteepened trunk signal has not propagated entirely up all downstream tributaries, as indicated by $k_{sn} < 400$ in upper reaches. **(b)** Longitudinal profile of trunk channel and tributaries with knickpoint indicated. **(c)** D∞ slope map (Tarboton, 2005) displaying steeper, more variable topography downstream of knickpoint, indicated by warmer colors and greater average slope (S_{ave}) and standard deviation. **(d)** Curvature colored by $\pm 3\sigma$ range with positive values concave (valleys) and negative values convex (hilltops). Note the planar slopes separating ridges and valleys and the increase in concavity near valley heads, indicating the shift from hillslope to fluvial processes.

sors) for comparison with the least-squares approach. This method balances goodness of fit for the piece-wise fit profile with model complexity (number of parameters and segments) to provide an m/n at the minimum corrected Akaike information criterion (AICc) (Akaike, 1974; Hurvich and Tsai, 1989). A SD (uncertainty) of this minimum AICc is also provided, over which AICc values falling within the SD range indicate other plausible m/n values (Mudd et al., 2014). Sensitivity tests were performed by varying fitting parameters, with final parameters reported in the Supplement.

3.4.2 Hillslope geomorphic metrics

In addition to channel profile analysis, signals of denudation and uplift may also be inferred from hillslope morphology as determined by geomorphic metrics, including characteristic hillslope length, local relief, slope angles, and curvature. These parameters allow the exploration of empirical geomorphic transport laws, which aid in topographic modeling over geologic timescales (see Dietrich et al., 2003). In particular, the accurate sampling of slope angles and curvatures allows patterns of erosion to be mapped from topography alone, thus playing key roles in geomorphic studies focused on the topographic expression of tectonic-climatic forcing (e.g., DiBiase et al., 2010; Hurst et al., 2012). Here, we test the newest generation of satellite-derived DEMs for assessing slope and curvature as well as the hillslope-to-valley transition marked by inflections in plots of slope, curvature, and drainage area to examine differences related not only to the channel knickpoint, but also to the resolution and quality of the DEM. We compared results between the high-quality 30 m SRTM-C, 30 m and 12 m TanDEM-X, 10 m CoSSC TDX, and 5 m AW3D5 DEMs. We did not include the ASTER DEMs in hillslope analyses because of elevation noise prevalent in these 30 m DEMs apparent in vertical uncertainty reporting (>6 m) and visual inspection. Despite low vertical uncertainty (<3 m), we also excluded the newly released 30 m AW3D30 because of unknown preprocessing steps taken to produce this DEM, which may have compounded high-frequency noise present in the 5 m AW3D5.

Since hillslopes represent a diffusive environment where flow is multidirectional, we calculated drainage area and slope at every pixel in the Quebrada Honda using the D∞ algorithm, allowing dispersive flow (Tarboton, 2005). Curvature was calculated using the Laplacian of elevation (e.g.,

Tarolli and Dalla Fontana, 2009):

$$C = \nabla^2 z = \left(\frac{\delta^2 z}{\delta x^2} + \frac{\delta^2 z}{\delta y^2}\right), \quad (2)$$

where concavity (valleys and channels) is denoted by $C > 0$, convexity (hillslopes and ridges) is denoted by $C < 0$, and planar slopes are denoted by $C = 0$ (Fig. 2d). Distributions of slope and curvature separated upstream and downstream of the knickpoint were visualized as box plots displaying medians, 25–75th percentile ranges, first and 99th percentile cutoffs, and all outlier measurements.

Filtering is a common step to smooth DEM noise before deriving geomorphic metrics, as the calculation of slope via steepest-descent algorithms is greatly affected by elevation errors in neighboring pixels (e.g., Raaflaub and Collins, 2006). However, here we use the D∞, which is less susceptible to these effects, as slope is divided between two cells. Often filtering is also done to reduce noise associated with shorter timescale geomorphic features, like tree throw, from high-resolution lidar DEMs (Grieve et al., 2016a). During initial tests we experimented with median, diffusion (Passalacqua et al., 2010b), and Wiener filtering (Wiener, 1949) prior to slope and curvature calculations. However, since all smoothing techniques were found to reduce the variability in slope and curvature measurements and blur sharper features of interest such as ridge crests and valley bottoms, we instead chose to measure derivatives from the raw elevation data.

To explore the influence of the oversteepened trunk reach on hillslope morphology, we combined measures of curvature (Laplacian-calculated), slope (D∞-calculated), and drainage area (also D∞-calculated) at every DEM pixel in the Quebrada Honda to explore differences between the gentle upstream and steep downstream catchment. We are particularly interested in differences at the hillslope-to-valley transition demarcated by the first inflection in plots of slope binned by area, occurring at a critical drainage area where channel heads are thought to initiate (Montgomery and Foufoula-Georgiou, 1993; Zhang and Montgomery, 1994; Ijjasz-Vasquez and Bras, 1995; Tarolli and Dalla Fontana, 2009). We generated plots of logarithmically binned contributing area versus median slope (area–slope), logarithmically binned area versus median curvature (area–curvature), and linearly binned curvature versus median slope (curvature–slope) – all separated upstream and downstream of the knickpoint. For area–slope plots the gradient at the graphical rollover in binned area is recorded along with this area bin. We also attempted rollover identification using 2-D kernel density estimates (Botev et al., 2010) to identify the densest concentrations of slope and area values demarcating the approximate rollover, but we found similar results to the graphical approach. Following the method of Roering et al. (2007), we divide the rollover drainage area by DEM resolution to approximate the characteristic horizontal hillslope length (L_{H}), providing an additional check on DEM applicability to geomorphology. This method relies on the assumption that DEM resolution is equivalent to unit contour width, which may be an oversimplification. Despite this caveat, the resolution, or unit contour width, serves as a constant for division, and differences between values will not alter the trend of the results. We use the horizontal definition of L_{H} since the difference between horizontal and downslope L_{H} (as measured by Grieve et al., 2016a, b, c) should be minimal, except for very high slope angles. Area–curvature and curvature–slope plots are used to visualize the slope and area trends related to curvature, particularly around the zero curvature planar inflection point in the landscape (Roering et al., 1999).

3.5 Two-dimensional fourier analysis

In a final step, we employ a two-dimensional discrete Fourier transform (2-D DFT) to quantify high-frequency noise in select datasets using 8 km by 14 km DEM clips centered on the Quebrada Honda. This common signal processing tool (Priestley, 1981) relies on the transformation of elevation matrices from the spatial to the frequency domain, providing information about the amplitude and periodicity of landscape features. Prior work using the 2-D DFT on gridded topography has focused on artifact identification (Arrell et al., 2008), landscape organization and scaling (e.g., Perron et al., 2008 and references therein), the identification of landslides (Booth et al., 2009), and the length scales of biotic influence on topography (Roering et al., 2010). We follow the methods outlined in Perron et al. (2008) and take the 2-D DFT of a rectangular elevation matrix, $z(x, y)$, with $N_x \times N_y$ measurements spaced evenly by Δx and Δy (Priestly, 1981; Perron et al., 2008; Booth et al., 2009):

$$Z\left(k_x, k_y\right) = \sum_{m=0}^{N_x-1} \sum_{n=0}^{N_y-1} z(m\Delta x, n\Delta y) e^{-2\pi i\left(\frac{k_x m}{N_x} + \frac{k_y n}{N_y}\right)}, \quad (3)$$

where k_x and k_y are wave numbers and m and n are indices of z. This transformation outputs an array with the amplitudes of the frequency components, from which the power spectrum can be calculated using the DFT periodogram:

$$P_{\mathrm{DFT}}\left(k_x, k_y\right) = \frac{1}{N_x^2 N_y^2} \left|Z(k_x, k_y)\right|^2. \quad (4)$$

The P_{DFT} array is a measure of the variance of z and has units of amplitude squared (square meters). To enhance visualization, this array is plotted against radial frequency in one dimension as wavelength (frequency^{-1}) versus mean-squared amplitude. Here, the wavelength represents the spatial scale (in meters) of the amplitude fluctuations, and can thus be converted to pixel steps given the DEM resolution. A linear regression through logarithmically spaced wavelength bins (at the mean value) is used as the background spectrum to normalize the mean-squared amplitude, as opposed to the randomly generated surfaces in Perron et al. (2008). Use of the median of the wavelength bins in the linear regression provided comparable results to the mean, with the R^2 value in all cases >0.98. Although this regression is somewhat skewed by the longest wavelength (lowest-frequency) values (Fig. S10), we are interested only in the high-frequency noise, which is effectively normalized here.

Through these steps, we achieve a 1-D plot of wavelength versus unitless spectral power that highlights large-amplitude outliers at specific low wavelengths for the 30 m SRTM-C, 12 m TanDEM-X, and 5 m AW3D5 DEMs. To test whether the DEM noise had an orientation-dependent spatial structure requiring a 2-D plotting scheme, we carried out the same analysis on each elevation matrix rotated 90° and found comparable results in the 1-D plots. Further details of this methodology, including topographic detrending and windowing functions to reduce spectral leakage, can be found in Perron et al. (2008). To quantitatively compare the 1-D graphical results of the 2-D DFT analysis between the 12 m TanDEM-X and 5 m AW3D5, we rely on the two-sample Kolmogorov–Smirnov (KS) test (Massey, 1951) and quantile–quantile (QQ) plots. These statistical techniques allow the direct comparison of normalized spectral power distributions from samples with different sizes (different resolutions) without the need for resampling of the data to a common resolution (e.g., 10 m), which introduces biases in the spectral analysis depending on the resampling scheme (e.g., bilinear, cubic, spline).

4 Results

4.1 Elevation accuracy

Vertical uncertainties, measured as the mean ± SD of differences between DEM elevation and rasterized dGPS height, for all DEMs are summarized in Table 2.

Histograms of the vertical uncertainty distributions are plotted for the 30 m (Fig. 3) and higher-resolution DEMs (Fig. 4). The 30 m SRTM-C and TanDEM-X both demonstrated a very low SD and smooth appearance upon visual inspections, with the TanDEM-X having the lowest SD (2.42 m) and narrowest distribution (Fig. 3a) of all 30 m DEMs. Conversely, despite a low SD, visual inspection of numerous artifacts in the AW3D30 with no landscape representation demonstrated its lower quality. The improvement in quality through weighted stacking of ASTER L1A stereopair DEMs versus the low-quality ASTER GDEM2 is apparent in the reduction of the SD from 11.42 m for a single L1A DEM to 6.93 m for the stack, although uncertainty distributions for all ASTER DEMs extend beyond the ±30 m outlier cutoff (Fig. 3b).

For the higher-resolution DEMs, we note the narrow uncertainty distributions with very few ±30 m outliers (Fig. 4). The 5 m AW3D5 has the lowest vertical SD of any DEM at 1.64 m. Importantly, the 12 m TanDEM-X DEM has a similarly low SD of 1.97 m but covers a much larger area of dGPS measurements (~50 000 km^2 for the TanDEM-X versus ~580 km^2 for the AW3D5), as indicated by the order of magnitude difference in counts. The wider, double-peaked vertical uncertainty distributions for the 6 November 2012 and 25 August 2013 CoSSC TDX DEMs are caused by their coverage over variable terrain east of the Salar de Pocitos, where accurate DEM generation is complicated by radar shadowing and layover in steeper topography. Visual inspection of these two DEMs containing the full Quebrada Honda catchment revealed minor hillslope artifacts often coinciding with rocky outcrops and other steep and rough features, occurring in only a few areas representing a small (<0.5 km^2) portion of the catchment relative to the total area from which geomorphic metrics were derived (66 km^2). Since the 2013 CoSSC TDX DEM had noticeable striping from radar processing, the 10 m CoSSC TDX DEM from 6 November 2012 was selected for further geomorphic comparison.

In addition to histograms, the vertical uncertainty distributions with respect to elevation, slope, and aspect of the topography are plotted for our highest-quality datasets (Figs. 5–7), with additional plots in the Supplement (Figs. S4–S9). On each plot the dearth of dGPS measurements on slopes > 30° is noted, as the majority of measurements were taken from low-gradient roads and flat salars. For our 30 m datasets, a narrow uncertainty range across all terrain attributes is apparent for the SRTM-C (Fig. 5), TanDEM-X (Fig. S7), and AW3D30 (Fig. S4). Conversely, the ASTER GDEM2 (Fig. S5) had the largest error distribution across all attributes, with even low-slope (0–3°) errors extending above ±10 m. Prevalent noise in the ASTER GDEM2 is also demonstrated by the greater number of measurements (> 15 000) taken from a greater-than-realistic landscape representation with only 6000–7000 measurements recorded on > 10° slopes. Furthermore, the ASTER GDEM2 appears to have a slight aspect-related bias, with an amplitude of ~5 m repeating at approximately ENE, SE, WSW, and NNW (Fig. S5c), which is not evident in any other 30 m DEMs. We note that this aspect bias is removed in the ASTER stack (Fig. S6); however, this stacked 30 m DEM still has large error bars and covers only the Pocitos Basin and thus far fewer measurements than the full ASTER GDEM2. For the SRTM-C, 30 m TanDEM-X, and AW3D30, error bars grow with slope and the smallest errors are consistently found in flat topography with slopes < 10°. At low slopes (0–3°), the 30 m TanDEM-X performs exceedingly well, with first and 99th percentile outlier cutoffs within ±5 m (Fig. S7b). The TanDEM-X performance is further enhanced in the 12 m version (Fig. 6), which has error bars mostly within ±5 m across the range of terrain attributes (with the exception of some limited high-slope measurements). Furthermore, the 12 m TanDEM-X does not have the aspect bias noted in the 10 m CoSSC TDX DEMs (Figs. S8–S9), which have a ~5–10 m bias repeating at approximately due N, E, S, and W, likely related to satellite slant angles. Unsurprisingly, from Table 2 and histogram plotting results, the 5 m AW3D5 has the small-

Table 2. Results of pixel-by-pixel DEM vertical accuracy (DEM minus dGPS). Mean and standard deviation before filtering denoted in parentheses, with a value of n/a if there were no outliers filtered.

Dataset	Mean (m)	Standard deviation (m)	Number of post-filtered rasterized measurements[a]	Reduction after ±30 m filtering (%)
30 m SRTM-C	2.18 (2.33)	3.33 (13.74)	64 782	0.02
30 m AW3D30	1.59 (1.66)	2.81 (16.19)	63 413	0.03
30 m ASTER GDEM2	−0.15 (0.02)	9.48 (17.65)	63 308	2.30
30 m ASTER Stack[b]	4.56 (4.58)	6.93[c] (7.00)	15 506	0.12
30 m TanDEM-X	−1.29 (−1.12)	2.42 (14.57)	55 791	0.02
12 m TanDEM-X	−1.41 (−1.31)	1.97 (11.16)	108 029	0.02
10 m CoSSC TDX (7 February 2011)	1.99 (2.36)	2.02 (21.26)	28 982	0.03
10 m CoSSC TDX (6 November 2012)[d]	1.32 (n/a)	3.83 (n/a)	22 182	0.00
10 m CoSSC TDX (25 August 2013)	2.94 (n/a)	3.22 (n/a)	22 175	0.00
5 m AW3D5	2.40 (n/a)	1.64 (n/a)	14 306	0.00

[a] After ±30 m outlier filtering.
[b] Generated for Pocitos Basin by weighted stacking of eight manually generated ASTER L1A DEMs.
[c] Compare with 11.42 and 10.06 m SD for single L1A DEM and ASTER GDEM2, respectively, clipped to same area.
[d] CoSSC TDX DEM selected for geomorphometric analysis.

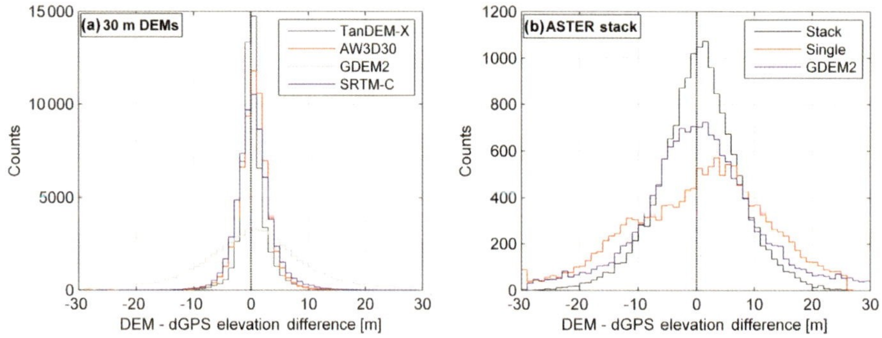

Figure 3. Vertical uncertainties for **(a)** global 30 m DEMs and **(b)** ASTER 30 m DEMs. Plots have been normalized by mean offsets, with statistics reported in Table 2. Note the order of magnitude difference in counts, as panel **(b)** covers only the Pocitos Basin (~ 2500 km^2), whereas panel **(a)** spans all dGPS measurements (~ 50 000 km^2) stretching over a 4000 m elevation range.

est vertical uncertainty error bars of any DEM tested, falling almost entirely within ±5 m for all bins over all terrain attributes (Fig. 7). We emphasize that this stacked DEM was purchased for only a small area (580 km^2), but covers highly variable topography around the Nevado Queva and Quebrada Honda (Fig. 1b).

4.2 Geomorphometric analysis

Based on the results of elevation validation and visual inspection of the datasets, we selected the 30 m SRTM-C, 30 m and 12 m TanDEM-X, 10 m CoSSC TDX from 6 November 2012, and 5 m AW3D5 for geomorphometric analysis of the Quebrada Honda (Fig. 2). These edited (SRTM-C, TanDEM-X, and AW3D5) and single-pair radar (CoSSC TDX) DEMs, all released in the past 3 years, have the highest potential for future studies seeking to derive geomorphic metrics in large regions without lidar.

4.2.1 Channel profile analysis

The results of m/n estimation from least-squares R^2 maximization (Perron and Royden, 2013; Schwanghart and Scherler, 2014) and from piece-wise fitting (Mudd et al., 2014) on the Quebrada Honda trunk channel on three DEMs are summarized in Table 3. Example plots for both methods of the SRTM-C are found in the Supplement (Figs. S2–S3). Not included here are the 30 m and 12 m TanDEM-X, as m/n results were similar to those listed. For all DEMs tested and listed in Table 1, the resulting m/n from least-squares fitting was 0.49–0.53 (all with $R^2 > 0.95$), representing the same range as those DEMs listed in Table 3, regardless of DEM vertical accuracy or resolution. For those DEMs tested with the piece-wise fitting method, the m/n range was similar, although slightly higher at 0.53–0.57. While the least-squares technique takes only a few minutes to setup and run, the computationally intensive piece-wise fitting takes hours to days, although it provides a range of minimum AICc that denote

Table 3. m/n values using two chi plot methods on the Quebrada Honda trunk.

Dataset	SD of dGPS uncertainty (m)	Least squares[a] m/n	R^2	Piece-wise fitting[b] AICc minimum value ± SD	m/n at AICc minimum	Plausible values of m/n[c]
30 m SRTM-C	3.33	0.53	0.97	27.98 ± 0.50	0.55	0.55–0.57
10 m CoSSC TDX	3.83	0.49	0.97	28.89 ± 0.22	0.54	–[d]
5 m AW3D5	1.64	0.51	0.98	31.54 ± 0.21	0.54	0.53–0.56

[a] Peron and Royden (2013) and Schwanghart and Scherler (2014).
[b] Mudd et al. (2014).
[c] With corresponding AICc value falling within SD range of AICc minimum.
[d] No tested values of m/n fell within the AICc SD range.

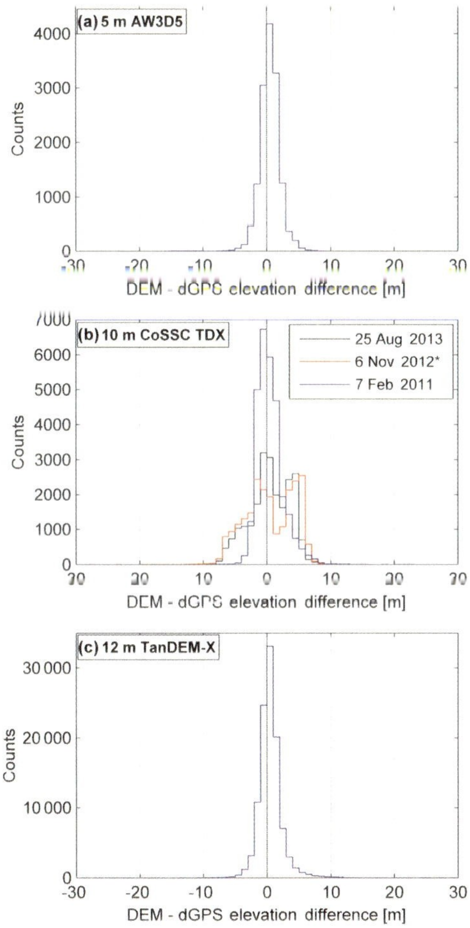

Figure 4. Vertical uncertainties for **(a)** 5 m AW3D5, **(b)** 10 m CoSSC TDX, and **(c)** 12 m TanDEM-X. Plots have been normalized by mean offsets, with statistics reported in Table 2. The star (*) in panel **(b)** denotes the 2012 CoSSC TDX DEM selected for geomorphometric analysis. Note the order of magnitude increase in counts for the 12 m TanDEM-X **(c)**, which covers nearly all dGPS measurements (~ 50 000 km^2) stretching over a 4000 m elevation range.

plausible m/n values. Interestingly, the 5 and 10 m DEMs had slightly lower SD for AICc, perhaps indicating the better fitting of the channel segments compared to the 30 m DEM.

4.2.2 Hillslope analysis

We turn to slope (D∞), drainage area (D∞), and curvature (Laplacian) geomorphic metrics calculated for every DEM pixel upstream and downstream of the knickpoint to further assess DEM quality at a finer scale than channel analysis. Box plots showing distributions for slope and curvature are presented in Fig. 8. We note that the quartile (boxes) and outlier (whiskers) ranges of slopes are similar regardless of DEM resolution (Fig. 8a). Only the outliers grow in number and spread with finer resolution, demonstrating in particular the greater slope variability measured by the highest-resolution AW3D5 DEM. In exception to this trend, more high-slope outliers are measured by the 12 m TanDEM-X compared with the 10 m CoSSC TDX. These higher slopes are caused by the better resolution of rocky outcrops and other steep features with very high slopes from the stacked TanDEM-X DEM versus the single-CoSSC TDX DEMs. For curvature distributions (Fig. 8b), the 30 m SRTM-C and TanDEM-X, 12 m TanDEM-X, and 10 m CoSSC TDX have very narrow quartile and outlier ranges compared with the wide distribution of the 5 m AW3D5. Similar to the slope results, the higher-resolution DEMs measure far more curvature outliers compared to the 30 m SRTM-C and TanDEM-X, whose full distributions extend only to approximately ±0.05 m^{-1}. Regarding differences in distributions related to the catchment knickpoint, median slopes downstream are consistently 0.1–0.2 m m^{-1} greater in magnitude than upstream (Fig. 8a). Conversely, the downstream curvatures have a slightly narrower range compared to the upstream curvatures, especially noticeable in the smaller downstream quartile range from the AW3D5 (Fig. 8b). This is likely caused by the fact that the upstream area (48.7 km^2) covers more than twice as much area as downstream (17.3 km^2), thus measuring more pixels and leading to a greater range in curvature measured when averaged over the entire subcatchment area.

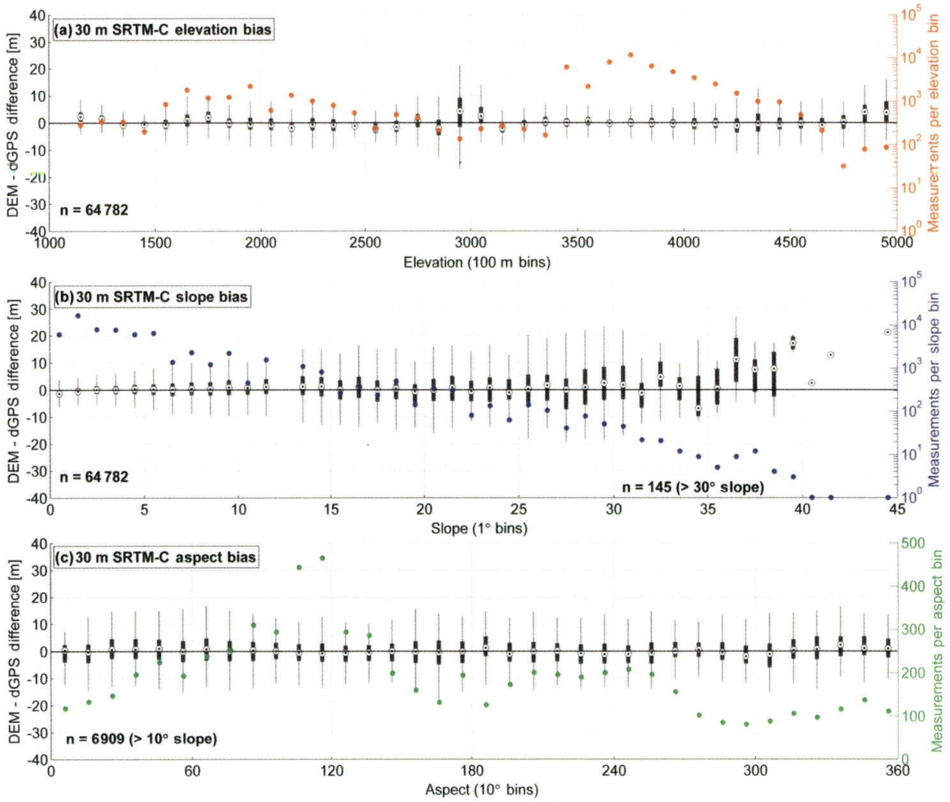

Figure 5. **(a)** Elevation, **(b)** slope (eight-connected neighborhood calculated), and **(c)** aspect (eight-connected neighborhood calculated) vertical uncertainties from 30 m SRTM-C. Median elevation difference (black circles) with 25–75th percentile range (boxes) and first and 99th percentile outlier cutoff (whiskers) plotted for each bin on left axis. Number of measurements indicated (n) with measurements per bin plotted as colored circles on right axis. For aspect **(c)**, only measurements on slopes > 10° are used; thus, n is reduced by an order of magnitude. Elevation differences are normalized by mean offset. We note the dearth of slope measurements > 30° **(b)**.

Higher downstream slopes are reflected in area–slope plots, with median slopes typically 0.1–0.2 m m^{-1} greater in magnitude downstream given the same contributing area (Fig. 9), regardless of DEM resolution. Moving from 30 to 5 m, the values for median local slope (and mean slope at the rollover) change only slightly, again indicating the similar slope values measured across DEMs. Differences in slope given the same drainage area are most pronounced in the 10^4–10^5 m^2 drainage area range, with results converging somewhat as drainage area trends towards very high values, until diverging once more in the 5–12 m DEMs at drainage areas > 10^6 m^2 (Fig. 9c–e). All DEMs demonstrate a broad changeover region from hillslope to fluvial processes, indicated by a wide first inflection (Fig. 9). Therefore, the graphical selection of the drainage area rollover is very approximate, and upstream and downstream rollover values do not show significant variation. Conversely, we note that this rollover drainage value does decrease with finer resolution of the DEM, indicating a resolution-dependent bias for this area–slope analysis, likely related to the ability of each DEM to represent the most upstream reaches of a catchment (< 10^4 m^2), which may only cover 4–5 pixels in the 30 m DEMs (900 m^2 per pixel) versus hundreds in the 5 m DEMs (25 m^2 per pixel). Nevertheless, when dividing this rollover drainage area by resolution to assess L_H, we find similar values across datasets. Downstream and upstream L_H, respectively, are 163 and 116 m (30 m SRTM-C), 116 and 116 m (30 m TanDEM-X), 76 and 94 m (12 m TanDEM-X), 130 and 83 m (10 m CoSSC TDX), and 119 and 73 m (5 m AW3D5). For the 10–30 m TerraSAR-X/TanDEM-X data presented here (Fig. 9b–d), there is a noticeable flattening of the downstream median slopes following the initial increase, indicating that for a 1 to 2 order-of-magnitude range of drainage areas, slope values are very similar in the steeper downstream sub-catchment.

Area–slope plots are complemented by area–curvature (Fig. 10) and curvature–slope (Fig. 11) plots to further illustrate differences related to the channel knickpoint and specific DEMs. The larger curvature variability captured in the 5 m data is again demonstrated with a larger range of curvature plotted (Figs. 10e and 11e) and larger percentile ranges about the median in each bin (Fig. 10e). Little difference in area–curvature (Fig. 10) is noticeable upstream or downstream in the 10–30 m DEMs. For all DEMs, there appears

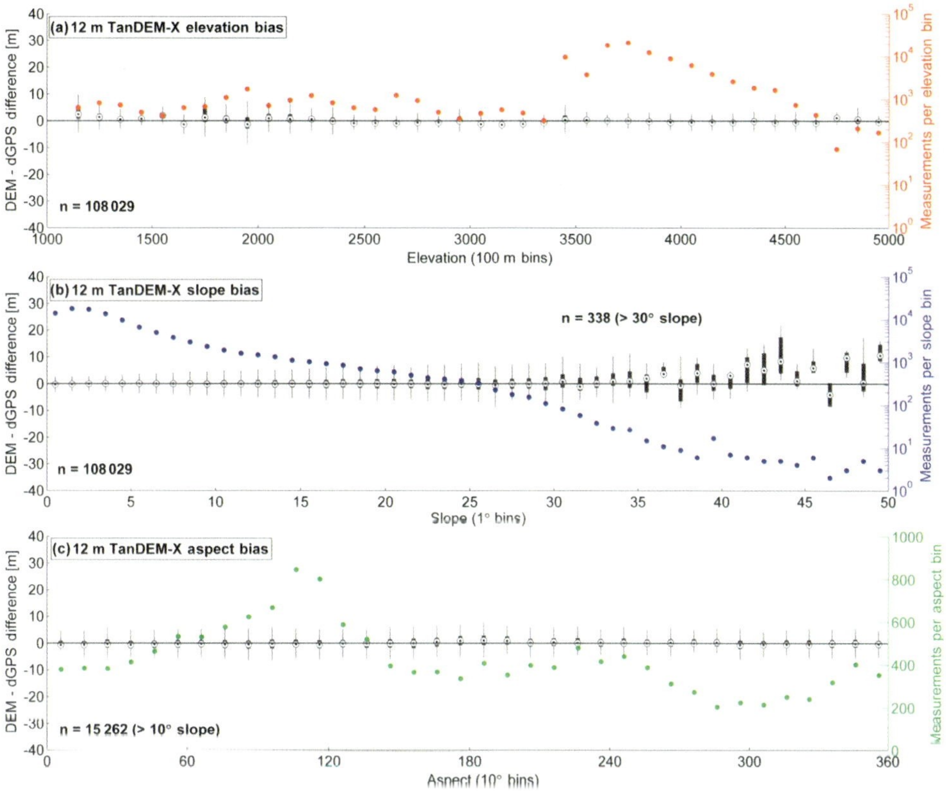

Figure 6. (a) Elevation, (b) slope, and (c) aspect vertical uncertainty bias for the 12 m TanDEM-X

to be a scaling break just below the zero-curvature planar inflection point in the concave hillslope realm, perhaps indicating the changeover between diffusive and advective processes resolvable in all DEMs (although again occurring at smaller drainage areas with finer resolution). Plots of slope binned by curvature again indicate similar slopes measured from 30 to 5 m resolution (Fig. 11). From this analysis, we note that slopes at high convex (negative) and concave (positive) curvatures are similar upstream and downstream, with greater differences on more planar slopes.

4.3 Two-dimensional fourier analysis

The calculation of the 2-D DFT for the 30 m SRTM-C, 12 m TanDEM-X, and 5 m AW3D5 provides a quantitative assessment of topographic wavelengths and noise of the DEMs. As detailed in Sect. 3.5., the normalized spectral power is plotted against wavelength for these three datasets at their native resolution (Fig. 12), with the non-normalized 1-D power spectra calculated from Eq. (4) and regression lines used for normalization shown in the Supplement (Fig. S10). In Fig. 12, a 99.9th percentile upper envelope of normalized power displays the similar trends in the distributions, with a distinct spike in normalized power at a wavelength of ~ 500 m and smaller secondary peaks above (at ~ 800 m) and below (at ~ 250 m) corresponding to ridge-and-valley structures in the area of the Quebrada Honda. However, what we are more interested in is the presence of a number of high peaks at approximately 10, 15, 20, and 40 m in the AW3D5 plot (with less pronounced expression at 25–35 m), causing a difference of 1 order of magnitude in normalized spectral power scaling (Fig. 12c). These wavelengths correspond to 2–8 pixel steps in the AW3D5 data, indicating a significant high-frequency component in this optical DEM that is not found in the 12 or 30 m radar-derived datasets.

Statistical tests help further quantify the effect of the high-power spikes in the 5 m AW3D5 data, given the overall similar shape of the spectral power distribution to the 12 m TanDEM-X. The two-sample KS test, which measures the difference in the cumulative distribution function (CDF) for each dataset, rejects the null hypothesis that both samples are taken from the same distribution at the 99 % confidence interval with a p value of 0. These results suggest that the power spectra distributions of AW3D5 and TanDEM-X are significantly different despite similar elevation validation and geomorphic metric results. To better explore how these differences relate to normalized spectral power spikes shown in Fig. 12, we plot the sample quantiles against one another (Fig. 13a). Noted are the 99th, 99.9th, 99.99th, and 99.999th quantiles, representing, respectively, normalized spectral powers of 11.1, 18.4, 33.6, and 106.1 for the AW3D5 data and 7.1, 12.5, 21.3, and 42.1 for the TanDEM-X

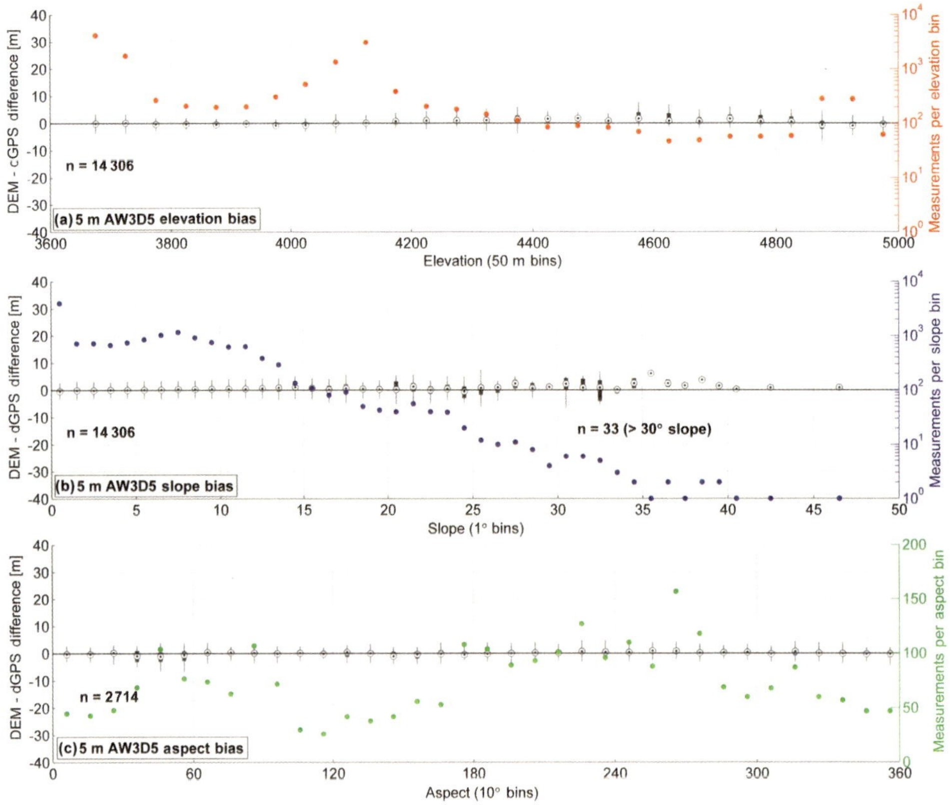

Figure 7. (**a**) Elevation, (**b**) slope, and (**c**) aspect vertical uncertainty bias for the 5 m AW3D5.

data. Nonlinear excursion from an approximately linear trend towards higher and higher normalized spectral power for the AW3D5 is caused by only a small percentage of DFT elements greater than the 99.99th quantile. For the AW3D5, the values above this quantile represent only 840 DFT elements (of a total $\sim 8.4 \times 10^6$) and only 210 for the TanDEM-X (of a total $\sim 2.1 \times 10^6$). Additionally, the inset normalized CDF plot (Fig. 13b) shows that while the median values (~ 0.5 normalized CDF) correspond between the datasets, there is the greatest diversion only at very high and very low spectral powers. Figure 13b provides an additional explanation of why the KS test is unable to reject the null hypothesis. Despite following a similar trend for over 99 % of the distribution, high-frequency (low-wavelength) noise in 2–8 pixel steps causes significant differences in the 5 m AW3D5.

5 Discussion

5.1 Elevation validation

5.1.1 The 30 m DEMs

The low quality of the ASTER GDEM2 is readily apparent in the wide uncertainty distribution (Fig. 3a), leading to a > 2 % outlier reduction in measurements used to assess uncertainty, but a SD remaining near 10 m (Table 2). For over 228 000 Australian National Gravity Database station heights with < 1 m vertical accuracy, Rexer and Hirt (2014) found similar results for the GDEM2, with SD ranging from 7.7 m in flat terrain to 11.29 m in mountainous regions. Baade and Schmullius (2016) also found vertical errors of 12–20 m for the GDEM2, using over 10 000 high-accuracy dGPS measurements. Other studies have reported vertical accuracies of 3–9 m for the GDEM2, but these are often determined with fewer (< 100) high-accuracy control points compared with our study using over 300 000 dGPS measurements (e.g., Mukherjee et al., 2013; Athmania and Achour, 2014; Bagnardi et al., 2016). Our results further confirm that in a mountainous, non-vegetated region the GDEM2 falls short of the reported vertical accuracy of 8.86 m (Tachikawa et al., 2011), even when ignoring gross outliers. In addition to the largest vertical uncertainty, the ASTER GDEM2 displays the largest uncertainties with respect to each topographic characteristic (elevation, slope, and aspect) (Fig. S5). An increase in uncertainty is apparent with increasing slopes, indicating overprediction of elevation for the ASTER GDEM2 at higher slopes; however, we also note the decrease in number of measurements at slopes > 30° (Fig. S5b). The GDEM2 experiences a clear aspect bias with an amplitude of ~ 5 m (Fig. S5c), which is lower than the ~ 50 m aspect bias reported in far-north glaciated terrain by Nuth and Kääb (2011). The ASTER stack generated for the Pocitos Basin shows im-

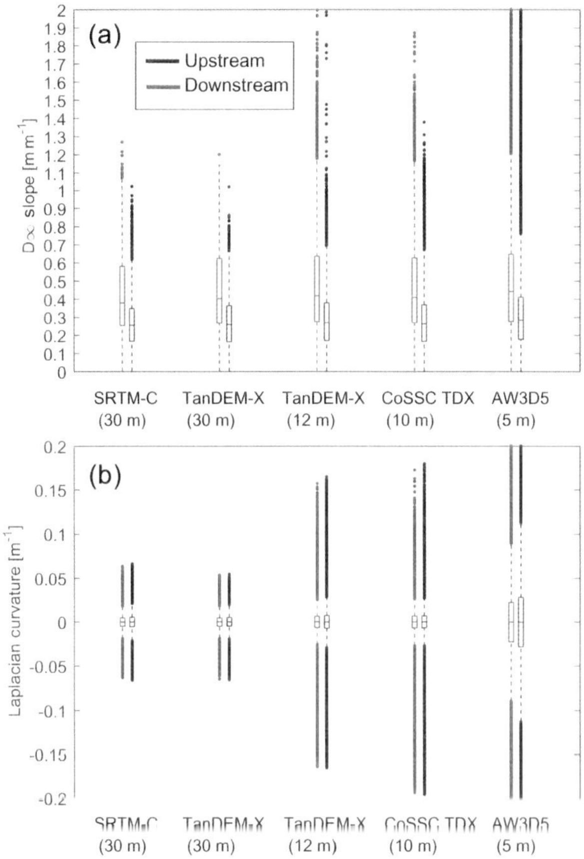

Figure 8. Slope **(a)** and curvature **(b)** box plots separated upstream (blue) and downstream (red) for five DEMs. Center line is median, boxes are 25–75th percentile range, dashed whiskers extend to first and 99th percentiles, and all outliers are plotted as points. Note that in cases where the outliers extend out of range of the plots, the points are truncated.

provement, with SD reduced to 6.93 m and only 0.12 % outlier reduction (Fig. 3b), as well as elimination of the aspect bias noted in the GDEM2 (Fig. S6c). This stack also represents an improvement over individual ASTER L1A stereopair DEMs with reported accuracies of 7–60 m (e.g., Toutin and Cheng, 2001; Hirano et al., 2003; Kääb, 2005; Nuth and Kääb, 2011). Despite this, the ASTER stack was deemed of insufficient quality for geomorphometry after visual inspection revealed remaining noise on hillslopes and channel elevation profiles, complicating slope and curvature measurements.

The SRTM-C, 30 m TanDEM-X, and AW3D30 have narrow vertical uncertainty distributions, with SDs of 3.33, 2.42, and 2.81 m, respectively, and < 0.04 % reduction in measurements from ±30 m outlier removal (Table 2). While elevation accuracy has not been previously reported for the AW3D30 or 30 m TanDEM-X, our results indicate that these datasets exceed mission specifications of < 5 m for the AW3D30 (Tadono et al., 2014) and < 10 m for the TanDEM-X (Wessel, 2016; Baade and Schmullius, 2016). Most elevation accuracy reporting for the SRTM DEMs has centered on the 30 m X-band and 90 m C-band products (e.g., Rexer and Hirt, 2014; Mukherjee et al., 2013; Kolecka and Kozak, 2014), and not the 2014 globally released (previously only USA) 30 m C-band DEM used here. In exception to this, Baade and Schmullius (2016) report vertical accuracy of 8–9 m for the 30 m SRTM-C, including outliers. Our SRTM-C (filtered) results are in close agreement with the 3.64 m accuracy found using 19 high-accuracy ground measurements for a steep volcano (Bagnardi et al., 2016) and less than the 8 m accuracy versus a control DEM on another volcano (Kervyn et al., 2008). Hofton et al. (2006) report a vertical SD of 2–7 m for low-vegetation regions in the USA for the SRTM-C versus high-accuracy lidar data. For the 30 m SRTM-C our results exceed the 6.2 m vertical accuracy found by Rodriguez et al. (2006) for dGPS tracks across South America.

These three high-quality 30 m DEMs exhibit no apparent biases with respect to elevation or aspect, and all show smaller ranges of uncertainties than the ASTER GDEM2. This is especially pronounced in the TanDEM-X, with the narrowest uncertainty ranges plotted (Fig. S7). Vertical uncertainty at higher slopes for the SRTM-C show overestimation of elevation, in agreement with the findings of Shortridge and Messina (2011). Conversely, the AW3D30 (Fig. S4b) and TanDEM-X (Fig. S7b) indicate lower uncertainties (but still increasing) at these higher slopes. Previous studies suggesting SRTM-C biases related to slope and aspect (e.g., Berthier et al., 2006, 2007; Van Niel et al., 2008; Shortridge and Messina, 2011) cannot be discounted by our findings, but we expect lower uncertainties with respect to slope in our non-glaciated, vegetation-free study area, where effects like radar penetration (e.g., Rignot et al., 2001; Becek, 2008; Gardelle et al., 2012) are minimal. Radar-associated biases are unexplored for TanDEM-X and are not apparent in our vegetation-free study area. These effects are also absent from the AW3D30 as this DEM was generated through optical methods by stacking of ALOS PRISM tri-stereopairs.

Results of 30 m global DEM elevation validation indicate the high quality of height information from SRTM-C, TanDEM-X, and AW3D30. ASTER GDEM2 is a far noisier dataset, which complicates geomorphic analyses requiring accurate slope and curvature calculations (e.g., Kervyn et al., 2008; Fisher et al., 2013; Pipaud et al., 2015). This noise is persistent, although slightly reduced, in the manually generated ASTER stack. Despite its low SD, visual inspection of the AW3D30 revealed its inadequacy for assessing geomorphic metrics. In addition to step-like artifacts on hillslopes, likely caused by resampling at the Japan Aerospace Exploration Agency (JAXA), this dataset also had numerous holes and hillslope artifacts caused by errors in optical DEM generation. Similar to the optically generated ASTER DEMs, these errors are caused by low contrast and cloud cover that hinder stereogrammetric methods. The 30 m TanDEM-X performed best in terms of agreement with dGPS measurements

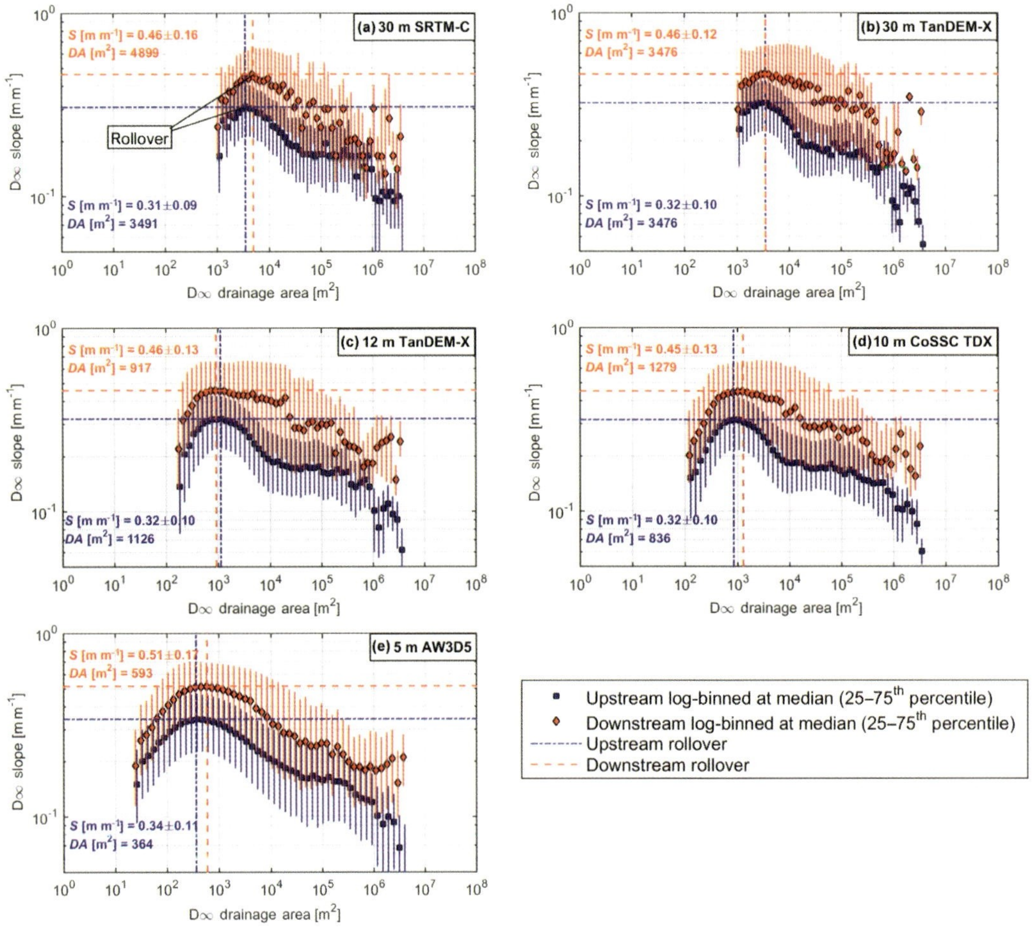

Figure 9. Logarithmically binned drainage area versus median slope with 25–75th percentile range of slopes indicated for **(a)** SRTM-C 30 m, **(b)** TanDEM-X 30 m, **(c)** TanDEM-X 12 m, **(d)** CoSSC TDX 10 m, and **(e)** AW3D5 5 m DEMs. Analysis separated downstream (red diamonds) and upstream (blue squares) of the knickpoint. Rollover drainage area demarcating approximate hillslope-to-valley transition, after which all subsequent slope values are less, is marked by vertical dashed line with slope at this bin marked by horizontal dashed line. Mean $\pm$ SD of slope (S) at the rollover drainage area bin (DA) recorded upstream and downstream. Plots are truncated to provide better visualization at very high drainage areas with low slopes.

and limited biases with respect to elevation, slope, and aspect. As the SRTM-C performed comparably well in terms of elevation accuracy, both of these 30 m datasets were selected for geomorphometric analysis.

5.1.2 The 5–12 m DEMs

Similar to the AW3D30, the vertical uncertainty for the AW3D5 exceeds the mission standard of <5 m (Tadono et al., 2014). The low SDs of 2.02, 3.83, and 3.22 m for our three CoSSC TDX DEMs are in close agreement with reported vertical accuracies of 5.74 m versus ground control points (Bagnardi et al., 2016), 3.57 m versus lidar data (Du et al., 2015), and <2 m versus laser altimetry (Rossi et al., 2016) for interferometrically generated CoSSC TDX DEMs with resolutions of 5–12 m. Wider, bimodal uncertainty distributions for the CoSSC TDX DEMs covering the Quebrada Honda and Nevado Queva (2012 and 2013 DEMs in Fig. 4b) are likely related to radar shadowing and layover in steeper terrain. Aspect biases for these single-CoSSC radar DEMs (Figs. S8–S9) were removed in the stacked 12 m TanDEM-X relying on descending and ascending orbits, which also had a lower SD of 1.97 m, again exceeding mission standards (Wessel, 2016; Baade and Schmullius, 2016).

Good vertical accuracy performance is seen in the stacked 5 m AW3D5 and stacked 12 m TanDEM-X product, with both datasets having narrow vertical uncertainty ranges plotted across terrain attributes (Figs. 6–7). While interferometrically generated single-CoSSC TDX DEMs (the same data used to generate the stacked TanDEM-X DEMs) also performed well in terms of vertical accuracy, a single stereogrammetrically generated ALOS PRISM tri-stereopair DEM (the same data used to generate the stacked AW3D DEMs) performed poorly and was not included in further

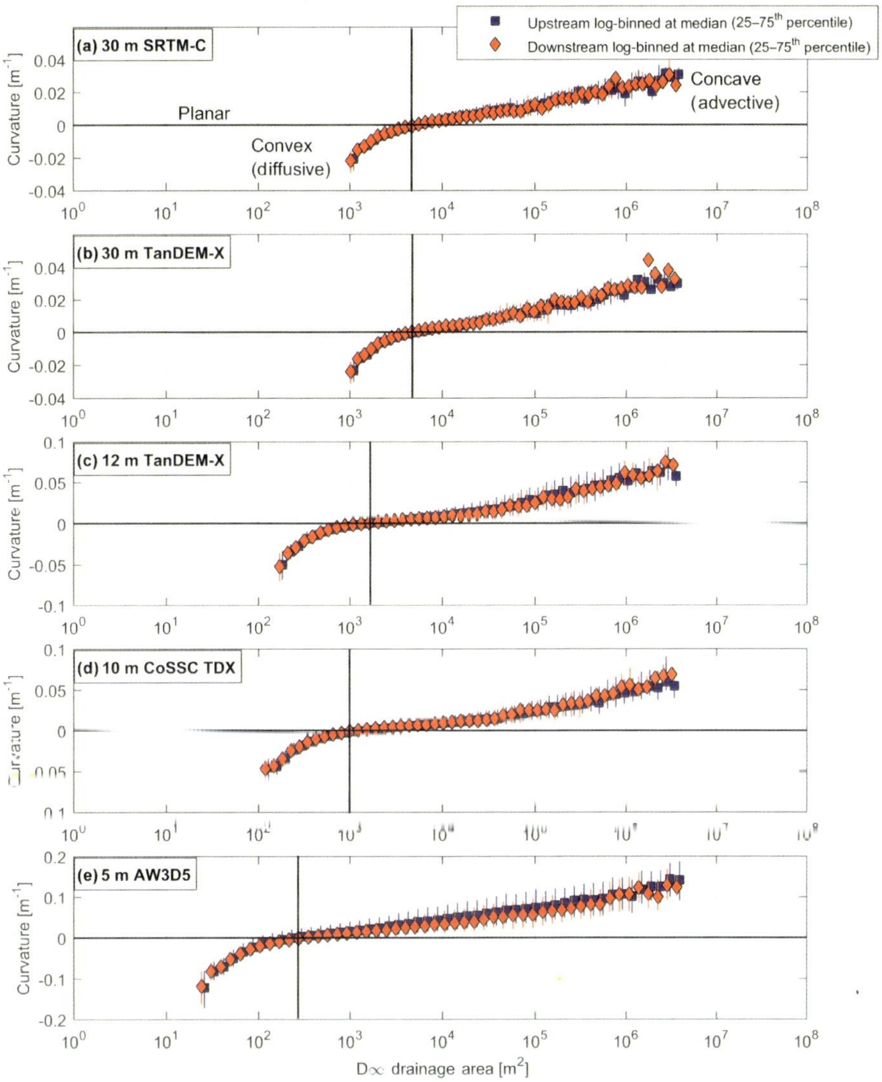

Figure 10. Logarithmically binned drainage area versus median curvature with 25–75th percentile range of curvatures indicated for **(a)** SRTM-C 30 m, **(b)** TanDEM-X 30 m, **(c)** TanDEM-X 12 m, **(d)** CoSSC TDX 10 m, and **(e)** AW3D5 5 m DEMs. Analysis separated downstream (red diamonds) and upstream (blue squares) of the knickpoint. Greater variability in curvature (larger error bars and greater *y*-axis range) is measured in the 5 m data. Vertical line marks approximate changeover from diffusive hillslope processes (convex curvature) to advective fluvial processes (concave curvature).

analysis (see Supplement). In conjunction with the improvement seen in our ASTER stack, these results indicate the importance of stacking multiple DEMs from the same data source to improve quality of the final product. This point is emphasized by the elimination of the aspect bias in the stacked 12 m TanDEM-X. The higher vertical accuracy and more realistic landscape representation of the single-CoSSC TDX radar DEM versus the single ALOS PRISM tri-stereopair DEM points to the greater potential of radar to accurately represent topography (e.g., the high-quality, radar SRTM-C versus the lower-quality, optical ASTER GDEM2).

Elevation accuracy for the higher-resolution DEMs is similar to the high-quality 30 m DEMs. The close agreement in vertical uncertainty (all < 3.5 m) between the highest-quality datasets (30 m SRTM-C, 30 and 12 m TanDEM-X, 10 m CoSSC TDX, and 5 m AW3D5) necessitates our geomorphic metric comparisons to better understand the limitations related not only to resolution but also to sensor. Our data show that for a number of DEMs, accurate elevation data are negligibly influenced by resolution at 5–30 m (Vaze et al., 2010), making differences in DEM quality for deriving geomorphic metrics unapparent from the pixel-by-pixel dGPS comparisons and SD metric.

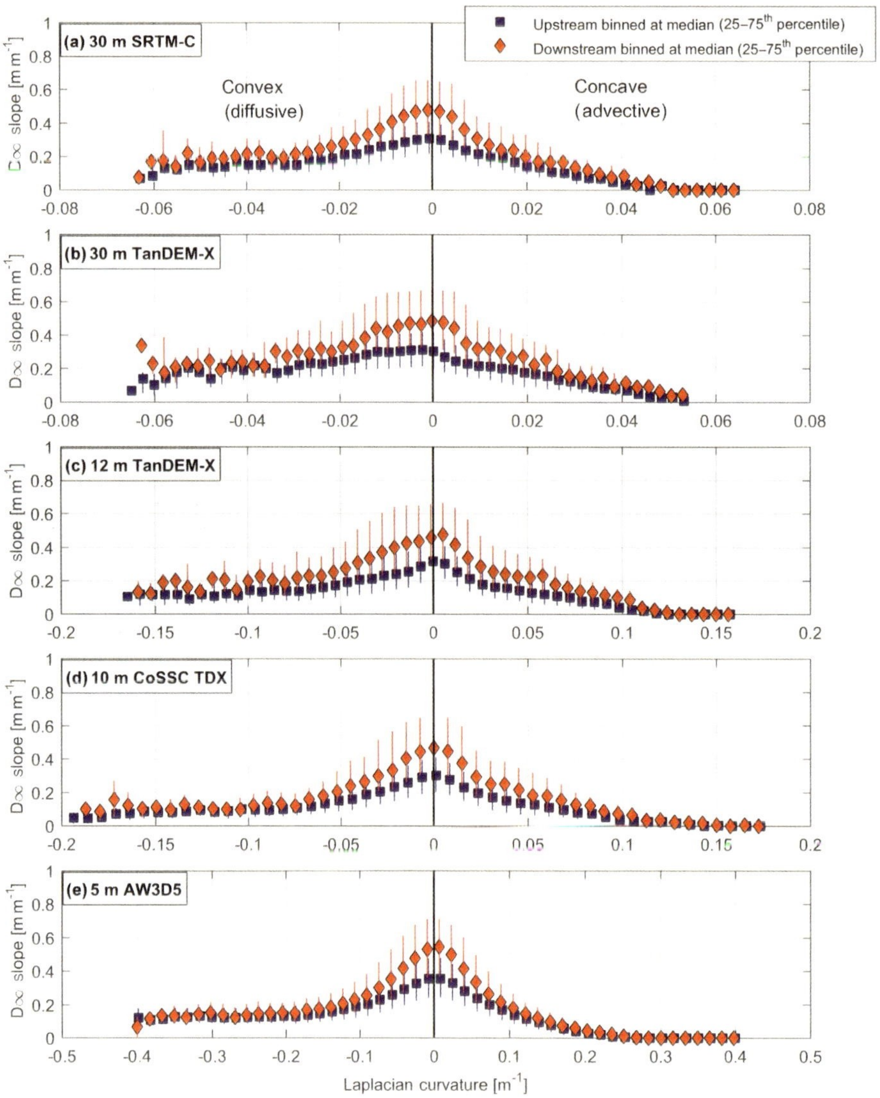

Figure 11. Linearly binned curvature versus median slope with 25–75th percentile range of slopes indicated for **(a)** SRTM-C 30 m, **(b)** TanDEM-X 30 m, **(c)** TanDEM-X 12 m, **(d)** CoSSC TDX 10 m, and **(e)** AW3D5 5 m DEMs. Analysis separated downstream (red diamonds) and upstream (blue squares) of the knickpoint. Greater variability in curvature is measured in the higher-resolution data as indicated by the growing x-axis range. Vertical line marks planar slopes dividing diffusive hillslope processes (convex curvature) to advective fluvial processes (concave curvature).

5.2 Geomorphometric validation

5.2.1 Channel profiles

The m/n values for the Quebrada Honda trunk correspond well across the datasets (30 m SRTM-C, 10 m CoSSC TDX, and 5 m AW3D5) and between the chi plot methods (Table 3). This is despite the fact that the knickpoint causes the channel to plot nonlinearly in chi space using the least-squares method (Fig. S2), whereas the piece-wise method allows exact fitting (Fig. S3). These values (0.49–0.57) fall well within the range of reported m/n values in a variety of other settings (e.g., Wobus et al., 2006; Kirby and Whipple, 2012; Harel et al., 2016). Testing on the 30 m DEMs revealed similar m/n values regardless of the elevation noise. For instance, the ASTER GDEM2, which had the largest vertical uncertainty and noisiest appearance, returned $m/n = 0.53$ with $R^2 = 0.97$ using the least-squares method, which is identical to the SRTM-C results. The only difference for the higher-resolution datasets is a slightly lower SD (uncertainty) of minimum AICc for piece-wise fitting: ~ 0.5 for the 30 m versus ~ 0.2 for the 5 and 10 m DEMs. Conversely, the coefficients of determination (R^2) from least-squares fitting are nearly identical for all three DEMs.

Differences in m/n values between the datasets are likely caused by differences in channel lengths for the area threshold channel delineation method or by minor differences in

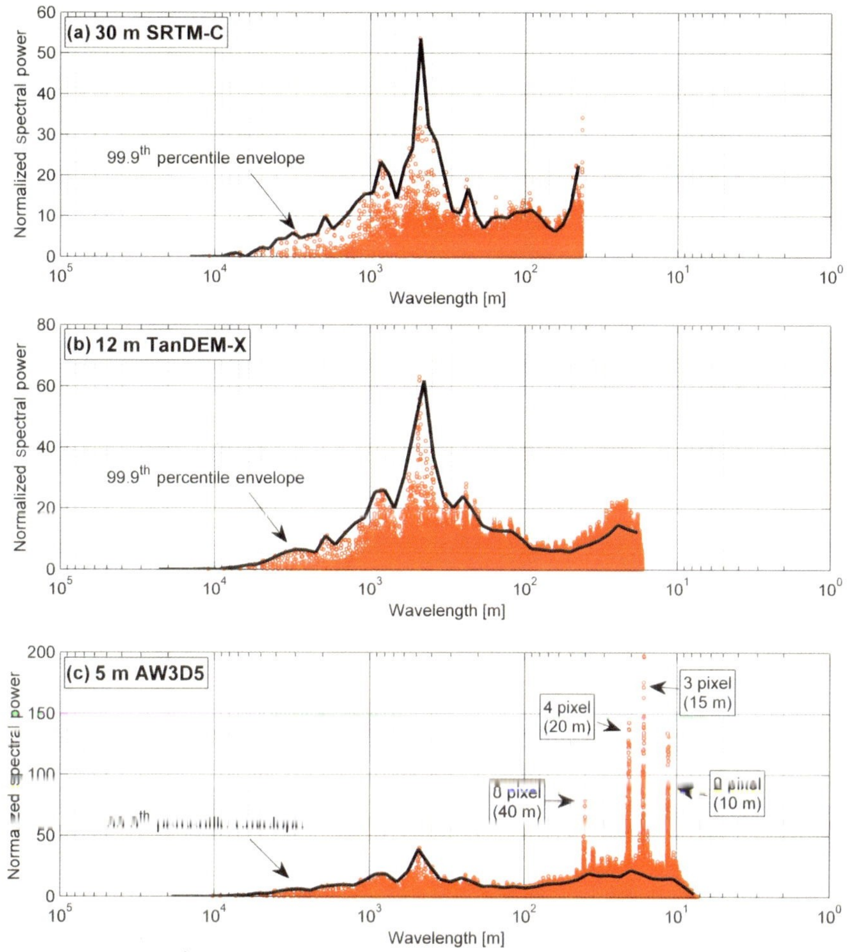

Figure 12. One-dimensional normalized power spectra for **(a)** 30 m SRTM-C, **(b)** 12 m TanDEM-X, and **(c)** 5 m AW3D5 plotted against wavelength (frequency^{-1}). Wavelength here is equivalent to spatial resolution in pixels. The 99.9th percentile envelope is calculated from 50 logarithmically spaced bins to highlight the peak in spectral power at ~500 m and the high-frequency (low-wavelength) spikes in the AW3D5 data. These spikes correspond to 2–8 pixel (10–40 m wavelength) steps in this 5 m DEM and cause a difference of 1 order of magnitude in y scaling for plot **(c)**.

exact channel placement downstream in the valley bottom. Nonetheless, the m/n values calculated using either chi plot method (least-squares or piece-wise fitting) are comparable regardless of DEM resolution (or noise, as indicated by the ASTER GDEM2 results), indicating the ability of all satellite-derived DEMs tested to resolve the valley bottom in our steep test catchment. This result only holds for relatively simple channel shapes, like the Quebrada Honda, whereas the inclusion of tributaries and more complex settings may warrant further testing and the preferred use of the statistically robust piece-wise fitting method (Mudd et al., 2014). Consideration of different channel lengths and changes in m/n may be an important factor when using the ASTER GDEM2 for chi plot analysis, as this dataset has demonstrated excessive channel foreshortening over long stretches (Fisher et al., 2013). Regardless, these results indicate that channel profile m/n analysis for mapping k_{sn} at $>1\ km^2$ scales, where minor differences related to channel head placement can be ignored (e.g., Grieve et al., 2016c), is readily achieved for open-access 30 m DEMs.

5.2.2 Hillslopes

The large increase in slope and curvature variability (outlier ranges; Fig. 8) with finer resolution – also demonstrated in Pipaud et al. (2015) – can be explored in a map view of curvature colored by a $\pm 3\sigma$ range for each DEM (Fig. 14). As the second derivative of elevation, curvature was selected for map view plots to highlight variability in elevation and slope (first derivative), as elevation errors propagate to higher derivatives. While the curvature signals of large ridges and narrow valleys are readily identified, although low in magnitude on the 30 m DEMs (Fig. 14a–b), many more features become apparent at higher resolutions. The 12 m TanDEM-X (Fig. 14c) and 10 m CoSSC TDX (Fig. 14d) DEMs appear

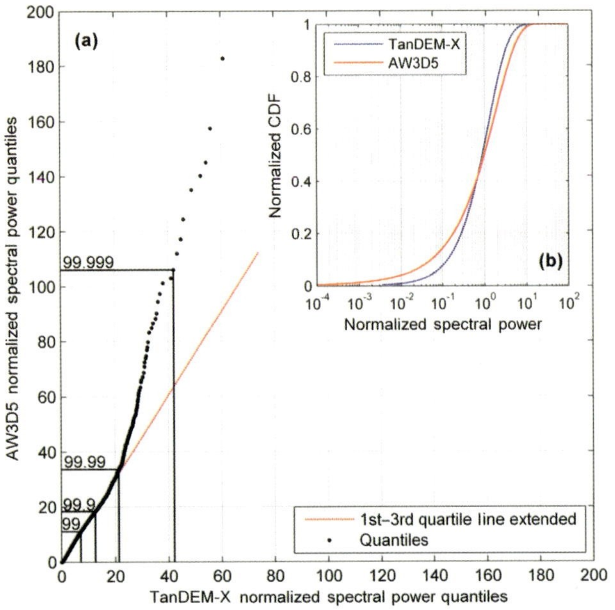

Figure 13. Statistical analysis of 1-D normalized spectral power shown in Fig. 12, with **(a)** QQ plot of 12 m TanDEM-X versus 5 m AW3D5 and **(b)** normalized CDF of normalized spectral power. Quantiles are plotted from 0.1 to 99.9998 in 0.0001 step sizes. The 99th, 99.9th, 99.99th, and 99.999th quantiles are noted for both datasets in panel **(a)** and the linear trend connecting the first and third quartiles is projected to display the diversion in the trend above the 99.99th quantile. We note the steeper than one to one trend in the data, demonstrating the higher spectral powers that dominate the AW3D5 signal.

similar (and have a similar $\pm 3\sigma$ range), although in the 10 m CoSSC TDX we note some striping becoming apparent in the second derivative of elevation, likely from interferometric processing of this single radar pair (Pipaud et al., 2015). For the 12 m TanDEM-X, the hillslopes appear smooth, separated by high-magnitude peaks at ridge crests and valley bottoms. The 5 m AW3D5 (Fig. 14e) shows the greatest variability, with sharp ridges and narrow valleys becoming obscured by other high curvatures (and thus high slopes) measured across the landscape. The cause of this may be the large number of rocky outcrops visible throughout the area in the 2.5 m panchromatic, nadir ALOS PRISM optical data (Fig. 14f). However, previous work suggests that large outliers in curvature can indicate DEM error and should be carefully considered (Sofia et al., 2013; Pipaud et al., 2015).

In Fig. 15, we explore this variability further with a map view of a hillshade image for the 5 m AW3D5 and 12 m TanDEM-X, in an area with less rocky outcrops, alongside the same area as viewed on the 2.5 m PRISM scene (the same data used to generate the AW3D5 DEM). While ridges and valleys are similar and the debris flow gully in the center can be identified on each DEM, there is a clear difference in smoothness between the optical (AW3D5) and radar (TanDEM-X) datasets. This is exactly the noise identified in our 2-D DFT analysis, which demonstrated spikes in spectral power in 2–8 pixel steps in the AW3D5 (Fig. 12c). Furthermore, it is this high-frequency, low-wavelength noise that causes the greater number of slope outliers and higher variability in curvature measurements, despite representing only a small (0.01 %) fraction of the power spectrum (Fig. 13). In other environments, it could be the case that this noise is caused by animal burrowing activity, tree throw, or some other high-frequency geomorphic process (e.g., Roering et al., 2010); however, it is clear in the optical data that no such processes are operating in this smooth, highly diffusive, vegetation-free environment. Rather, this error is from optical DEM generation and stacking. Such high-frequency noise was also present in the optical ASTER stack and GDEM2, as well as in the ALOS PRISM tri-stereopair, RapidEye, and SPOT6 DEMs, all manually generated via stereogrammetry but not used for further analysis (see Supplement for details on these datasets). Conversely, the radar-derived 12 m TanDEM-X provides a much more realistic representation of the landscape, despite a coarser resolution.

One important distinction to make with slope and curvature measurements is the window size used for calculation as differences may not only be related to the different sensors but also to the different resolutions. By using an equally sized nine-cell window (3×3) across the datasets, we are measuring different length scales. Numerous authors (e.g., Albani et al., 2004; Sofia et al., 2013) point this effect out with regard to elevation error propagation. To test this, we bilinear resampled the 5 m AW3D5 to 10 and 30 m and examined the slope and curvature distributions compared to the 12 m TanDEM-X and 30 m SRTM-C. By resampling the 5 m data to the coarser resolutions, we are essentially changing the length scale over which the derivatives are calculated (still in a nine-cell window). Our results show that with coarsening resolution the 5 m AW3D5 still shows high-frequency noise, particularly with respect to curvature, but overall they become increasingly similar to TanDEM-X and SRTM-C (Fig. S11). This result indicates that higher-resolution data captures more information even when measuring over the same length scale as coarser data. However, we demonstrate with the Fourier analysis that, in this case, the additional information is just sensor-related noise.

Although curvature measurements may differ widely depending on DEM resolution (and quality), it is clear from our analysis that slope measurements are less sensitive. All DEMs display a broad changeover between dominantly hillslope and dominantly fluvial processes, indicated by a wide area of inflection at the first rollover in area–slope plots, occurring at comparable slope values across datasets (Fig. 9). This stands in contrast to other studies using lidar DEMs, which show a very narrow changeover at DEM resolutions of 1–5 m (Tarolli and Dalla Fontana, 2009; Tseng et al., 2015). However, these studies were based in much wetter environments, where fluvial processes may act to increase this contrast. It is likely the case that this hillslope-to-valley

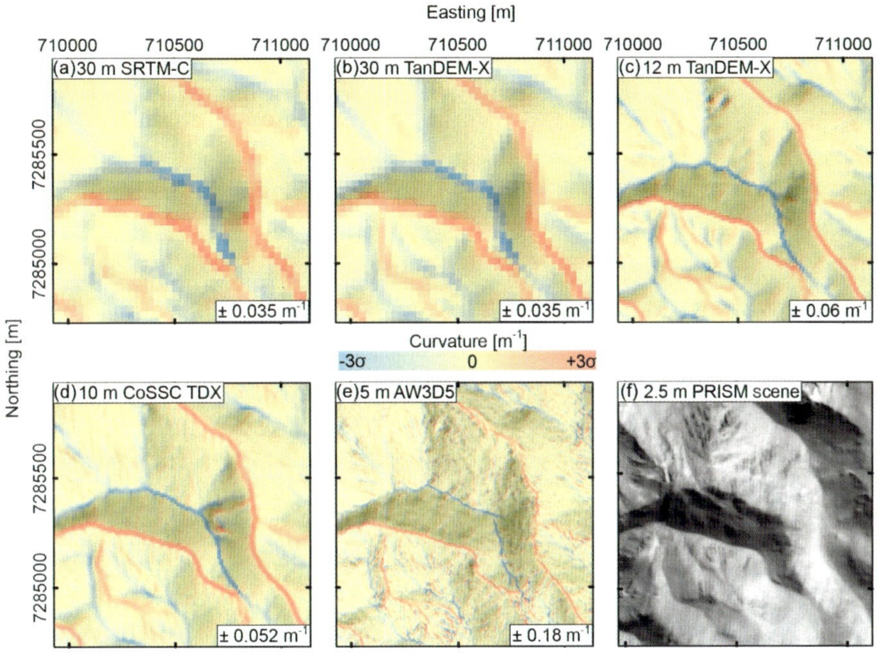

Figure 14. Map view of curvature from a section of the Quebrada Honda overlain on hillshade for **(a)** 30 m SRTM-C, **(b)** 30 m TanDEM-X, **(c)** 12 m TanDEM-X, **(d)** 10 m CoSSC TDX, and **(e)** 5 m AW3D5. Curvatures colored by a $\pm 3\sigma$ range to emphasize the high values, with the color bar range noted for each DEM in the lower right. Note the striping present in the CoSSC TDX **(d)** and the large variability measured in the AW3D5 **(e)**, which may be explained by the rocky outcrops apparent in the ALOS PRISM optical data for this area **(f)**.

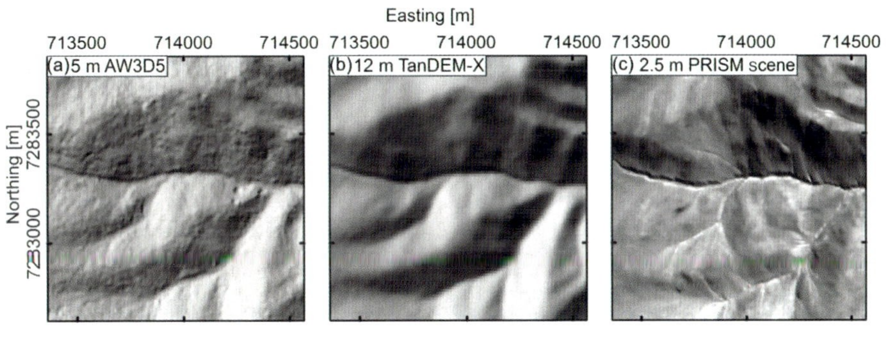

Figure 15. Hillshade view from a section of the Quebrada Honda with few rocky outcrops for **(a)** 5 m AW3D5 and **(b)** 12 m TanDEM-X, alongside **(c)** 2.5 m ALOS PRISM optical scene (nadir view). Ridges, valleys, and central debris flow gully are well represented by both DEMs; however, the high-frequency noise throughout the 5 m data does not correspond to pit and mound topography in the optical scene.

changeover is highly dependent on environmental conditions and typically occurs over a much larger range, even in the same conditions, due to local complexities at small ($\sim 1\,\text{km}^2$) scales. The more pronounced difference in slope values in the 10^4–$10^5\,\text{m}^2$ drainage area range (Fig. 9), may indicate the increased influence of landsliding and other mass-wasting processes in the steeper downstream catchment (Tseng et al., 2015). The flattening of downstream slope values in the TanDEM-X data in this range (Fig. 9b–d) also points to a change in geomorphic processes to more mass wasting in the steeper downstream catchment, further indicating the high quality of this data at resolutions of 10–30 m. In area–curvature and curvature–slope plots (Figs. 10 and 11), we show a greater range of curvature measured with finer resolution, yet slope values in Fig. 11 remain comparable regardless of resolution. Furthermore, Fig. 11 demonstrates that the 0.1–0.2 $\text{m}\,\text{m}^{-1}$ magnitude difference in slope measurements upstream and downstream of the knickpoint (Fig. 9) peaks on near-planar hillslopes (zero curvature). Hence, while valley bottoms (highly concave) and hilltops (highly convex) may have small differences in slope and curvature upstream and downstream of the knickpoint in the Quebrada Honda, the intervening hillslopes represent much of the erosional signal in the topography in this landscape.

Interestingly, the horizontal L_H measured in area–slope plots differs little upstream, downstream, and between datasets. Agreement in L_H between DEMs indicates that even the coarser (and noisier) DEMs are capable of measuring this key landscape metric. With a mean of 109 ± 26 m (1σ), L_H results are within the range (though on the high side) reported in the literature using this technique (e.g., Montgomery and Foufoula-Georgiou, 1993; Roering et al., 2007; Tarolli and Dalla Fontana, 2009; DiBiase et al., 2012; Grieve et al., 2016a). The higher L_H compared with other studies may be caused by the fact that the arid Central Andean Plateau, with low precipitation and little fluvial erosion (Bookhagen and Strecker, 2008, 2012), emphasizes diffusive hillslope processes. This, in turn, leads to longer hillslope measurements compared to studies in less-arid regions (e.g., Roering et al., 2007).

Given the presence of a knickpoint, we expect longer hillslopes in the gently sloped upstream catchment, whereas our data demonstrate equally long hillslopes in the steeper downstream section. This result may be caused by the graphical selection of a rollover value in area–slope plots, which have known deficiencies in assuming uniform basin shape at these scales (Grieve et al., 2016a). However, it could be that upstream and downstream L_H are in fact similar. In that case, steeper slopes downstream given the same curvature and drainage area (Figs. 9 and 11) would be compensated for by greater relief rather than longer hillslopes in order to maintain steady state of hillslopes with respect to changing base levels caused by the migrating knickpoint. Relief measured across DEMs in a 100 m radius (consistent with the measured L_H) is ~ 35 m greater downstream, indicating that this hypothesis is a possible explanation for the similar L_H.

The relative steady state of the upstream and downstream portions of the catchment would be surprising given the presence of a migrating knickpoint, which should affect erosion rates. However, in this dominantly diffusive environment there is little advection via fluvial erosion to disturb the landscape. This hypothesis allows hillslopes to adjust to changing base levels as the knickpoint migrates primarily during short pluvial periods on the Puna de Atacama Plateau (e.g., Trauth et al., 2000; Bookhagen et al., 2001; Luna et al., 2017). Although the hillslopes may adjust quickly to the changing base level caused by a migrating knickpoint, hilltop curvature should keep pace with erosion for much longer (Hurst et al., 2012). Despite this, it has been suggested that at resolutions ≥ 5 m, DEMs are unable to capture the fine variability of these curvature measurements (Clubb et al., 2016). Recent work by Grieve et al. (2016c) suggests that for high-accuracy DEMs, the resolution chosen for geomorphometric analysis should depend on how broad features such as ridges and valleys are and whether that resolution can capture variations in these regions.

Attempts to examine this hilltop signal with 27 manually selected subbasins from the 5 m AW3D5 and 12 m TanDEM-X data were complicated by transient landscape features in the Quebrada Honda. While a number of steep subbasins can be found downstream of the knickpoint, some of these basins appear to have downstream steepened sections and upstream gently sloped sections. This is indicated by channel steepness values decreasing moving up-valley from the subbasin outlets downstream of the knickpoint (Fig. 2a). It is possible that the trunk knickpoint signal has yet to reach the higher portions of these catchments given the low rate of fluvial erosion on the Puna de Atacama Plateau. Thus, only a limited number of sharper (more convex) hilltop measurements adjusted to the knickpoint are available to analyze subbasin transient adjustment, which is further limited by the number of measurements available (number of pixels making up a hilltop) as resolution decreases. Furthermore, many of these hilltops are not soil-mantled (many rocky outcrops are visible in optical data) and are thus unusable for hilltop curvature measurements (Hurst et al., 2012).

Taken together, results of our study indicate that despite having low vertical uncertainties, DEMs at resolutions of 5–30 m may not all be sufficient for deriving hillslope-scale geomorphic metrics, especially curvature. This is particularly true for optical DEMs (ASTER and AW3D5), which suffer from high-frequency noise that complicates slope and curvature assessments. Conversely, radar-derived DEMs provide a more realistic representation of the landscape, even in the case of single-CoSSC TDX DEMs, in agreement with Pipaud et al. (2015). While stacking further improves DEM quality (e.g., TanDEM-X DEMs), all DEMs were capable of accurate slope and hillslope length measurements regardless of resolution (or of noise in the case of the AW3D5). Conversely, differences in curvature measurements (particularly outliers) were distinctly dependent on resolution. These results are in agreement with Grieve et al. (2016c), but we emphasize that apparent differences in extracted geomorphic metrics at different resolutions are not only the result of the resolution tested – as in the resampled lidar DEMs of Grieve et al. (2016c) – but also the original sensor biases and multistep processing required to get the data in usable, gridded format. Our results lead us to conclude that the potential of high-accuracy, satellite-derived DEMs in geomorphometry and tectonic geomorphology for exploring transient responses and equilibrium adjustment of large (10–100 km^2) catchments remains high; however, 1–5 m lidar data with realistic landscape representation may still be necessary to assess fine-scale differences in hillslope processes.

6 Conclusions

We were able to determine the elevation accuracy of the current generation of global 30 m digital elevation models (DEMs) by using a 307 509-measurement differential GPS (dGPS) dataset from the high-elevation, vegetation- and cloud-free southern Central Andean Plateau (Puna de Atacama). Results indicate the high quality of the SRTM-C, TanDEM-X, and ALOS World 3D 30 m DEMs, with fewer no-data voids and fewer step-like hillslope artifacts in the SRTM-C and TanDEM-X DEMs. Furthermore, we have demonstrated the ASTER products' lower quality even after weighted stacking of eight meticulously generated raw L1A DEMs. We extended this analysis to the 12 m TanDEM-X DEM and demonstrated the very high quality of this dataset. In addition, we assessed the vertical accuracy of a single manually generated CoSSC TerraSAR-X/TanDEM-X 10 m DEM and the 5 m ALOS World 3D. With the exception of the ASTER data, all DEMs had vertical accuracy below 4 m. Our dGPS dataset also allowed us to explore the terrain elevation, slope, and aspect vertical biases of these DEMs. For our 30 m DEMs, we found little evidence of error biases in the SRTM-C, TanDEM-X, and ALOS World 3D and only minor aspect biases with the ASTER GDEM2. The 12 m TanDEM-X performed exceptionally well across all terrain attributes over a large area. The 5 m ALOS World 3D performed comparably well to the TanDEM-X, although over a much smaller area. More future measurements from higher slopes (> 30°) would allow a fuller assessment of the previously described DEM errors on steeper topography but are limited by dGPS measuring capabilities in the field. Both the 12 m TanDEM-X and 5 m ALOS World 3D had vertical accuracies below 2 m. The 10 m single-CoSSC TerraSAR-X/TanDEM-X DEMs displayed wide error distributions and some aspect bias, indicating the necessity of stacked scenes for accurate DEM generation (i.e., multiple TanDEM-X pairs).

Having assessed the accuracy of the DEMs, we chose the highest-quality datasets (30 m SRTM-C, 30 and 12 m TanDEM-X, 10 m single-CoSSC TerraSAR-X/TanDEM-X, and 5 m ALOS World 3D) for a geomorphometric analysis of channel profiles and hillslopes in a large (66 km^2) catchment with a clear river knickpoint. We show that chi plot analysis of m/n values provides comparable results regardless of DEM resolution or chi plot method when applied to the trunk channel alone in this simple setting. Regarding hillslope analyses, the 5 m ALOS World 3D (25 m^2 per pixel), 10 m single-CoSSC TerraSAR-X/TanDEM-X (100 m^2), 12 m TanDEM-X (144 m^2), and 30 m SRTM-C and TanDEM-X (900 m^2) demonstrate similar slopes and characteristic hillslope lengths. The primary difference in geomorphic analysis came in the measurement of curvature, which is shown to be highly resolution (and noise) dependent. Despite low vertical uncertainty, the ALOS World 3D data were found to be excessively noisy in 2–8 pixel steps via a Fourier frequency analysis. This high-frequency, low-wavelength noise is caused by difficulties in optical DEM generation via stereogrammetry, whereas the interferometric radar-derived TanDEM-X DEM had much smoother, more realistic landscape representation with no spurious high-frequency noise.

We demonstrate that the newer generation of 5–12 m DEM products can be useful in assessing hillslope parameters at larger scales and lower costs than lidar but may still be insufficient for fine-scale analysis, for example of hilltop curvatures or channel-head location. Caution should be taken by geomorphologists when using noisy optically generated DEMs, when even single non-stacked radar DEMs (e.g., single-CoSSC TerraSAR-X/TanDEM-X) may provide more realistic landscape representation. DEMs acquired by remote sensing technology onboard satellites are reaching better and higher potential for geomorphic analyses, including landslide detection and identification of smaller-scale features. Despite this improvement, the variability of mountainous landscapes and identification of transient responses in the realm of tectonic geomorphology benefits not only from vertical accuracy but also DEM resolution and the ability to derive higher-order derivatives.

Author contributions. B. Bookhagen provided field equipment and training as well as financial support and supervision of analysis. B. Purinton carried out analyses, produced figures, and prepared the manuscript under B. Bookhagen's guidance.

Competing interests. The authors declare that they have no conflict of interest.

Acknowledgements. CoSSC TerraSAR-X/TanDEM-X radar data were received through proposal ID XTI_GEOL6727 granted to B. Bookhagen, and TanDEM-X DEMs were received through proposal ID DEM_CALVAL1028 granted to B. Purinton, both from the DLR. RapidEye scenes were received through RapidEye Science Archive (RESA) proposal ID 00195 granted to B. Purinton. The authors thank the DFG Graduate School StRATEGy (IGK2018) and NEXUS funded through the MWFK Brandenburg, Germany. Ricardo Alonso, Manfred Strecker, Lisa Luna, and Patrick Lanouette are thanked for help with field work in March 2015. We particularly thank Kanayim Teshebaeva for help in generating the interferometric TerraSAR-X/TanDEM-X DEMs and Wolfgang Schwanghart for development of the terrain analysis toolbox and productive conversations throughout preparation of the paper. Simon Mudd, Stuart Grieve, and the anonymous reviewer are thanked for suggested improvements on the original manuscript.

References

Akaike, H.: A New Look at the Statistical Model Identification, IEEE Trans. Autom. Control, 19, 716–723, doi:10.1109/TAC.1974.1100705, 1974.

Albani, M., Klinkenberg, B., Andison, D. W., and Kimmins, J. P.: The choice of window size in approximating topographic surfaces from Digital Elevation Models, Int. J. Geogr. Inf. Sci., 18, 577–593, 2004.

Allmendinger, R. W., Jordan, T. E., Kay, S. M., and Isacks, B. L.: The Evolution of the Altiplano-Puna Plateau of the Central Andes, Annu. Rev. Earth Pl. Sc., 25, 139–174, 1997.

Arrell, K., Wise, S., Wood, J., and Donoghue, D.: Spectral filtering as a method of visualising and removing striped artefacts in digital elevation data, Earth Surf. Proc. Land., 33, 943–961, 2008.

Athmania, D. and Achour, H.: External Validation of the ASTER GDEM2, GMTED2010 and CGIAR-CSI- SRTM v4.1 Free Access Digital Elevation Models (DEMs) in Tunisia and Algeria, Remote Sensing, 6, 4600–4620, 2014.

Baade, J. and Schmullius, C.: TanDEM-X IDEM precision and accuracy assessment based on a large assembly of differential GNSS measurements in Kruger National Park, South Africa, ISPRS J. Photogramm., 119, 496–508, 2016.

Bagnardi, M., González, P. J., and Hooper, A.: High-resolution digital elevation model from tri-stereo Pleiades-1 satellite imagery for lava flow volume estimates at Fogo Volcano, Geophys. Res. Lett., 43, 6267–6275, 2016.

Becek, K.: Investigation of elevation bias of the SRTM C-and X-band digital elevation models, Int. Arch. Photogramm. Remote Sens. Spatial Inf. Sci., 37, 105–110, 2008.

Berthier, E. and Toutin, T.: SPOT5-HRS digital elevation models and the monitoring of glacier elevation changes in North-West Canada and South-East Alaska, Remote Sens. Environ., 112, 2443–2454, 2008.

Berthier, E., Arnaud, Y., Vincent, C., and Rémy, F.: Biases of SRTM in high-mountain areas: Implications for the monitoring of glacier volume changes, Geophys. Res. Lett., 33, L08502, doi:10.1029/2006GL025862, 2006.

Berthier, E., Arnaud, Y., Kumar, R., Ahmad, S., Wagnon, P., and Chevallier, P.: Remote sensing estimates of glacier mass balances in the Himachal Pradesh (Western Himalaya, India), Remote Sens. Environ., 108, 327–338, 2007.

Bookhagen, B. and Strecker, M. R.: Orographic barriers, high-resolution TRMM rainfall, and relief variations along the eastern Andes, Geophys. Res. Lett., 35, L06403, doi:10.1029/2007GL032011, 2008.

Bookhagen, B. and Strecker, M. R.: Spatiotemporal trends in erosion rates across a pronounced rainfall gradient: Examples from the southern Central Andes, Earth Planet. Sc. Lett., 327–328, 97–110, 2012.

Bookhagen, B., Haselton, K., and Trauth, M. H.: Hydrological modelling of a Pleistocene landslide-dammed lake in the Santa Maria Basin, NW Argentina, Palaeogeogr. Palaeocl., 169, 113–127, 2001.

Booth, A. M., Roering, J. J., and Perron, J. T.: Automated landslide mapping using spectral analysis and high-resolution topographic data: Puget Sound lowlands, Washington, and Portland Hills, Oregon, Geomorphology, 109, 132–147, 2009.

Botev, Z. I., Grotowski, J. F., and Kroese, D. P.: Kernel density estimation via diffusion, Ann. Stat., 38, 2916–2957, 2010.

Burbank, D. W., Leland, J., Fielding, E., Anderson, R. S., Brozovic, N., Reid, M. R., and Duncan, C.: Bedrock incision, rock uplift and threshold hillslopes in the northwestern Himalayas, Nature, 379, 505–510, 1996.

Clubb, F. J., Mudd, S. M., Milodowski, D. T., Hurst, M. D., and Slater, L. J.: Objective extraction of channel heads from high-resolution topographic data, Water Resour. Res., 50, 4283–4304, 2014.

Clubb, F. J., Mudd, S. M., Attal, M., Milodowski, D. T., and Grieve, S. W. D.: The relationship between drainage density, erosion rate, and hilltop curvature: Implications for sediment transport processes, J. Geophys. Res.-Earth, 121, 1724–1745, 2016.

DiBiase, R. A., Whipple, K. X., Heimsath, A. M., and Ouimet, W. B.: Landscape form and millennial erosion rates in the San Gabriel Mountains, CA, Earth Planet. Sc. Lett., 289, 134–144, 2010.

DiBiase, R. A., Heimsath, A. M., and Whipple, K. X.: Hillslope response to tectonic forcing in threshold landscapes, Earth Surf. Proc. Land., 37, 855–865, 2012.

Dietrich, W. E., Bellugi, D. G., Sklar, L. S., Stock, J. D., Heimsath, A. M., and Roering, J. J.: Geomorphic Transport Laws for Predicting Landscape form and Dynamics, in: Prediction in Geomorphology, Wiley-Blackwell, Washington, D.C., USA, 103–132, 2003.

Du, Y.-N., Feng, G.-C., Li, Z.-W., Zhu, J.-J., and Peng, X.: Generation of high precision DEM from TerraSAR-X/TanDEM-X, Chin. J. Geophys, 58, 1634–1644, 2015.

Farr, T. G., Rosen, P. A., Caro, E., Crippen, R., Duren, R., Hensley, S., Kobrick, M., Paller, M., Rodriguez, E., Roth, L., Seal, D., Shaffer, S., Shimada, J., Umland, J., Werner, M., Oskin, M., Burbank, D., and Alsdorf, D.: The Shuttle Radar Topography Mission, Rev. Geophys., 45, RG2004, doi:10.1029/2005RG000183, 2007.

Fisher, G. B., Bookhagen, B., and Amos, C. B.: Channel planform geometry and slopes from freely available high-spatial resolution imagery and DEM fusion: Implications for channel width scalings, erosion proxies, and fluvial signatures in tectonically active landscapes, Geomorphology, 194, 46–56, 2013.

Fisher, P. F. and Tate, N. J.: Causes and consequences of error in digital elevation models, Prog. Phys. Geog., 30, 467–489, 2006.

Forte, A. M., Whipple, K. X., Bookhagen, B., and Rossi, M. W.: Decoupling of modern shortening rates, climate, and topography in the Caucasus, Earth Planet. Sc. Lett., 449, 282–294, 2016.

Gardelle, J., Berthier, E., and Arnaud, Y.: Impact of resolution and radar penetration on glacier elevation changes computed from DEM differencing, J. Glaciol., 58, 419–422, 2012.

Grieve, S. W. D., Mudd, S. M., and Hurst, M. D.: How long is a hillslope?, Earth Surf. Proc. Land., 41, 1039–1054, 2016a.

Grieve, S. W. D., Mudd, S. M., Hurst, M. D., and Milodowski, D. T.: A nondimensional framework for exploring the relief structure of landscapes, Earth Surf. Dynam., 4, 309–325, doi:10.5194/esurf-4-309-2016, 2016b.

Grieve, S. W. D., Mudd, S. M., Milodowski, D. T., Clubb, F. J., and Furbish, D. J.: How does grid-resolution modulate the topographic expression of geomorphic processes?, Earth Surf. Dynam., 4, 627–653, doi:10.5194/esurf-4-627-2016, 2016c.

Harel, M.-A., Mudd, S. M., and Attal, M.: Global analysis of the stream power law parameters based on worldwide ^{10}Be denudation rates, Geomorphology, 268, 184–196, 2016.

Hengl, T.: Finding the right pixel size, Comput. Geosci., 32, 1283–1298, 2006.

Hirano, A., Welch, R., and Lang, H.: Mapping from ASTER stereo image data: DEM validation and accuracy assessment, ISPRS J. Photogramm., 57, 356–370, 2003.

Hofton, M., Dubayah, R., Blair, J. B., and Rabine, D.: Validation of SRTM Elevations Over Vegetated and Non-vegetated Terrain Using Medium Footprint Lidar, Photogramm. Eng. Rem. S., 72, 279–285, 2006.

Höhle, J. and Höhle, M.: Accuracy assessment of digital elevation models by means of robust statistical methods, ISPRS J. Photogramm., 64, 398–406, 2009.

Holzer, N., Vijay, S., Yao, T., Xu, B., Buchroithner, M., and Bolch, T.: Four decades of glacier variations at Muztagh Ata (eastern Pamir): a multi-sensor study including Hexagon KH-9 and Pléiades data, The Cryosphere, 9, 2071–2088, doi:10.5194/tc-9-2071-2015, 2015.

Hooshyar, M., Wang, D., Kim, S., Medeiros, S. C., and Hagen, S. C.: Valley and channel networks extraction based on local topographic curvature and k -means clustering of contours, Water Resour. Res., 52, 8081–8102, 2016.

Howard, A. D., Dietrich, W. E., and Seidl, M. A.: Modeling fluvial erosion on regional to continental scales, J. Geophys. Res.-Sol. Ea., 99, 13971–13986, 1994.

Hurst, M. D., Mudd, S. M., Walcott, R., Attal, M., and Yoo, K.: Using hilltop curvature to derive the spatial distribution of erosion rates, J. Geophys. Res.-Earth, 117, F02017, doi:10.1029/2011JF002057, 2012.

Hurvich, C. M. and Tsai, C.-H.: Regression and time series model selection in small samples, Biometrika, 76, 297–307, 1989.

Ijjasz-Vasquez, E. J. and Bras, R. L.: Scaling regimes of local slope versus contributing area in digital elevation models, Geomorphology, 12, 299–311, 1995.

Jaber, W. A., Floricioiu, D., Rott, H., and Eineder, M.: Surface elevation changes of glaciers derived from SRTM and TanDEM-X DEM differences, in: 2013 IEEE International [illegible] IGARSS, 21–26 July 2013, Melbourne, Australia, IEEE, 1893–1896, doi:10.1109/IGARSS.2013.6723173, 2013.

Jarvis, A., Reuter, H. I., Nelson, A., and Guevara, E.: Hole-filled SRTM for the globe Version 4, CGIAR-CSI SRTM 90 m Database, available at: http://srtm.csi.cgiar.org (last access: 1 September 2016), 2008.

Kääb, A.: Combination of SRTM3 and repeat ASTER data for deriving alpine glacier flow velocities in the Bhutan Himalaya, Remote Sens. Environ., 94, 463–474, 2005.

Kervyn, M., Ernst, G. G. J., Goossens, R., and Jacobs, P.: Mapping volcano topography with remote sensing: ASTER vs. SRTM, Int. J. Remote Sens., 29, 6515–6538, 2008.

Kirby, E. and Whipple, K. X.: Expression of active tectonics in erosional landscapes, J. Struct. Geol., 44, 54–75, 2012.

Kolecka, N. and Kozak, J.: Assessment of the Accuracy of SRTM C- and X-Band High Mountain Elevation Data: a Case Study of the Polish Tatra Mountains, Pure Appl. Geophys., 171, 897–912, 2014.

Lague, D.: The stream power river incision model: evidence, theory and beyond, Earth Surf. Proc. Land., 39, 38–61, 2014.

Li, Z.: On the measure of digital terrain model accuracy, Photogramm. Rec., 12, 873–877, 1988.

Liu, Z., Peng, M., and Di, K.: A continuative variable resolution digital elevation model for ground-based photogrammetry, Comput. Geosci., 79, 71–79, 2014.

Luna, L. V., Bookhagen, B., Niedermann, S., Rugel, G., Scharf, A., and Merchel, S.: Glacial chronology and production rate cross-calibration of six cosmogenic nuclide/mineral systems from the central southern Andean Plateau (Puna, 24° S, 4000–5000 m), Earth Planet. Sc. Lett., in review, 2017.

Massey, F. J.: The Kolmogorov-Smirnov Test for Goodness of Fit, J. Am. Stat. Assoc., 46, 68–78, 1951.

METI/NASA/USGS: ASTER Global DEM Validation Summary Report, Tech. rep., METI/ERSDAC, NASA/LPDAAC, USGS/EROS, available at: https://lpdaac.usgs.gov/sites/default/files/public/aster/docs/ASTER_GDEM_Validation_Summary_Report.pdf (last access: 12 April 2017), 2009.

Montgomery, D. R. and Foufoula-Georgiou, E.: Channel network source representation using digital elevation models, Water Resour. Res., 29, 3925–3934, 1993.

Montgomery, D. R., Balco, G., and Willett, S. D.: Climate, tectonics, and the morphology of the Andes, Geology, 29, 579–579, 2001.

Mudd, S. M., Attal, M., Milodowski, D. T., Grieve, S. W. D., and Valters, D. A.: A statistical framework to quantify spatial variation in channel gradients using the integral method of channel profile analysis, J. Geophys. Res.-Earth, 119, 138–152, 2014.

Mukherjee, S., Joshi, P. K., Mukherjee, S., Ghosh, A., Garg, R. D., and Mukhopadhyay, A.: Evaluation of vertical accuracy of open source Digital Elevation Model (DEM), Int. J. Appl. Earth Obs., 21, 205–217, 2013.

Neckel, N., Braun, A., Kropácek, J., and Hochschild, V.: Recent mass balance of the Purogangri Ice Cap, central Tibetan Plateau, by means of differential X-band SAR interferometry, The Cryosphere, 7, 1623–1633, doi:10.5194/tc-7-1623-2013, 2013.

Neelmeijer, J., Motagh, M., and Bookhagen, B.: High-Resolution Digital Elevation Models from Single-Pass TanDEM-X Imagery Over Mountainous Regions: A Case Study for Inylchek Glacier, Central Asia, ISPRS J. Phtogramm., in review, 2017.

Nuth, C. and Kääb, A.: Co-registration and bias corrections of satellite elevation data sets for quantifying glacier thickness change, The Cryosphere, 5, 271–290, doi:10.5194/tc-5-271-2011, 2011.

Olen, S. M., Bookhagen, B., and Strecker, M. R.: Role of climate and vegetation density in modulating denudation rates in the Himalaya, Earth Planet. Sc. Lett., 445, 57–67, 2016.

Ouimet, W. B., Whipple, K. X., and Granger, D. E.: Beyond threshold hillslopes: Channel adjustment to base-level fall in tectonically active mountain ranges, Geology, 37, 579–582, 2009.

Pandey, P. and Venkataraman, G.: Comparison of DEMs derived from TanDEM-X and SRTM-C for Himalayan terrain, in: 2013 IEEE International Geoscience and Remote Sensing Symposium – IGARSS, 21–26 July 2013, Melbourne, Australia, IEEE, 322–325, doi:10.1109/IGARSS.2013.6721157, 2013.

Passalacqua, P., Tarolli, P., and Foufoula-Georgiou, E.: Testing space-scale methodologies for automatic geomorphic feature extraction from lidar in a complex mountainous landscape, Water Resour. Res., 46, W11535, doi:10.1029/2009WR008812, 2010a.

Passalacqua, P., Trung, T. D., Foufoula-Georgiou, E., Sapiro, G., and Dietrich, W. E.: A geometric framework for channel network extraction from lidar: Nonlinear diffusion and geodesic paths, J. Geophys. Res.-Earth, 115, F01002, doi:10.1029/2009JF001254, 2010b.

Passalacqua, P., Belmont, P., Staley, D. M., Simley, J. D., Arrowsmith, J. R., Bode, C. A., Crosby, C., DeLong, S. B., Glenn, N. F., Kelly, S. A., Lague, D., Sangireddy, H., Schaffrath, K., Tarboton, D. G., Wasklewicz, T., and Wheaton, J. M.: Analyzing high resolution topography for advancing the understanding of mass and energy transfer through landscapes: A review, Earth-Sci. Rev., 148, 174–193, 2015.

Paul, F. and Haeberli, W.: Spatial variability of glacier elevation changes in the Swiss Alps obtained from two digital elevation models, Geophys. Res. Lett., 35, L21502, doi:10.1029/2008GL034718, 2008.

Perron, J. T. and Royden, L.: An integral approach to bedrock river profile analysis, Earth Surf. Proc. Land., 38, 570–576, 2013.

Perron, J. T., Kirchner, J. W., and Dietrich, W. E.: Spectral signatures of characteristic spatial scales and nonfractal structure in landscapes, J. Geophys. Res., 113, F04003, doi:10.1029/2007JF000866, 2008.

Perroy, R. L., Bookhagen, B., Asner, G. P., and Chadwick, O. A.: Comparison of gully erosion estimates using airborne and ground-based LiDAR on Santa Cruz Island, California, Geomorphology, 118, 288–300, 2010.

Pike, R. J., Evans, I. S., and Hengl, T.: Geomorphometry: A Brief Guide, chap. 1, in: Geomorphometry: Concepts, Software, Applications, edited by: Hengl, T. and Reuter, H. I., Elsevier, Amsterdam, the Netherlands, 3–30, doi:10.1016/S0166-2481(08)00001-9, 2009.

Pipaud, I., Loibl, D., and Lehmkuhl, F.: Evaluation of TanDEM-X elevation data for geomorphological mapping and interpretation in high mountain environments – A case study from SE Tibet, China, Geomorphology, 246, 232–254, 2015.

Priestley, M. B.: Spectral Analysis and Time Series, Academic Press, New York, USA, 1981.

Purinton, B.: DEM Fourier noise analysis code for MATLAB, GitHub repository, available at: https://github.com/bpurinton/DEM_fourier_noise and https://github.com/UP-RS-ESP, last access: 14 January 2017.

Raaflaub, L. D. and Collins, M. J.: The effect of error in gridded digital elevation models on the estimation of topographic parameters, Environ. Modell. Softw., 21, 710–732, 2006.

Racoviteanu, A. E., Manley, W. F., Arnaud, Y., and Williams, M. W.: Evaluating digital elevation models for glaciologic applications: An example from Nevado Coropuna, Peruvian Andes, Global Planet. Change, 59, 110–125, 2007.

Rankl, M. and Braun, M.: Glacier elevation and mass changes over the central Karakoram region estimated from TanDEM-X and SRTM/X-SAR digital elevation models, Ann. Glaciol., 57, 273–281, 2016.

Reuter, H. I., Hengl, T., Gessler, P., and Soille, P.: Preparation of DEMs for Geomorphometric Analysis, chap. 4, in: Geomorphometry: Concepts, Software, and Applications, edited by: Hengl, T. and Reuter, H. I., Elsevier, Amsterdam, the Netherlands, 87–120, doi:10.1016/S0166-2481(08)00004-4, 2009.

Rexer, M. and Hirt, C.: Comparison of free high resolution digital elevation data sets (ASTER GDEM2, SRTM v2.1/v4.1) and validation against accurate heights from the Australian National Gravity Database, Aust. J. Earth Sci., 61, 213–226, 2014.

Rignot, E., Echelmeyer, K., and Krabill, W.: Penetration depth of interferometric synthetic-aperture radar signals in snow and ice, Geophys. Res. Lett., 28, 3501–3504, 2001.

Rodriguez, E., Morris, C. S., and Belz, J. E.: A Global Assessment of the SRTM Performance, Photogramm. Eng. Rem. S., 72, 249–260, 2006.

Roering, J. J., Kirchner, J. W., and Dietrich, W. E.: Evidence for nonlinear, diffusive sediment transport on hillslopes and implications for landscape morphology, Water Resour. Res., 35, 853–870, 1999.

Roering, J. J., Perron, J. T., and Kirchner, J. W.: Hillslope morphology and functional relationships between topographic relief and denudation, Earth Planet. Sc. Lett., 264, 245–258, doi:10.1016/j.epsl.2007.09.035, 2007.

Roering, J. J., Marshall, J., Booth, A. M., Mort, M., and Jin, Q.: Evidence for biotic controls on topography and soil production, Earth Planet. Sc. Lett., 298, 183–190, 2010.

Roering, J. J., Mackey, B. H., Marshall, J. A., Sweeney, K. E., Deligne, N. I., Booth, A. M., Handwerger, A. L., and Cerovski-Darriau, C.: "You are HERE": Connecting the dots with airborne lidar for geomorphic fieldwork, Geomorphology, 200, 172–183, 2013.

Rohrmann, A., Strecker, M. R., Bookhagen, B., Mulch, A., Sachse, D., Pingel, H., Alonso, R. N., Schildgen, T. F., and Montero, C.: Can stable isotopes ride out the storms? The role of convection for water isotopes in models, records, and paleoaltimetry studies in the central Andes, Earth Planet. Sc. Lett., 407, 187–195, 2014.

Rossi, C., Minet, C., Fritz, T., Eineder, M., and Bamler, R.: Temporal monitoring of subglacial volcanoes with TanDEM-X – Application to the 2014–2015 eruption within the Bárðarbunga volcanic system, Iceland, Remote Sens. Environ., 181, 186–197, 2016.

Scherler, D., Bookhagen, B., Wulf, H., Preusser, F., and Strecker, M. R.: Increased late Pleistocene erosion rates during fluvial aggradation in the Garhwal Himalaya, northern India, Earth Planet. Sc. Lett., 428, 255–266, 2015.

Schwanghart, W. and Scherler, D.: Short Communication: TopoToolbox 2 – MATLAB-based software for topographic analysis and modeling in Earth surface sciences, Earth Surf. Dynam., 2, 1–7, doi:10.5194/esurf-2-1-2014, 2014.

Shelef, E. and Hilley, G. E.: Impact of flow routing on catchment area calculations, slope estimates, and numerical simulations of landscape development, J. Geophys. Res.-Earth, 118, 2105–2123, 2013.

Shortridge, A. and Messina, J.: Spatial structure and landscape associations of SRTM error, Remote Sens. Environ., 115, 1576–1587, 2011.

Smith, B. and Sandwell, D.: Accuracy and resolution of shuttle radar topography mission data, Geophys. Res. Lett., 30, 1467, doi:10.1029/2002GL016643, 2003.

Snyder, N. P., Whipple, K. X., Tucker, G. E., and Merritts, D. J.: Landscape response to tectonic forcing: Digital elevation model analysis of stream profiles in the Mendocino triple junction region, northern California, Geol. Soc. Am. Bull., 112, 1250–1263, 2000.

Sofia, G., Pirotti, F., and Tarolli, P.: Variations in multiscale curvature distribution and signatures of LiDAR DTM errors, Earth Surf. Proc. Land., 38, 1116–1134, 2013.

Sofia, G., Hillier, J. K., and Conway, S. J.: Frontiers in Geomorphometry and Earth Surface Dynamics: possibilities, limitations and perspectives, Earth Surf. Dynam., 4, 721–725, doi:10.5194/esurf-4-721-2016, 2016.

Tachikawa, T., Kaku, M., Iwasaki, A., Gesch, D., Oimoen, M., Zhang, Z., Danielson, J., Krieger, T., Curtis, B., Haase, J., Abrams, M., Crippen, R., and Carabajal, C.: ASTER Global Digital Elevation Model Version 2 – Summary of Validation Results, Tech. rep., Joint Japan-US ASTER Science Team, available at: http://www.jspacesystems.or.jp/ersdac/GDEM/ver2Validation/Summary_GDEM2_validation_report_final.pdf (last access: 12 April 2017), 2011.

Tadono, T., Ishida, H., Oda, F., Naito, S., Minakawa, K., and

Iwamoto, H.: Precise Global DEM Generation by ALOS PRISM, ISPRS Ann. Photogramm. Remote Sens. Spatial Inf. Sci., II-4, 71–76, 2014.

Tarboton, D. G., Bras, R. L., and Rodriguez-Iturbe, I.: On the extraction of channel networks from digital elevation data, Hydrological Processes, 5, 81–100, 1991.

Tarolli, P.: High-resolution topography for understanding Earth surface processes: Opportunities and challenges, Geomorphology, 216, 295–312, 2014.

Tarolli, P. and Dalla Fontana, G.: Hillslope-to-valley transition morphology: New opportunities from high resolution DTMs, Geomorphology, 113, 47–56, 2009.

Tarolli, P. and Tarboton, D. G.: A new method for determination of most likely landslide initiation points and the evaluation of digital terrain model scale in terrain stability mapping, Hydrol. Earth Syst. Sci., 10, 663–677, doi:10.5194/hess-10-663-2006, 2006.

Toutin, T. and Cheng, P.: DEM generation with ASTER stereo data, Earth Observation Magazine, 10, 10–13, 2001.

Trauth, M. H., Alonso, R. A., Haselton, K. R., Hermanns, R. L., and Strecker, M. R.: Climate change and mass movements in the NW Argentine Andes, Earth Planet. Sc. Lett., 179, 243–256, 2000.

Tseng, C.-M., Lin, C.-W., Fontana, G. D., and Tarolli, P.: The topographic signature of a major typhoon, Earth Surf. Proc. Land., 40, 1129–1136, 2015.

Van Niel, T. G., McVicar, T. R., Li, L., Gallant, J. C., and Yang, Q.: The impact of misregistration on SRTM and DEM image differences, Remote Sens. Environ., 112, 2430–2442, 2008.

Vaze, J., Teng, J., and Spencer, G.: Impact of DEM accuracy and resolution on topographic indices, Environ. Modell. Softw., 25, 1086–1098, 2010.

Wechsler, S. P.: Uncertainties associated with digital elevation models for hydrologic applications: a review, Hydrol. Earth Syst. Sci., 11, 1481–1500, doi:10.5194/hess-11-1481-2007, 2007.

Wessel, B.: TanDEM-X Ground Segment – DEM Products Specification Document, EOC, DLR, Oberpfaffenhofen, Germany, Public Document TD-GS-PS-0021, Issue 3.1, available at: https://tandemx-science.dlr.de/, last access: 1 December 2016.

Whipple, K. X. and Tucker, G. E.: Dynamics of the stream-power river incision model: Implications for height limits of mountain ranges, landscape response timescales, and research needs, J. Geophys. Res.-Sol. Ea., 104, 17661–17674, 1999.

Wiener, N.: Extrapolation, interpolation, and smoothing of stationary time series, vol. 7, MIT press, Cambridge, MA, USA, 1949.

Wilson, J. P.: Digital terrain modeling, Geomorphology, 137, 107–121, 2012.

Wobus, C., Whipple, K. X., Kirby, E., Snyder, N., Johnson, J., Spyropolou, K., Crosby, B., and Sheehan, D.: Tectonics from topography: procedures, promise, and pitfalls, in: Tectonics, Climate, and Landscape Evolution, edited by: Willett, S. D., Hovius, N., Brandon, M. T., and Fisher, D. M., Geological Society of America Special Paper, 398, 55–74, doi:10.1130/2006.2398(04), 2006.

Zhang, W. and Montgomery, D. R.: Digital elevation model grid size, landscape representation, and hydrologic simulations, Water Resour. Res., 30, 1019–1028, 1994.

11

Complex coastlines responding to climate change: do shoreline shapes reflect present forcing or "remember" the distant past?

Christopher W. Thomas[1], **A. Brad Murray**[2], **Andrew D. Ashton**[3], **Martin D. Hurst**[4], **Andrew K. A. P. Barkwith**[4], **and Michael A. Ellis**[4]

[1]British Geological Survey, Lyell Centre, Edinburgh, EH14 4AP, Scotland, UK
[2]Division of Earth and Ocean Sciences, Nicholas School of the Environment and Earth Sciences and Center for Nonlinear and Complex Systems, Duke University, Durham, North Carolina 27708, USA
[3]Department of Geology and Geophysics, Woods Hole Oceanographic Institution, Woods Hole, Massachusetts 02543, USA
[4]British Geological Survey, Nicker Hill, Keyworth, Nottingham, NG12 5GG, England, UK

Correspondence to: Christopher W. Thomas (cwt@bgs.ac.uk)

Abstract. A range of planform morphologies emerge along sandy coastlines as a function of offshore wave climate. It has been implicitly assumed that the morphological response time is rapid compared to the timescales of wave climate change, meaning that coastal morphologies simply reflect the extant wave climate. This assumption has been explored by focussing on the response of two distinctive morphological coastlines – flying spits and cuspate capes – to changing wave climates, using a coastline evolution model. Results indicate that antecedent conditions are important in determining the evolution of morphologies, and that sandy coastlines can demonstrate hysteresis behaviour. In particular, antecedent morphology is particularly important in the evolution of flying spits, with characteristic timescales of morphological adjustment on the order of centuries for large spits. Characteristic timescales vary with the square of aspect ratios of capes and spits; for spits, these timescales are an order of magnitude longer than for capes (centuries vs. decades). When wave climates change more slowly than the relevant characteristic timescales, coastlines are able to adjust in a quasi-equilibrium manner. Our results have important implications for the management of sandy coastlines where decisions may be implicitly and incorrectly based on the assumption that present-day coastlines are in equilibrium with current conditions.

1 Introduction

Many recent studies demonstrate how distinctive rhythmic planform coastline shapes on scales of kilometres to hundreds of kilometres (Fig. 1) can develop on wave-dominated sandy coasts due to morphodynamic interactions with wave climates (Ashton and Murray, 2006a, b; Ashton et al., 2001; Falqués and Calvete, 2005; Falqués et al., 2000; Kaergaard and Fredsoe, 2013a, b, c; van den Berg et al., 2012; Hurst et al., 2015; Nienhuis et al., 2013). Typically, previous studies assume (implicitly or explicitly) that coastline shapes chiefly reflect the contemporary wave climate (generally with specific regard to the angular distribution of wave angles), adjusting very rapidly to any changes. That is, intrinsic timescales for change in morphology are shorter than the timescales characterizing shifts in wave climate. Such rapid morphological change, relative to the rate of change in the wave climate, reflects a quasi- or dynamic-equilibrium morphological response to changing climate forcing: put simply, the rate of change in the morphology is much faster than the rate of change in the wave climate.

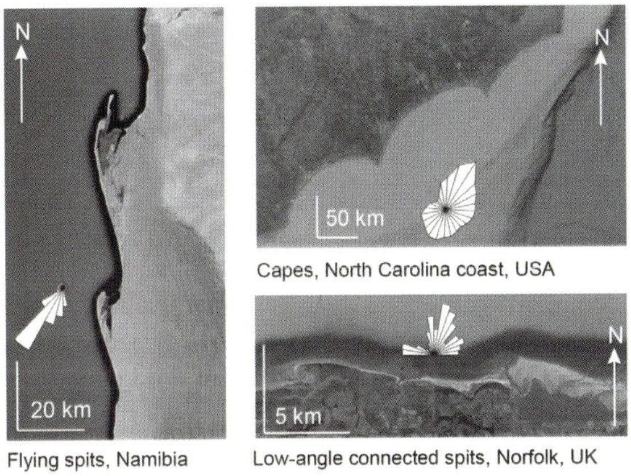

Figure 1. Diverse large-scale morphologies developed on open sandy coastlines under the influence of dominantly high-angle offshore waves ($U > 0.5$). Spits develop under an asymmetric distribution of wave angles ($A \sim 0.7$); nearly symmetric capes develop where the wave angles are distributed symmetrically ($A \sim 0.5$). Representative wave angle distributions are taken from wave buoy data available for these examples. Namibia: after Bosman and Joubert (2008). North Carolina capes, USA: WIS station 509, after Ashton and Murray (2006b). North Norfolk, UK: data from Blakeney Overfalls Waverider III buoy, Cefas WaveNet.

Changes in storm patterns in future climates will likely yield different wave climates (WASA Group, 1998; Hemer et al., 2013; Storlazzi et al., 2015; Wolf and Woolf, 2006) with concomitant changes in coastline behaviour. Moore et al. (2013) demonstrated temporal shifts in coastline shape on the Carolina capes caused by observed shifts in wave climate, supporting the prediction that even slight differences in wave climate can be associated with different morphologies (cf. Ashton and Murray, 2006a, b). Similarly, Allard et al. (2008) related spit growth and morphological change to wave climate variations on the Arçay Spit, France. Changes in wave climates will be manifest in altered patterns of shoreline change, including zones of enhanced erosion and/or accretion (Barkwith et al., 2014; McNamara et al., 2011; Slott et al., 2006). However, the assumption of quasi-equilibrium coastline change, with shoreline shape reflecting the present wave climate, has not been previously examined.

In an analogous study, Nienhuis et al. (2013) have shown how inherited delta morphology can influence the evolution of the coastline when sediment supply is perturbed: the shape of a delta can reflect not just present or recent sediment supply, but can depend strongly on the long-term history of sediment input. Might the influence of former morphology on coastline evolution also be true for sandy coastlines when wave climates are perturbed?

In this paper, we present the results of a study exploring the two key questions that arise from the above summary, specifically, the influence of former morphology of sandy coastlines on their subsequent evolution as wave climates change with regard to approaching wave angles (focussing on rates and spatial scales of response), and the degree to which coastlines can respond to changing wave climate in a quasi-equilibrium manner.

To address these questions, we have used the one-line Coastline Evolution Model (CEM; Ashton and Murray, 2006a) to examine (a) whether, and under what conditions, coastline shapes exhibit quasi-equilibrium responses to changes in the angular distribution of approaching waves; and (b) whether there are conditions under which coastlines might retain long-term memory of previous wave-climate forcing. Such hysteresis would greatly complicate forecasting and identification of coastline behaviours to be expected with changing wave climate.

We focussed on two end-member coastal morphologies, flying spits and cuspate capes (Fig. 1). These morphologies emerge as self-organized structures in response to high-angle wave instability that results when waves approach the coast at highly oblique angles, causing shoreline undulations to grow (Ashton et al., 2001; Falqués and Calvete, 2005). Their value from an academic perspective notwithstanding, our interest in these particular morphologies extends beyond their fascinating self-organized complexity. These morphologies are themselves important in many coastal regions because they often shelter fragile but highly diverse and dynamic shallow marine and estuarine ecosystems upon which human and animal communities depend, and also shelter socially and economically important coastal infrastructure. The fate of capes and spits under changing climate is thus of material concern to humanity and needs to be better understood to aid the development of appropriately informed coastal management policies.

In our experiments, we generated spit and cape coastlines using appropriate wave angle distributions (Ashton and Murray, 2006a). We then changed the wave angle distributions to be more diffusive in character, such that complex, high-amplitude perturbations (as exemplified here by spits and capes) are smoothed. Changes were made either instantaneously or gradually over a period of time. Outputs from the modelling were used to quantify how coastal morphologies responded to these changes.

In this initial investigation, we have focussed specifically on changes in the distributions of approaching wave angle, as these distributions fundamentally control gross coastal morphology at and above kilometre scales. Changes in other wave properties (height, period) control the rates at which changes occur, so are important in coastline evolution. However, we wished only to study the effects of changing wave angle distribution. The decision to move linearly from an unstable to diffusive wave angle distribution was motivated by wishing to understand the degree of stability of complex morphologies under changing wave angle distributions, hence the simplicity of our experiments. More complex experiments involving nonlinear or oscillating changes in wave climate were not warranted at this stage, but could be explored in future work. The timescales for change in those experiments where the wave angle distribution was changed gradually were to some extent arbitrary, but guided by timescales over which climate is known to be changing (decades–centuries) under the influence of anthropogenic activity.

2 Background and methods

2.1 Alongshore gradients and shoreline planform change

Gradients in alongshore sediment flux, generated by wave-driven currents, cause shoreline erosion or progradation on sandy coastlines. Assuming conservation of mass in the shoreface, the temporal change in coastline location in relation to flux gradients is given by

$$\frac{\partial y}{\partial t} = -\frac{1}{D_{\mathrm{sf}}}\frac{\partial Q_{\mathrm{s}}}{\partial x}, \tag{1}$$

where y (L) is the cross-shore position at long-shore position x (L) at time t, and Q_{s} ($\mathrm{L}^3\,\mathrm{T}^{-1}$) is the alongshore sediment flux; D_{sf} (L) is shoreface depth – that over which erosion or accretion extends (Ashton and Murray, 2006a).

Critically, Q_{s} is chiefly dependent on the angle between offshore wave crests and the shoreline (Ashton et al., 2001; Ashton and Murray, 2006a, b, Fig. 2b; Falqués and Calvete, 2005; see Fig. 2a), here expressed in terms of the commonly used CERC equation (Komar, 1971, 1998; Ashton and Murray, 2006a), but recast in terms of offshore waves (Ashton and Murray, 2006a):

$$Q_{\mathrm{s}} = K_2 H_0^{\frac{12}{5}} T^{\frac{1}{5}} \cos^{\frac{6}{5}}(\varnothing_0 - \theta)\sin(\varnothing_0 - \theta). \tag{2}$$

K_2 ($\mathrm{m}^{3/5}\,\mathrm{s}^{-6/5}$) is a constant, H_0 is *significant* offshore wave height in metres and T is wave period in seconds; $\varnothing_0$ is the offshore wave angle (representing waves at the offshore limit of approximately shore-parallel contours), and θ the local coastline angle (both in degrees), both relative to some general coastline orientation.

The angle dependency of Q_{s} is such that sediment flux is maximized at offshore wave approach angles of $\sim 45°$ with respect to the shoreline. Waves approaching from offshore at angles $< 45°$ ("low-angle" waves) will smooth a coastline, such that a straight coastline is in a stable equilibrium. Waves approaching at angles of $45°$ or greater ("high-angle" waves) induce instability along a coastline and protuberances tend to grow. How coastline morphology evolves depends on the relative degrees of influence of high-angle and low-angle waves in the wave climate, as well as the degree of asymmetry in the wave angle distribution (Ashton and Murray, 2006a).

2.2 Net sediment flux and shoreline instability and diffusivity

For a nearly straight coastline, coastline evolution can be described by a linear diffusion equation, where the diffusivity is either positive (stable) or negative (unstable) (Ashton and Murray, 2006a, their Eq. 8). Every approaching wave condition contributes to sediment transport and the consequent evolution of the coastline. The overall effect of a wave climate on a coastline can be determined from the net diffusivity, μ_{net} ($\mathrm{m\,s}^{-2}$), calculated from the sum of the individual diffusivities induced by each wave condition (analogous to the net alongshore sediment flux), and from a dimensionless "stability index", Γ, that measures the competition between stability and instability (Ashton and Murray, 2006b). We use these indices to quantify the behaviour and state of a coastline under the influence of a particular wave climate.

For each wave at each location along a coastline, individual alongshore sediment flux values are calculated using Eq. (2), and individual diffusivity (μ) values are calculated with respect to the local coastline orientation using the shoreline diffusivity equation obtained by Ashton and Murray (2006a, their Eq. 8; 2006b):

$$\mu = \frac{K_2}{D} T^{\frac{1}{5}} H_0^{\frac{12}{5}} \left(\cos^{\frac{1}{5}}(\varnothing_0 - \theta)\right. \tag{3}$$
$$\left.\left[\left(\frac{6}{5}\right)\sin^2(\varnothing_0 - \theta) - \cos^2(\varnothing_0 - \theta)\right]\right).$$

Summing over individual fluxes and diffusivities at each coastline location over n wave events gives the net flux and diffusivity, respectively:

$$Q_{\mathrm{s,net}} = \sum_{i=1}^{n} Q_{\mathrm{s},i}\Delta t_i \Big/ \sum_{i=1}^{n}\Delta t_i \tag{4}$$

$$\mu_{\mathrm{net}} = \sum_{i=1}^{n} \mu_i \Delta t_i \Big/ \sum_{i=1}^{n}\Delta t_i \tag{5}$$

(Ashton and Murray, 2006b, their Eq. 5). The "stability index", Γ, is given by

$$\Gamma = \sum_{i=1}^{n} \mu_i \Delta t_i \Big/ \sum_{i=1}^{n} |\mu_i|\Delta t_i, \tag{6}$$

where Δt_i (s) is the time step (Ashton and Murray, 2006b, their Eq. 6).

Γ ranges between 1 for a fully low-angle wave climate and -1 for a fully high-angle climate. Mapping Γ along a coastline for different wave climates elucidates the nature of the response of a coastline to those wave climates (cf. Ashton and Murray, 2006b). We have used calculations of Γ to quantify the behaviour of coastlines characterized by capes and spits that have grown under an anti-diffusive wave climate in contrast to those formed under a different, more diffusive wave climate. We subsequently investigated how μ and Γ changed during the transition from anti-diffusive to diffusive wave climates in order to reveal how domains of erosion and deposition would change along a coast with changing wave climate.

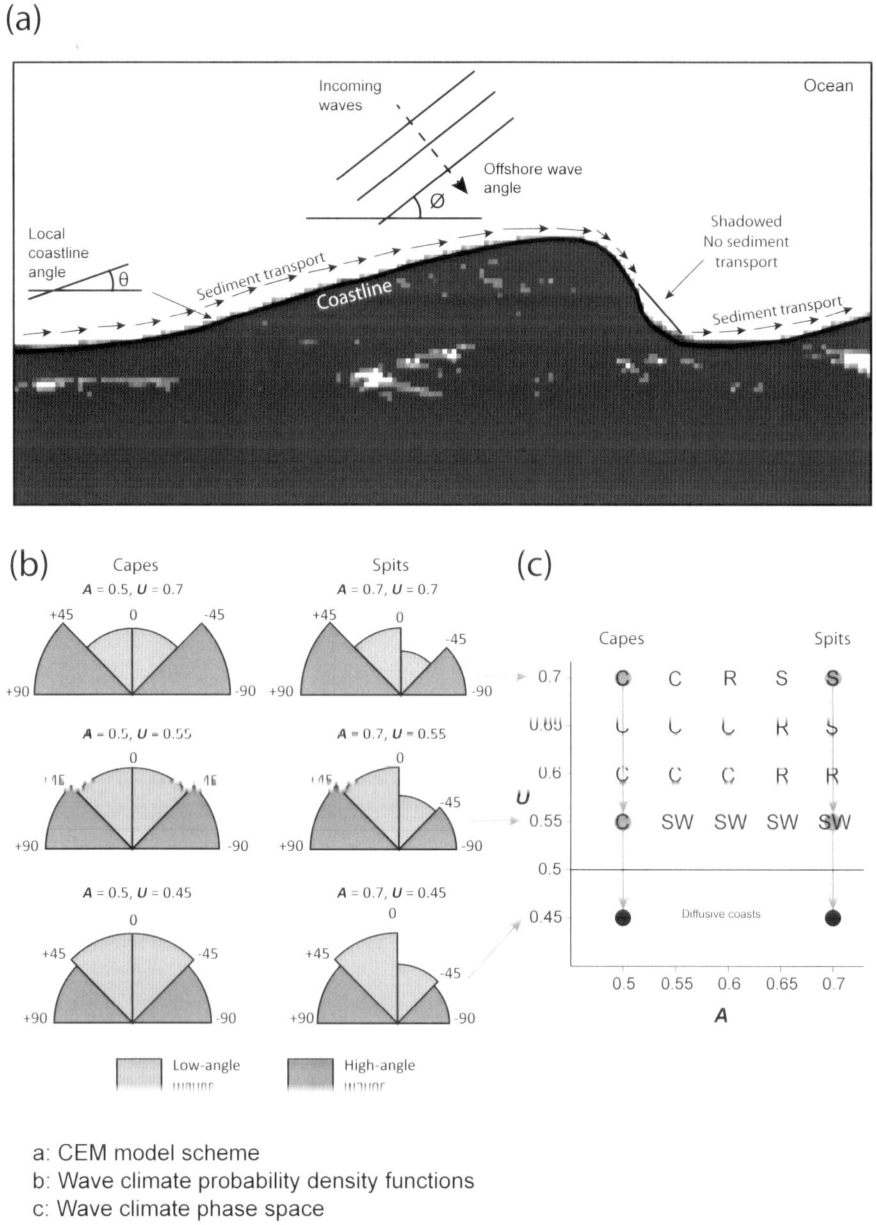

Figure 2. (a) The CEM model scheme. The coastline is discretized into cells. Local shoreline angles, θ, at each cell are calculated according to the schema outlined by Ashton and Murray (2006a). Sediment transport is determined by the effective wave angle ($Ø$–θ). Arrows indicate net flux direction under waves incoming from the left; arrow lengths qualitatively indicate the flux. Sand is not transported through cells which are in shadow for a particular wave. Barrier width is maintained by a simple overwash scheme described by Ashton and Murray (2006a). **(b)** Rose diagrams of the four-bin probability density functions for wave climates used in the modelling, based on parameters U and A. **(c)** Wave climate trajectories used in the simulations defined by the U and A phase space, after Ashton and Murray (2006a). Wave climate asymmetry is defined by A; U defines the proportion of offshore high-angle waves ($> 45°$). C: cuspate capes; R: recurved, connected spits; S: flying spits; SW: large-scale sand waves.

2.3 Coastal evolution modelling

2.3.1 Coastline planform evolution

The CEM uses Eqs. (1) and (2) to explore coastline planform behaviour numerically. The details of the theory and its implementation in the CEM are discussed extensively elsewhere (Ashton et al., 2001; Ashton and Murray, 2006a, b), so here we summarize the model domain and setup. In this study, the model domain consisted of 2000×300 cells alongshore and cross-shore respectively; each cell is 100×100 m. The model was driven by synthetic offshore wave angle distributions, as described in Sect. 3.1.2 below. Model runs used daily time steps, with a new offshore wave angle being chosen at random from the wave angle distribution at each model day.

Wave angle distributions define the relative influences on alongshore transport of all the waves approaching from angles falling within each angle bin (Ashton and Murray, 2006b). Observed (or hindcast) wave records can be transformed into such angular distributions, additionally weighting each wave by height and period ($H_0^{\frac{12}{5}} T^{\frac{1}{5}}$; Eq. 2), and summing all the wave occurrences in each bin (Ashton and Murray, 2006b). However, in the model experiments reported here, we used simplified synthetic wave distributions. These generic distributions could represent either systematic differences in wave height as a function of approach angle, or in the frequency with which waves approach from different angles (or a combination of the two). The angular distribution of wave influences on alongshore transport determines the emergent dynamic-equilibrium coastline morphology (Ashton and Murray, 2006b); we discuss a relevant measure of dynamic equilibrium below. Rhythmic coastline features retain a self-similar shape under a constant wave climate, even though the scale of the shape increases slowly through time (Ashton and Murray, 2006a; Ashton et al., 2001). The effective average wave height (Ashton and Murray, 2006b) and period only influence the timescales for coastline development.

Because the principle interest of this study is in the effect of changing wave angle distribution on planform change, the effects of variations in wave period T and height H_0 can be folded into angular distributions of wave influences on alongshore sediment transport (Ashton and Murray, 2006b); in this study, T and H_0 were fixed at 10 s and 1 m in all runs.

Using linear wave theory, each offshore wave is refracted progressively over shore-parallel contours until depth-limited breaking occurs (e.g. see Hurst et al., 2015, Appendix A). At this point, the standard breaking-wave version of the CERC equation (e.g. Komar, 1998) is used to calculate the sediment flux as a function of the angle between the locally determined coastline orientation (Fig. 2a) and the wave angle at breaking and the breaking wave height. The coastline position is evolved based on the calculated gradients in flux, assuming conservation of mass in the shoreface (Eq. 1).

2.3.2 Coastline instability and diffusivity

Net flux and diffusivity data generated by the CEM can be used to explore the sensitivity of a coastline to change under existing and modified wave climates, and the processes by which any change would occur, as discussed in Sect. 2.2. In this study, the CEM was used to capture the *potential* net flux and diffusivity data for particular wave climates operating on a coastline at particular points in time. To do this, spits and capes were grown for a specific time using wave climates with appropriate U and A values, as described below in Sect. 3.1.2. Coastline evolution was then paused and 10 000 sample wave angles drawn randomly from the selected wave climate distribution were run over the extant coastline. The average potential net flux and net diffusivity can then be calculated at each location (cell) along the coastline. This characterizes the *potential* change in the response (either unstable or diffusive) of the coastline to that wave climate, at any point along its length, and given its current morphological state

3 Experiments with changing wave climate

3.1 Experimental design

3.1.1 Instantaneous and gradual wave climate change

We set up experiments by growing either flying spits or cuspate capes ("capes" from here on) from an initially straight coast (with small white noise perturbations) over an initial fixed period of time. In most experiments, this initial period was 250 model years. This time frame allows these morphologies to attain length scales commensurate with those observed along real coastlines. We then subjected these model coastlines either to a gradual change in wave climate, or an instantaneous change. In experiments with gradual wave climate change, the initial wave climate was evolved linearly towards the new state over an arbitrary period of 100 years. These experiments were used to explore the influence of pre-existing morphology on the nature and rate of response of a coastline to changing wave climate.

In experiments involving instantaneous change, the wave climate is transformed to the target state immediately following the period of initial growth; in these experiments, we also used initial periods of 50 and 125 years to provide additional data that we could use to determine characteristic timescales for change. This allowed us to explore the possibility of scaling relationships between time and the rate of change of length scales, and the degree to which a quasi-equilibrium response in morphology is possible for given rates of wave climate change.

In both cases, coastline morphology is in dynamic equilibrium with the initial wave climate just before the wave-climate transition begins. As the wave climate changes, the coastline progressively approaches a new morphological state, settling into dynamic equilibrium with the final wave climate. We characterized coastline morphology using the aspect ratio (cross-shore extent / alongshore wavelength) of coastline features: previous work has shown that for equilibrium coastline shapes, aspect ratio varies with wave climate (Ashton and Murray, 2006a). As a coastline continues to evolve under a constant wave climate, the scale of coastline features grows (Ashton and Murray, 2006a; Ashton et al., 2001). However, aspect ratio remains constant even as length scales increase (Ashton and Murray, 2006a). Hence,

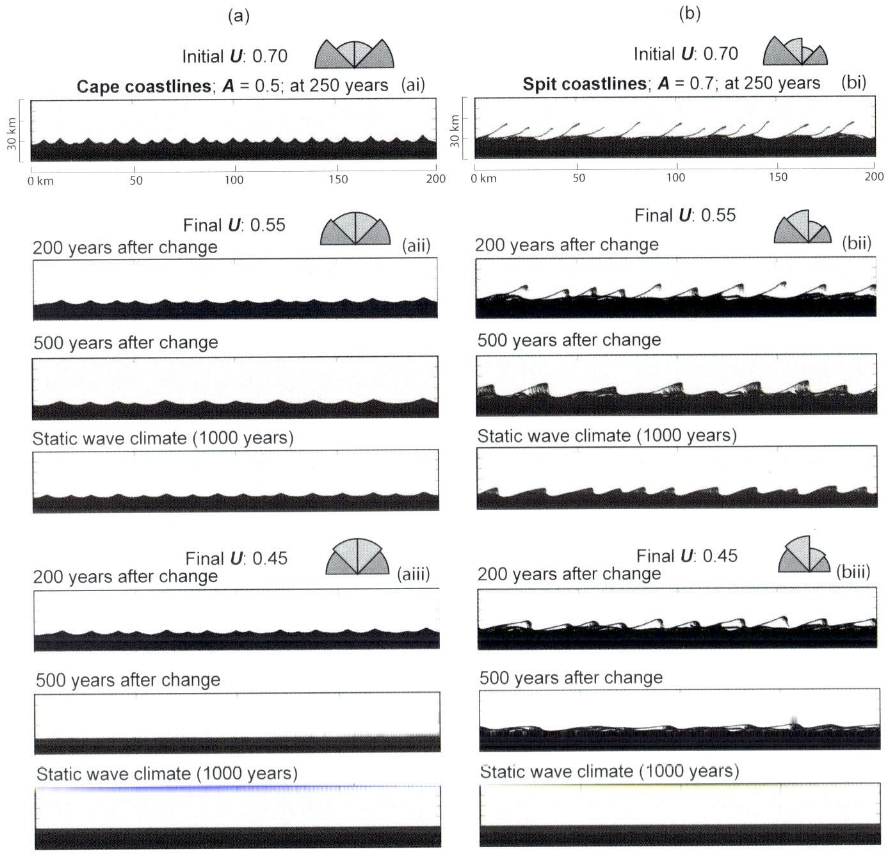

Figure 3. Snapshots of simulated coastline morphologies evolved under changing wave climate. See Section 3.1.3 onwards for discussion. **(ai–iii)** Capes evolved under a symmetric wave climate. **(bi–iii)** Spits evolved under an asymmetric wave climate. Horizontal and vertical scales are the same for all panels, representing a domain 30 × 200 km. **(ai, bi)** Coastlines evolved for 250 years under initial conditions of $U = 0.7$ for $A = 0.5$ (capes) and 0.7 (spits). **(aii, bii)** The U values of the changed wave climate show the coastline morphologies evolved 200 and 500 years *after* the wave climate is changed at 250 years, and the morphologies evolved over 1000 years under static wave climates with the same U; A values as the changed wave climate (note that in these examples, the wave climate is changed gradually over 100 years). The greater length scales in the morphologies of the coastlines altered by the changed wave climates are clear. The persistence of cross-shore amplitude for a spit coastline evolving under U of 0.45 is particularly evident. Model scales are the same in all panels. PDFs of the wave climates are included for comparison.

aspect ratio is an appropriate measure reflecting the degree of dynamic equilibration with respect to a particular wave climate.

3.1.2 Experimental wave climates

Model simulations were driven by wave approach angles drawn from a probability density function (PDF) defined by two parameters (Ashton and Murray, 2006a): U, the fraction of waves approaching from angles $> 45°$ (representing the fraction of wave influences on alongshore transport from these angles) (Fig. 2b) and A, the fraction of waves approaching from the left (CEM convention); the wave climate is asymmetric when $A > 0.5$. When $U > 0.5$, the model coastline experiences instability and perturbations will grow. The PDFs used in our modelling are shown in Fig. 2b.

Ashton and Murray (2006a, their Fig. 9) mapped different coastline shapes that emerge for different values of U and A. We have explored two trajectories across the (U, A) parameter space in our experiments, one for capes and one for spits (Fig. 2c). In both cases, the trajectory is towards a diffusive wave climate, under which perturbations are smoothed.

3.1.3 Experimental conditions

Model capes are generated over 250 years with U at 0.7, and A at 0.5 (Fig. 3a); flying spits are generated with U at 0.7 and with A set to 0.7 (Fig. 3bi). Subsequently, U is changed from 0.7 to 0.55 (moderately anti-diffusive), or to 0.45 (diffusive) while holding A constant (0.5 for capes, 0.7 for spits); U is changed either instantaneously, or gradually over 100 years. The new wave climate is then held constant for a further 650 years. Total model run times are 900 years for models in which the wave climate is changed instantaneously, and 1000 years for models in which the wave climate is changed gradually over 100 years. In addition, to investigate how response times vary with the spatial scale of the features, in separate experiments we generated capes and spits over 50 and 125 years, followed by instantaneous change in U to 0.45

4 Results

4.1 Changes in morphology under gradual wave climate change

Examples of the changes that occur in the planform morphology of our experimental coastlines during the model runs are shown in Fig. 3; note that these data are for experiments in which the wave climate is changed gradually over 100 years from the initial $U = 0.7$ to $U = 0.55$ or 0.45.

In Figs. 4 and 5 we plot the evolution of aspect ratio, wavelength and amplitude of coastal features (capes and/or spits) during the experiments. The data have been smoothed using a seven-point moving average window to aid clarity. Animations of the model simulations from which the results discussed in this study were derived are also included in the Supplement.

4.1.1 Capes

Cape morphology adjusts mostly in the 100-year period over which the wave climate changes (Fig. 4ai). Cape amplitude declines through erosion of the cape tips (Fig. 3aii, first and second panels). Rapidly declining aspect ratio (mean amplitude / mean wavelength) indicates quasi-equilibrium adjustment in amplitude, approaching that expected for $U = 0.55$ (Fig. 3aii, second and third panels; Fig. 4ai). For a diffusive wave climate the coast eventually becomes smooth (Figs. 3aiii; 4ai, iii).

4.1.2 Spits

Flying spits change much more slowly, exhibiting pronounced long-term memory (Fig. 4bi). The coast shape in dynamic equilibrium with $U = 0.55$ features small, relatively low-amplitude reconnected sand waves (Fig. 3bii, third panel; see Figs. 1c, 2c). However, in the experiment in which U changes from 0.7 to 0.55, the coastline morphology differs with regard to both shape and scale from that expected at $U = 0.55$ (Fig. 3bii, second and third panels), even several centuries after the wave-climate transition ends. More strikingly, when the wave climate becomes diffusive ($U = 0.45$), the experimental coastline retains significant undulations even after 750 years (twice the time it took to grow the flying spits) (see Fig. 3bii, second and third panels). Figure 4bi–iii show most of the adjustment occurs after the transition period. The aspect ratio converges toward the dynamic equilibrium value, but only over a timescale of many centuries.

4.1.3 Resultant length scales

The results show that for both capes and spits, the scale of both wavelength and amplitude of coastline features is larger than would be expected had the coastlines formed under the final wave climates ($U = 0.55$, 0.45; Figs. 3aii, bii; 4aii, iii, bii, iii). Under anti-diffusive wave climates, this scale "memory" is a permanent result of the path through wave-climate space. Shape "memory", on the other hand, fades over timescales that differ for capes and spits: spits retain a memory of shape over many centuries, capes only over a few decades (Fig. 4ai, bi).

4.2 Instantaneous wave climate change: characteristic timescales and temporal–spatial scaling

To derive characteristic timescales for change in morphology and to examine the relationship between temporal and spatial scales, further model experiments were run with initial wave climate conditions lasting 50, 125 and 250 years, followed by an instantaneous change in wave climate. The rates of morphological change in capes and spits are shown in Fig. 5a and b respectively, with reference to initial conditions run for 250 years.

Using the change in aspect ratio as the metric, we determined characteristic timescales for morphological change (i.e. first and second e-folding scales, $1/e$ and $1/e^2$) from the point at which the wave climate is changed to become diffusive ($U = 0.45$). For capes, first and second e-folding times are approximately 20 and 80 years (Table S1 in the Supplement), respectively; for spits they are approximately 90 and 320 years (Table S2). Comparing characteristic timescales of morphological change in aspect ratio for capes and spits generated over 50, 125 and 250 years, we find that diffusive scaling pertains, with timescale varying as the square of the characteristic wavelength (Tables S1, S2).

5 Discussion

5.1 Physical mechanisms of coastline morphological adjustment

Based on the results from the instantaneous change experiments, we can distinguish two modes in which coastlines can adjust to changing wave climate, exemplified by the cape and spit experiments, respectively: cape morphologies adjust to a zero net flux condition, whilst spit morphologies adjust to a condition in which there is a constant down-drift translation of the feature.

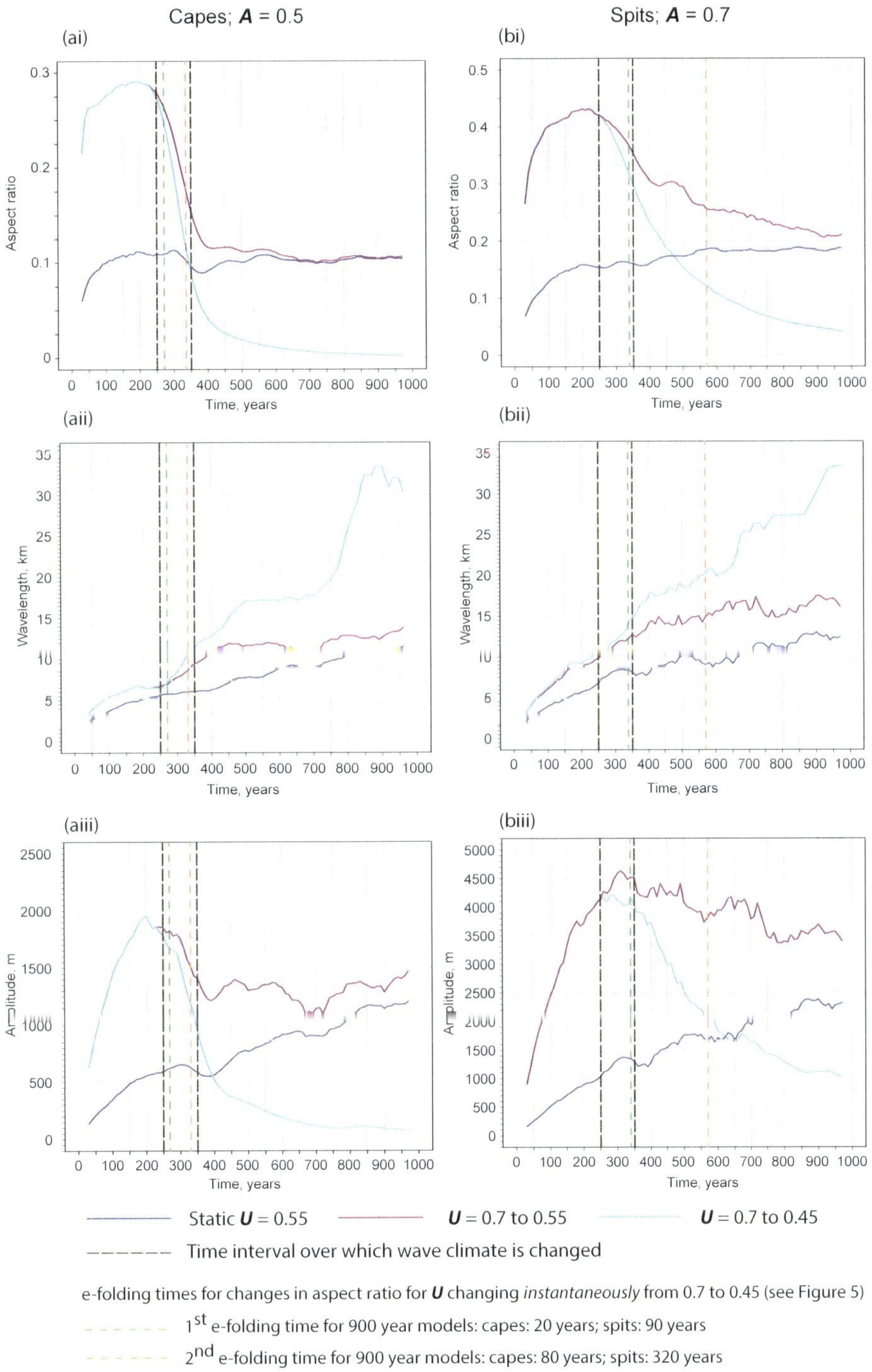

Figure 4. Time series of the evolution of aspect ratio (amplitude / wavelength), wavelength and amplitude for **(ai–iii)** capes and **(bi–iii)** spits under 100 years of *gradual* change in wave climate, from 250 years. In all panels, the dark blue line: results for the coastline evolved under a static wave climate with $U = 0.55$ for both spits and capes; light blue line: progressive change in U from 0.7 to 0.45 over 100 years; and purple line: progressive change in U from 0.7 to 0.55 over 100 years. The *e*-folding times are shown graphically for 900-year models (see Fig. 5). These relate *specifically* to change from $U = 0.7$ to $U = 0.45$ based on change in aspect ratio and are shown to provide additional context for timescales of gradual change. See text and Supplement. Note: data are smoothed by a seven-point moving average to aid clarity. The indication of change in morphology slightly before 250 years is an artefact of this smoothing.

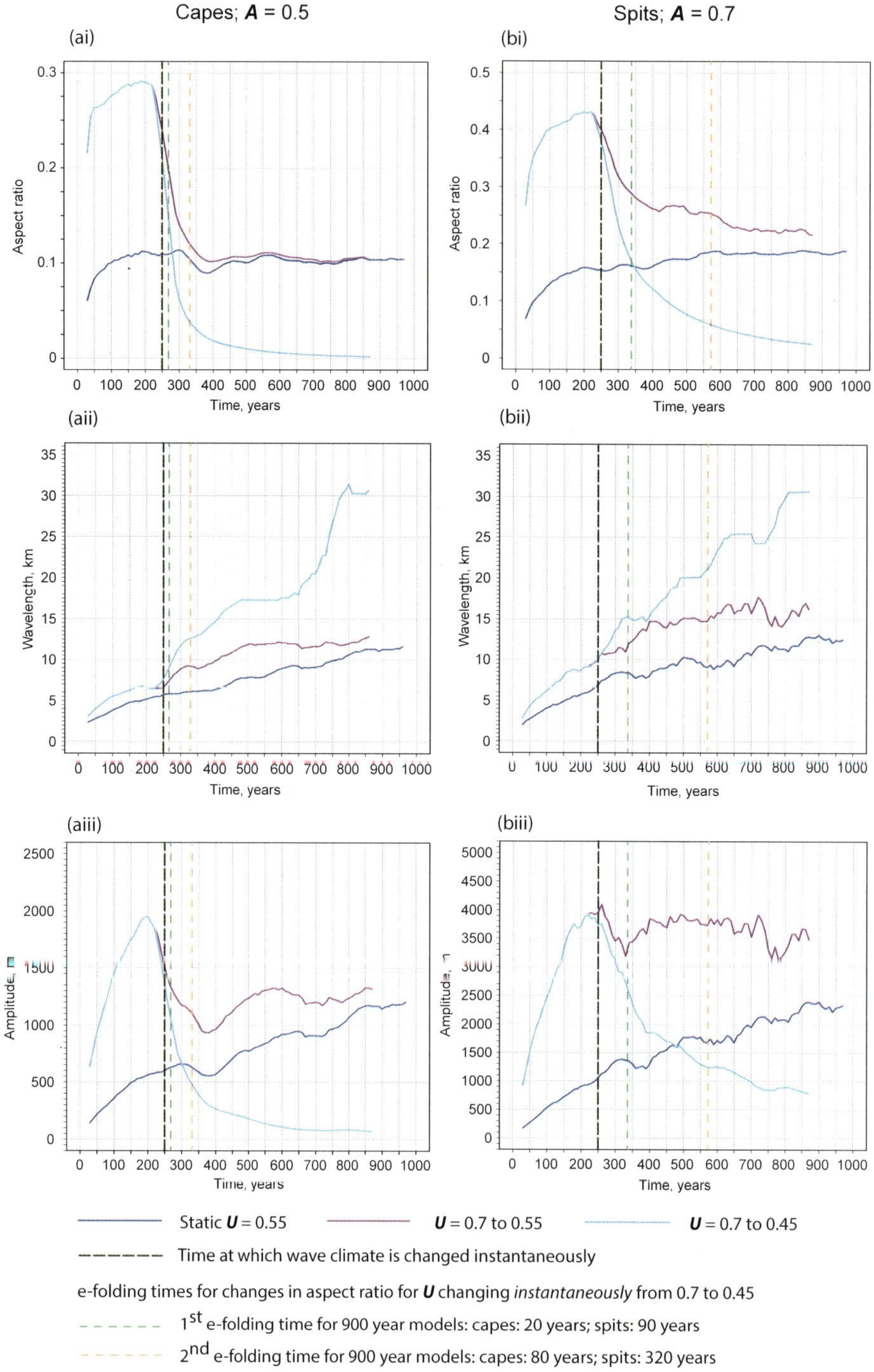

Figure 5. Time series of the evolution of aspect ratio (amplitude / wavelength), wavelength and amplitude for **(ai–iii)** capes and **(bi–iii)** spits under *instantaneous* change in wave climate at 250 years (900-year models). In all panels, dark blue line: the results for the coastline evolved under a static wave climate with $U = 0.55$ for both spits and capes; light blue line: U changing from 0.7 to 0.45; and purple line: U changing from 0.7 to 0.55. The e-folding times are shown graphically for the 900-year models. These relate *specifically* to change from $U = 0.7$ to $U = 0.45$, based on change in aspect ratio. See text and Supplement. Note: data are smoothed by a seven-point moving average to aid clarity; see Fig. 4.

5.1.1 Mechanisms of cape morphology adjustment

To understand how capes adjust, we begin by considering capes in dynamic equilibrium with a symmetric wave climate dominated by high-angle waves ($U > 0.5$; Figs. 2b, 3ai). Cape tips feel the full distribution of wave angles in the regional climate. In contrast, cape flanks and inter-cape bays are affected by wave climates that differ from the regional wave climate because of shadowing by adjacent capes (Ashton and Murray, 2006b; Murray and Ashton, 2013). The shadowing shelters the flanks from the higher angle waves in the distribution, the degree of shadowing depending on location in the inter-cape bays.

For a dynamic equilibrium to develop under a symmetric wave climate – a state in which the coastline adjusts very rapidly to small changes in wave climate at short temporal and small spatial scales – all local coastlines must adjust to orientations that produce little or no net sediment flux under the local wave climate. As high-angle waves become more dominant in the wave climate, the local coastlines adjusted to zero net flux lie at progressively greater angles, relative to the regional coastline orientation, towards the cape tips. Consequently, cape amplitudes increase with larger U. This behaviour is evident in the *potential* net flux ($Q_{s,net}$) and stability index (Γ) data calculated to explore their spatial variation along a coastline.

Examples of these characteristics are shown in Fig. 6a, c for a section of coastline with two sample capes. Figure 6a shows a cape pair from part of a model cape coastline grown for 250 years under a symmetric wave climate with $U = 0.7$. Dynamic equilibrium is indicated by the near-zero net flux condition across most of the inter-cape bay (Fig. 6a). In addition, both the net flux and stability index data (Fig. 6a, c) show that only the cape tips experience the anti-diffusive effects of the full wave climate, whereas the flanks and the bay experience the diffusive effects of the variably shadowed local wave climate.

As the wave climate shifts to one dominated by waves approaching more directly onshore (e.g. $U = 0.45$; Figs. 2b, 3aii, iii), orientations previously adjusted to the zero net flux condition become subject to significant change in net sediment fluxes, which are directed away from the cape tips (Fig. 6a). The resulting strong gradients in these fluxes cause cape tips to erode and bays to prograde. The fluxes are proportional to the maximum net flux divided by the alongshore length scale for the cape tip (some small fraction of the total cape wavelength). The change is strongly apparent in the stability index data (Fig. 6c; blue line). The inter-cape bay is no longer shadowed with respect to the dominant wave directions and becomes relatively less stable under the new wave climate. The very tips of the capes behave more diffusively but stability falls rapidly along the cape flanks.

5.1.2 Mechanisms of spit morphology adjustment

For spits, both the mode of emergence of the steady-state pattern and the mode of subsequent adjustment under a changing wave climate are more complicated. For an asymmetric wave climate, under which there is translation of finite-amplitude coastline features (Ashton and Murray, 2006a), a dynamic equilibrium is indicated by constant alongshore translation of the spits.

The tips of spits tend to propagate in the direction parallel to the shoreline orientation that produces the maximum sediment flux for the given regional wave climate (Ashton and Murray, 2006a; Ashton et al., 2016). However, the flanks and tips of spits also experience alongshore translation caused by erosion at their updrift ends. The updrift portion of each spit is shadowed by its updrift neighbour. Given that flying spit growth requires a wave climate in which the dominant wave angles are high and from the updrift direction, this shadowing is greatest at the updrift end of each spit. The shadowing decreases progressively downdrift, the spit coastline becoming more exposed to the waves approaching from the dominant direction. Gradients in net sediment flux arising from down-drift decline in wave shadowing effects tend to produce erosion, resulting in seaward concavity in the coastline (Fig. 6b, d). However, the development of concavity is limited because the increasing curvature of the coastline tends to result in accretion as the change in coastline geometry interacts with the locally experienced wave climate. Thus, the erosion tends to be balanced by coastline flattening induced by the local wave climates that result from wave shadowing. The concavity tends to be flattened and the locus of erosion propagates downdrift to the spit flank and tip. As long as different parts of the spit translate alongshore at different rates, the shape evolves (Ashton et al., 2016). However, when the shape and the gradients in net sediment flux adjust such that each segment of the spit translates alongshore at the same rate (given by the local cross-shore erosion rate / local coastline angle), the shape becomes persistent in a (dynamic) steady state; this steady state is characterized by episodic spit loss and subsequent rearrangements of the remaining spits.

When the wave climate shifts to one still asymmetric in character, but with an increased proportion of low-angle waves (e.g. see wave rose PDFs in Fig. 3 for $U = 0.55$ and 0.45), the net sediment flux will still be directed toward the tip of the spit (cf. capes where the flux is directed away from the tip). Thus, rather than contributing directly to a reduction in aspect ratio, the positive net flux continues to be captured at the spit tip. However, because the shoreline orientation experiencing the maximum net flux is different after the change in wave climate, the spit tip propagates in a new net direction. The degree of offshore propagation is reduced (indeed, it can be directed onshore when $U < 0.5$) and the alongshore component increases. In addition, shadow zones caused by the updrift neighbouring spits shift in location and intensity as spit morphologies change. The combined changes in flux

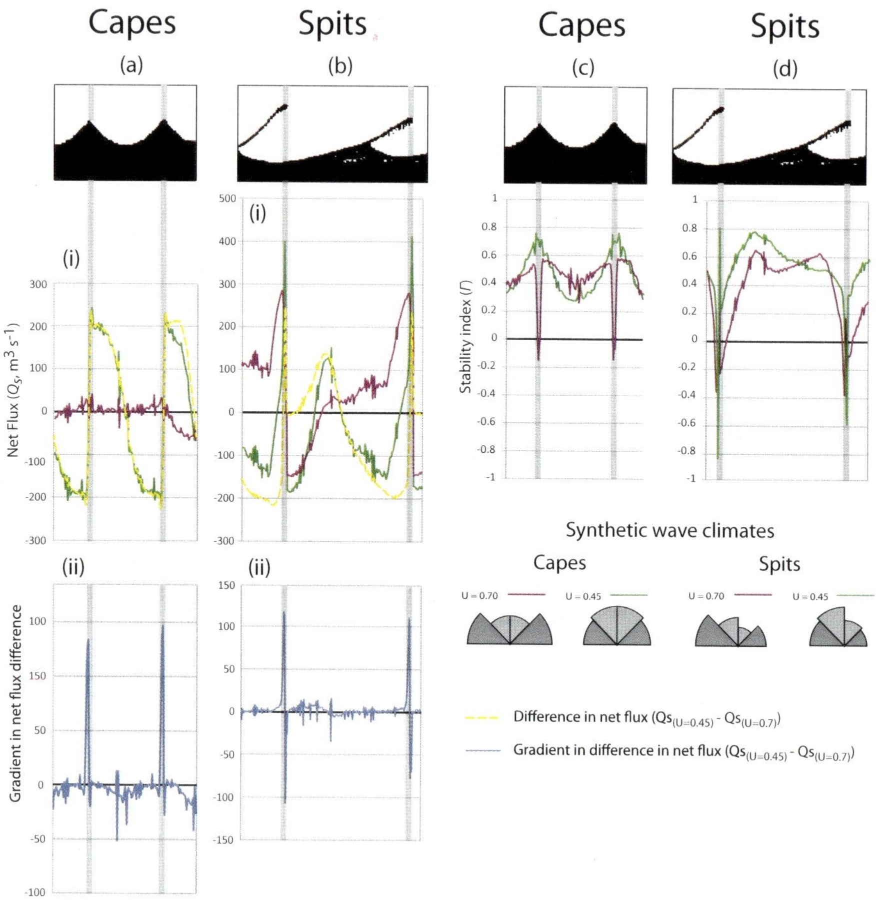

Figure 6. Detailed mapping of potential net flux and stability index across example coastline features for different wave climates impacting on static (unevolving) coastlines. Net flux data for **(ai)** capes and **(bi)** spits; blue line: gradients in the *difference* between the net fluxes for **(aii)** capes and **(bii)** spits. Stability index data for **(c)** capes and **(d)** spits. Purple line: $U = 0.7$; green line: $U = 0.45$; dashed yellow line: difference in potential net flux. Rose diagrams of the wave climate PDFs are included for comparison.

and transport direction at each end of a spit result in flux gradients that cause the spit shape to shift gradually to a more shore-parallel orientation, in the manner described above.

Thus, for flying spits, the flux gradients scale approximately with the maximum net sediment flux divided by the 23 total spit wavelength, rather than the small fraction of cape wavelength represented by cape tips. These two modes of adjustment – one in which gradients in net sediment flux occur over some small fraction of the wavelength, and one in which they occur over the whole wavelength – do not apply exclusively to capes and spits however. Once an adjusting flying spit reconnects with the down-drift coastline (Fig. 3bii, second panel), gradients in sediment flux occurring over a small fraction of the wavelength begin to exert an influence (Nienhuis et al., 2013). Furthermore, capes formed by slightly asymmetric wave climates (Ashton and Murray, 2006a), which produce a net alongshore sediment flux, also migrate alongshore and are, therefore, also subject to the constant alongshore-translation condition.

5.2 Limitations and implications

The experiments reported here involve only two types of coastline morphology and two types of wave climate change. This limited exploration motivates a wider, more systematic investigation of the responses of a broad range of morphologies to changes in wave climate. Although beyond the scope of this initial study, experiments like those depicted in Fig. 2c could be conducted for morphologies within the (U, A) phase space (Ashton and Murray, 2006a, their Fig. 9a), with different trajectories through that space. However, even the initial results presented here show that (explicit or implicit) assumptions common in previous analyses of coastline shapes (Ashton and Murray, 2006b; Kaergaard and Fredsoe, 2013b; Moore et al., 2013; Ribas et al., 2013; van den Berg et al., 2012; Idier and Falqués, 2014), or adjustments to wave climate change (Barkwith et al., 2014; McNamara et al., 2011; Slott et al., 2006) could be wrong. Coastline morphology should not be assumed to be in quasi-equilibrium with the current forcing; rather it could instead represent a legacy of past forcing conditions, possibly from many centuries ago.

The results reported here provide *guidelines* for critical timescales of wave climate change below which either capes or spits would fail to respond in a quasi-equilibrium fashion. These set limits on the mode of coastline adjustment that will occur when wave climate changes occur over timescales shorter than the characteristic timescale for morphological adjustment and/or the shoreline response will exhibit significant morphologic memory. In contrast, longer timescales for change permit a quasi-equilibrium response, in which morphological adjustment keeps pace with the change in wave climate. This scaling makes it possible to extrapolate the temporal limits on equilibrium adjustment described above, permitting distinction between quasi-equilibrium and long-term memory response of coastline features at arbitrary spatial scales.

These critical timescales depend not only on coastline morphology and the scale of coastline features (see also Hurst et al., 2015), but also on the characteristic wave heights and shoreface depths, which influence rates of coastline change in quantifiable ways. The timescales predicted by model experiments, such as those we present here, will be altered quantitatively when different wave heights, shoreface depths or alongshore-sediment flux relationships (or, indeed, empirical coefficient values) are used. However, such quantitative changes will not affect the relevance of comparing timescales for morphological response and wave climate change to understand or forecast whether coastlines will exhibit a quasi-equilibrium response or be influenced by a "memory" induced by preceding morphological states.

These results also have implications for management of potentially fragile sandy coasts. Management policies and plans are commonly underpinned by predictions of future shoreline erosion (or accretion) rates along developed coastlines that are based on shoreline change observed over previous (usually very few) decades. Observations accumulated over such relatively short timescales may not be sufficient to discern the true direction of morphological change, since these are of roughly the same order as the characteristic timescales indicative of limits to the potential for equilibrium morphological response, as calculated above. Furthermore, change in environmental drivers (weather patterns, storm frequency, etc.) may be poorly understood. Thus, wider analysis of environmental conditions and coastline response is likely needed for more informed decision making. Indeed, coastlines deemed to be under threat from climate change effects, and therefore requiring socio-economic as well as environmental management, should benefit from long-term monitoring of weather and geomorphology to understand what kind of intervention might be necessary, and to help preclude costly, non-beneficial or even damaging actions.

Our analysis has purposefully not considered changes in cross-shore sediment flux resulting from erosion related to sea-level rise (e.g. Moore et al., 2010; Wolinsky and Murray, 2009). However, the changes in coastline shape we have addressed can be superimposed on shoreline change associated with sea-level rise (e.g. Bruun, 1962; Walkden and Hall, 2005, 2011). Increases in the rate of sea-level rise generally cause increases in shoreline erosion, but the response to sea-level rise is approximately uniform along a sandy coastline at the scales of interest (e.g. Moore et al., 2010). In contrast, planform changes in coastline position arising from alongshore gradients in sediment flux indicate heterogeneous erosion and accretion along a shore: spatial patterns and magnitudes of accentuated shoreline erosion could be different in the future, compared to the recent past. This could result from quasi-equilibrium style adjustment to future changes in wave climate, analogous to that which Moore et al. (2013) found for cuspate capes over recent decades. However, our results illuminate the possibility that even without wave climate change occurring in the present or near future, coastline shape could be in the midst of a long-term readjustment to changes in wave climate that occurred in the past. Given the diffusive scaling of coastline response timescales with length scales, large coastal features inherited from the early Holocene may not yet have adjusted to current wave conditions. In such a case, zones of accentuated erosion would tend to shift in location and intensity over time, without any warning from identifiable changes in forcing. This possibility motivates future work to develop metrics in terms of alongshore patterns of local wave climates (e.g. Ashton and Murray, 2006b) or sediment fluxes that could identify coastlines that are in or out of (quasi-) equilibrium with the present directional wave climate.

Finally, the results of these model experiments might also have implications for geologic and paleo-climate interpretations. Our modelling results indicate that flying spits adjusting to changes in the wave angle distribution in a wave climate can leave behind records of the adjustment in the form of complex arrays of lagoons enclosed by beach ridges (e.g. see Fig. 3bii, second panel; 3biii, first and second panels). In natural settings, the lagoons, which can potentially extend far enough landward to be preserved, would fill with fine sediment over time. However, the morphological arrangement of such lagoons and associated bounding beach ridges, preserved in the geological record, could indicate the effects of changing wave climate on a spit coastline. Such complicated coastal plain deposits can form in other ways (including reworking of a relict, potentially crenulated coastline present at the beginning of the current sea-level high stand, and/or episodic fluvial or coastal sediment delivery, Nienhuis et al., 2015), but being aware of the possible influence of wave climate change on the morphology and structure of coastal hinterlands could inform reconstructions of coastal geographies and paleo-wave climates.

6 Conclusions

We have explored the degree to which complex sandy coastlines can adjust in a quasi-equilibrium manner to changing wave climate, and the degree to which past wave climates, as manifest in existing morphology, might influence subsequent coastline evolution during wave climate change. We conducted numerical modelling experiments in which capes and spits grown under some initial condition were subjected subsequently to a different wave climate. We find that the resulting evolution of complex coastlines (capes and spits) is influenced by previous morphology for a significant period of time – in some cases rather longer than indicative characteristic timescales of morphological change (few decades to few centuries) after wave climate change has occurred. Characteristic timescales for change show that spits respond at rates approximately an order of magnitude more slowly than capes, and these timescales depend on the spatial scale of the coastline feature (a diffusive scaling). In particular, quasi-equilibrium response cannot be assumed: such behaviour will depend on the rate of wave climate change, and how it compares with the characteristic timescale for morphological adjustment for a given coastline morphology. Significant changes in wave climate on decadal to centennial scales may result in hysteresis in the response of the coastline morphology. Thus, modern coastlines may be out of morphological equilibrium with respect to current wave conditions, reflecting instead the wave climates to which they have been subject in the past, and the resulting antecedent coastline morphology. The preservation of beach land forms in the coastal hinterlands behind modern sandy coasts may reflect this history and provide insights into past paleo-wave climate and coastal geography.

Author contributions. Christopher W. Thomas and A. Brad Murray planned the experimental design and undertook the modelling and research, and also prepared the manuscript, supported by Andrew D. Ashton and Martin D. Hurst. The original ideas for this work, and their subsequent development, grew out of initial discussions between Christopher W. Thomas and Andrew K. A. P. Barkwith; Andrew K. A. P. Barkwith subsequently provided further ideas, advice and support for the later modelling. Michael A. Ellis supported and helped shape the research through discussion and further ideas. All have contributed to the manuscript through additional or modified text and comments.

Acknowledgements. Christopher W. Thomas gratefully acknowledges support from the British Geological Survey and the Nicholas School of the Environment, Duke University, North Carolina, for a short sabbatical at Duke University in 2015, during which this work was largely undertaken. Eli Lazarus and an anonymous referee are thanked for constructive and insightful reviews. This work was funded by NERC national capability core funding to the British Geological Survey. The Coastline Evolution Model (CEM) can be downloaded from the Community Surface Dynamics Modeling System model repository.

References

Allard, J., Bertin, X., Chaumillon, E., and Pouget, F.: Sand spit rhythmic development: A potential record of wave climate variations? Arçay Spit, western coast of France, Mar. Geol., 253, 107–131, doi:10.1016/j.margeo.2008.05.009, 2008.

Ashton, A. D. and Murray, A. B.: High-angle wave instability and emergent shoreline shapes: 1. Modeling of sand waves, flying spits and capes, J. Geophys. Res., 111, F04011, doi:10.1029/2005JF000422, 2006a.

Ashton, A. D. and Murray, A. B.: High-angle wave instability and emergent shorleine shapes: 2. Wave climate analysis and comparisons to nature, J. Geophys. Res., 111, F04012, doi:10.1029/2005JF000423, 2006b.

Ashton, A. D., Murray, A. B., and Arnoult, O.: Formation of coastline features by large-scale instabilities induced by high-angle waves, Nature London, 414, 296–300, 2001.

Ashton, A. D., Nienhuis, J., and Ells, K.: On a neck, on a spit: controls on the shape of free spits, Earth Surf. Dynam., 4, 193–210, doi:10.5194/esurf-4-193-2016, 2016.

Barkwith, A., Thomas, C. W., Limber, P. W., Ellis, M. A., and Murray, A. B.: Coastal vulnerability of a pinned, soft-cliff coastline – Part I: Assessing the natural sensitivity to wave climate, Earth Surf. Dynam., 2, 295–308, doi:10.5194/esurf-2-295-2014, 2014.

Bosman, D. E. and Joubert, J. R.: Report on wave, wind and water level conditions at Walvis Bay, Technical Report Namibian Port Authority, WB 12-2008/1, 2008.

Bruun, P.: Sea-level rise as a cause of shore erosion, J. Waterway. Harbors div., 88, 117–132, 1962.

Falqués, A. and Calvete, D.: Large-scale dynamics of sandy coastlines: Diffusivity and instability, J. Geophys. Res.-Oceans, 110, C03007, doi:10.1029/2004jc002587, 2005.

Falqués, A., Coco, G., and Huntley, D. A.: A mechanism for the generation of wave-driven rhythmic patterns in the surf zone, J. Geophys. Res.-Oceans, 105, 24071–24087, doi:10.1029/2000jc900100, 2000.

Hemer, M. A., Fan, Y., Mori, N., Semedo, A., and Wang, X. L.: Projected changes in wave climate from a multi-model ensemble, Nat. Clim. Change, 3, 471–476, doi:10.1038/nclimate1791, 2013.

Hurst, M. D., Barkwith, A., Ellis, M. A., Thomas, C. W., and Murray, A. B.: Exploring the sensitivities of crenulate-bay shorelines to wave climates using a new vector-based one-line model, J. Geophys. Res.-Earth, 120, 2586–2608, doi:10.1002/2015JF003704, 2015.

Idier, D. and Falqués, A.: How kilometric sandy shoreline undulations correlate with wave and morphology characteristics: preliminary analysis on the Atlantic coast of Africa, Adv. Geosci., 39, 55–60, doi:10.5194/adgeo-39-55-2014, 2014.

Kaergaard, K. and Fredsoe, J.: Numerical modeling of shoreline undulations part 1: Constant wave climate, Coast. Eng., 75, 64–76, doi:10.1016/j.coastaleng.2012.11.006, 2013a.

Kaergaard, K. and Fredsoe, J.: Numerical modeling of shoreline undulations part 2: Varying wave climate and comparison with observations, Coast. Eng., 75, 77–90, doi:10.1016/j.coastaleng.2012.11.003, 2013b.

Kaergaard, K. and Fredsoe, J.: A numerical shoreline model for shorelines with large curvature, Coast. Eng., 74, 19–32, doi:10.1016/j.coastaleng.2012.11.011, 2013c.

Komar, P. D.: The mechanics of sand transport on beaches, J. Geophys. Res., 76, 713–721, 1971.

Komar, P. D.: Beach processes and sedimentation, 2nd Edn., Prentice Hall, 544 pp., 1998.

McNamara, D. E., Murray, A. B., and Smith, M. D.: Coastal sustainability depends on how economic and coastline responses to climate change affect each other, Geophys. Res. Lett., 38, L07401, doi:10.1029/2011GL047207, 2011.

Moore, L. J., List, J. H., Williams, S. J., and Stolper, D.: Complexities in barrier island response to sea level rise: Insights from numerical model experiments, North Carolina Outer Banks, J. Geophys. Res.-Earth, 115, F03004, doi:10.1029/2009JF001299, 2010.

Moore, L. J., McNamara, D. E., Murray, A. B., and Brenner, O.: Observed changes in hurricane-driven waves explain the dynamics of modern cuspate shorelines, Geophys. Res. Lett., 40, 5867–5871, doi:10.1002/2013GL057311, 2013.

Murray, A. B. and Ashton, A. D.: Instability and finite-amplitude self-organization of large-scale coastline shapes, Philos. T. R. Soc. A, 371, 2004, doi:10.1098/rsta.2012.0363, 2013.

Nienhuis, J. H., Ashton, A. D., Roos, P. C., Hulscher, S. J. M. H., and Giosan, L.: Wave reworking of abandoned deltas, Geophys. Res. Lett., 40, 5899–5903, doi:10.1002/2013GL058231, 2013.

Nienhuis, J. H., Ashton, A. D., and Giosan, L.: What makes a delta wave-dominated? Geology, 43, 511–514, doi:10.1130/g36518.1, 2015.

Ribas, F., Falqués, A., Van Den Berg, N., and Caballeria, M.: Modeling shoreline sand waves on the coasts of Namibia and Angola, Int. J. Sediment Res., 28, 338–348, doi:10.1016/S1001-6279(13)60044-X, 2013.

Slott, J. M., Murray, A. B., Ashton, A. D., and Crowley, T. J.: Coastline responses to changing storm patterns, Geophys. Res. Lett., 33, L18404, doi:10.1029/2006GL027445, 2006.

Storlazzi, C. D., Shope, J. B., Erikson, L. H., Hegermiller, C. A., and Barnard, P. L.: Future wave and wind projections for United States and United States-affiliated Pacific Islands, US Geological Survey Open-File Report, 2015-1001, 426 pp., doi:10.3133/ofr20151001, 2015.

van den Berg, N., Falqués, A., and Ribas, F.: Modeling large scale shoreline sand waves under oblique wave incidence, J. Geophys. Res.-Earth, 117, F03019, doi:10.1029/2011JF002177, 2012.

Walkden, M. J. A. and Hall, J. W.: A predictive Mesoscale model of the erosion and profile development of soft rock shores, Coast. Eng., 52, 535–563, 2005.

Walkden, M. J. A. and Hall, J. W.: A mesoscale predictive model of the evolution and management of a soft-rock coast, J. Coast. Res., 27, 529–543, 2011.

WASA Group: Changing Waves and Storms in the Northeast Atlantic?, B. Am. Meteorol. Soc., 79, 741–760, doi:10.1175/1520-0477(1998)079<0741:CWASIT>2.0.CO;2, 1998.

Wolf, J. and Woolf, D. K.: Waves and climate change in the north-east Atlantic, Geophys. Res. Lett., 33, L06604, doi:10.1029/2005GL025113, 2006.

Wolinsky, M. A. and Murray, A. B.: A unifying framework for shoreline migration: 2. Application to wave-dominated coasts, J. Geophys. Res., 114, F01009, doi:10.1029/2007JF000856, 2009.

Permissions

All chapters in this book were first published in ESD, by Copernicus Publications; hereby published with permission under the Creative Commons Attribution License or equivalent. Every chapter published in this book has been scrutinized by our experts. Their significance has been extensively debated. The topics covered herein carry significant findings which will fuel the growth of the discipline. They may even be implemented as practical applications or may be referred to as a beginning point for another development.

The contributors of this book come from diverse backgrounds, making this book a truly international effort. This book will bring forth new frontiers with its revolutionizing research information and detailed analysis of the nascent developments around the world.

We would like to thank all the contributing authors for lending their expertise to make the book truly unique. They have played a crucial role in the development of this book. Without their invaluable contributions this book wouldn't have been possible. They have made vital efforts to compile up to date information on the varied aspects of this subject to make this book a valuable addition to the collection of many professionals and students.

This book was conceptualized with the vision of imparting up-to-date information and advanced data in this field. To ensure the same, a matchless editorial board was set up. Every individual on the board went through rigorous rounds of assessment to prove their worth. After which they invested a large part of their time researching and compiling the most relevant data for our readers.

The editorial board has been involved in producing this book since its inception. They have spent rigorous hours researching and exploring the diverse topics which have resulted in the successful publishing of this book. They have passed on their knowledge of decades through this book. To expedite this challenging task, the publisher supported the team at every step. A small team of assistant editors was also appointed to further simplify the editing procedure and attain best results for the readers.

Apart from the editorial board, the designing team has also invested a significant amount of their time in understanding the subject and creating the most relevant covers. They scrutinized every image to scout for the most suitable representation of the subject and create an appropriate cover for the book.

The publishing team has been an ardent support to the editorial, designing and production team. Their endless efforts to recruit the best for this project, has resulted in the accomplishment of this book. They are a veteran in the field of academics and their pool of knowledge is as vast as their experience in printing. Their expertise and guidance has proved useful at every step. Their uncompromising quality standards have made this book an exceptional effort. Their encouragement from time to time has been an inspiration for everyone.

The publisher and the editorial board hope that this book will prove to be a valuable piece of knowledge for researchers, students, practitioners and scholars across the globe.

LIST OF CONTRIBUTORS

Jasper H. J. Candel, Cindy Quik and Jakob Wallinga
Soil Geography and Landscape Group, Wageningen University & Research, Wageningen, 6700AA, the Netherlands

Maarten G. Kleinhans and Wim Z. Hoek
Department of Physical Geography, Utrecht University, Utrecht, 3508TC, the Netherlands

Fiona J. Clubb and Simon M. Mudd
School of GeoSciences, University of Edinburgh, Drummond Street, Edinburgh, EH8 9XP, UK

David T. Milodowski
School of GeoSciences, University of Edinburgh, Crew Building, King's Buildings, Edinburgh, EH9 3JN, UK

Declan A. Valters
School of Earth, Atmospheric, and Environmental Science, University of Manchester, Oxford Road, Manchester, M13 9PL, UK

Louise J. Slater
Department of Geography, Loughborough University, Loughborough, LE11 3TU, UK

Martin D. Hurst
School of Geographical and Earth Sciences, East Quadrangle, University of Glasgow, Glasgow, G12 8QQ, UK

Ajay B. Limaye
Department of Earth Sciences and St. Anthony Falls Laboratory, University of Minnesota, Minneapolis, Minnesota, USA

Daniel Wujanz and Frank Neitzel
Institute of Geodesy and Geoinformation Science, Technische Universität Berlin, Berlin, Germany

Michael Avian
Austrian Institute of Technology GmbH, Tulln, Austria

Daniel Krueger
GFaI – Society for the Advancement of Applied Computer Science, Berlin, Germany

Bradley A. Weymer and Mark E. Everett
Department of Geology and Geophysics, Texas A&M University, College Station, Texas 77843, USA

Phillipe Wernette
Department of Geography, Texas A&M University, College Station, Texas 77843, USA

Chris Houser
Department of Earth and Environmental Sciences, University of Windsor, Windsor, Ontario N9B 3P4, Canada anow at: GEOMAR – Helmholtz Center for Ocean Research Kiel, Wischhofstraße 1-3, 24148 Kiel, Germany

Étienne Cossart and Mathieu Fressard
Université de Lyon (Jean Moulin, Lyon 3), UMR 5600 CNRS – Environnement Ville Société, Lyon, 69007, France

Sandro Rossato and Paolo Mozzi
Department of Geosciences, University of Padova, Padova, 35131, Italy

Anna Carraro, Giovanni Monegato and Fabio Tateo
Institute of Geosciences and Earth Resources (IGG) – National Research Council (CNR), Padova, 35131, Italy

Yoni Israeli and Simon Emmanuel
Institute of Earth Sciences, The Hebrew University of Jerusalem, Jerusalem, Israel

Tobias Müller and Marwan A. Hassan
Department of Geography, the University of British Columbia, Vancouver, BC, Canada

Eliisa S. Lotsari
Department of Geographical and Historical Studies, University of Eastern Finland, Joensuu, Yliopistokatu 2, 80101 Joensuu, Finland

Mikel Calle and Gerardo Benito
National Museum of Natural Sciences, Spanish Research Council (CSIC), Madrid, Calle de Serrano 117, 28006 Madrid, Spain

Antero Kukko, Harri Kaartinen, Juha Hyyppä and Petteri Alho
Department of Remote Sensing and Photogrammetry, Finnish Geospatial Research Institute, National Land Survey of Finland, Kirkkonummi, Geodeetinrinne 2, 02430 Masala, Finland

Antero Kukko and Hannu Hyyppä
Department of Built Environment, Aalto University, Espoo, Vaisalantie 8, 00076 Aalto, Finland

Eliisa S. Lotsari, Harri Kaartinen and Petteri Alho
Department of Geography and Geology, University of Turku, 20014 Turku, Finland

Benjamin Purinton and Bodo Bookhagen
Institute of Earth and Environmental Science, Universität Potsdam, Potsdam, Germany

Christopher W. Thomas
British Geological Survey, Lyell Centre, Edinburgh, EH14 4AP, Scotland, UK

A. Brad Murray
Division of Earth and Ocean Sciences, Nicholas School of the Environment and Earth Sciences and Center for Nonlinear and Complex Systems, Duke University, Durham, North Carolina 27708, USA

Andrew D. Ashton
Department of Geology and Geophysics, Woods Hole Oceanographic Institution, Woods Hole, Massachusetts 02543, USA

Martin D. Hurst, Andrew K. A. P. Barkwith and Michael A. Ellis
British Geological Survey, Nicker Hill, Keyworth, Nottingham, NG12 5GG, England, UK

Index

S

T

V

W